THE Naval Operations

SMARTbook

Guide to Designing, Planning & Conducting Maritime Operations

The Lightning Press
Norman M. Wade

The Lightning Press

2227 Arrowhead Blvd.
Lakeland, FL 33813
24-hour Voicemail/Fax/Order: 1-800-997-8827
E-mail: SMARTbooks@TheLightningPress.com
www.TheLightningPress.com

The Naval Operations & Planning SMARTbook
Guide to Designing, Planning & Conducting Maritime Operations

Copyright © 2010 Norman M. Wade

ISBN: 978-0-9824859-5-8

All Rights Reserved

No part of this book may be reproduced or utilized in any form or other means, electronic or mechanical, including photocopying, recording or by any information storage and retrieval systems, without permission in writing by the publisher. Inquiries should be addressed to The Lightning Press.

Notice of Liability

The information in this SMARTbook and quick reference guide is distributed on an "As Is" basis, without warranty. While every precaution has been taken to ensure the reliability and accuracy of all data and contents, neither the author nor The Lightning Press shall have any liability to any person or entity with respect to liability, loss, or damage caused directly or indirectly by the contents of this book. If there is a discrepancy, refer to the source document. This SMARTbook does not contain classified or sensitive information restricted from public release.

"The views presented in this publication are those of the author and do not necessarily represent the views of the Department of Defense or its components."

The Lightning Press would like to recognize and thank Darrell Ames for his work on this book and the series of SMARTbooks over the years, Jim Brown for his layout and design assistance, and Jason Salata for his naval warfighting focus and perspective in making this book a reality.

SMARTbook is a trademark of The Lightning Press.

Photos. Courtesy Department of Defense and the Military Services.

Printed and bound in the United States of America.

Preface

The *Navy Operations & Planning SMARTbook* outlines the Navy operational-level fundamentals, command, control, and organization. It is also a bridge between the theory of operational art and the practical specific guidance that Navy commanders and staffs require to accomplish their mission. It is prepared to complement existing **joint and Navy doctrine** and provides a general guide to the application of command at the **operational level of war** and the staff organization and functionality required to support the operational commander.

The **U.S. Navy** is an instrument of national power, employed to prevent conflict and, if necessary, prevail in war. It is organized, trained and equipped primarily to fight at and from the sea and to influence events on land. Unlike the other components of the joint force, the **maritime component** routinely conducts operations across all of the domains, described as air, land, maritime, space, and the information environment.

The **JFMCC** is the JFC's maritime warfighter. The JFMCC's forces/capabilities may consist of subordinate commanders and forces from any Service and may include multinational forces to accomplish the portion of joint operations that occurs predominately in the maritime domain, defined as "the oceans, seas, bays, estuaries, islands, coastal areas, and the airspace above these, including the littorals."

Design is a methodology for applying critical and creative thinking to understand, visualize, and describe complex, ill-structured problems and develop approaches to solve them. **Military planning**, and by extension Navy planning, is the process by which a commander visualizes an end state and then determines the most effective ways by which to reach the end state. Through the Navy Planning Process (NPP), a commander can effectively plan for and execute operations, ensure that the employment of forces is linked to objectives, and integrate naval operations seamlessly with the actions of a joint force. **Preparation** helps the JFMCC transition between planning and execution. **Execution** is putting a plan into action by applying combat power to accomplish the mission and using situational understanding to assess progress and make execution and adjustment decisions. Execution combines continued planning, preparation, and assessment with the challenges of a dynamic adversary and the fog of war. Assessment is a key portion of the commander's decision cycle. **Assessment** is the continuous monitoring and evaluation of the current situation and the progress of an operation. Based on their assessment, commanders direct adjustments, thus ensuring the operation remains focused on accomplishing the mission.

SMARTbooks - The Essentials of Warfighting!

Recognized as a doctrinal reference standard by military professionals around the world, SMARTbooks are designed with all levels of Soldiers, Sailors, Airmen, Marines and Civilians in mind.

SMARTbooks can be used as quick reference guides during actual tactical combat operations, as study guides at military education and professional development courses, and as lesson plans and checklists in support of training. Serving a generation of warfighters, military reference SMARTbooks have become "mission-essential" around the world. Visit www.TheLightningPress.com for complete details!

Preface-1

References

The following references were used to compile The Navy Operations & Planning SMARTbook. All references are unclassified, considered public domain, available to the general public, and designated as "approved for public release; distribution is unlimited." The Navy Operations & Planning SMARTbook does not contain classified or sensitive material restricted from public release.

Joint Publications (JPs)

JP 3-0 (Chg 1)	Feb 2008	Joint Operations (with Change 1)
JP 3-08	Mar 2006	Interagency, IGO and NGO Coordination (I & II)
JP 3-16	Mar 2007	Multinational Operations
JP 3-32 (Chg 1)	Aug 2006	Command and Control for Joint Maritime Operations
JP 5-0	Dec 2006	Joint Operation Planning

Naval Doctrine Publications (NDPs)

NDP1	Mar 1994	Naval Warfare
NDP4	Feb 2001	Naval Logistics
NDP5	Jan 1996	Naval Planning

Navy Warfare Publications (NWPs)

NWP 1-14M	Jul 2007	The Commander's Handbook on the Law of Naval Operations
NWP 3-32	Oct 2008	Maritime Operations at the Operational Level of War
NWP 3-56 (REV A)	Aug 2001	Composite Warfare Commander's Manual
NWP 5-01	Jan 2007	Navy Planning

Navy Tactics, Techniques, and Procedures (NTTPs)

NTTP 3-32.1	Oct 2008	Maritime Operations Center

Training Circulars (TCs) and other Publications

JWFC	Sept 2010	Pamphlet 10, Design in Military Operations: A Primer for Joint Warfighters
NWC MCCH	Feb 2010	NWC Maritime Component Commander Handbook
NWC 3153K	Jul 2009	Joint Military Operations Reference Guide
NWC 4020A	Sept 2008	On Naval Warfare. Dr. Milan Vego, Joint Military Operations Department
NWC NP32	2008	Newport Papers 32: Major Naval Operations. Milan Vego, Center for Naval Warfare Studies
TCH-TSC	2009	Tactical Commander's Handbook for Theater Security Cooperation
NSW Press Kit	2010	Naval Special Warfare Command Fact File

Table of Contents

Chap 1: Maritime Forces, Organization & Capabilities

I. U.S. Navy 1-1
    I. Mission and Purpose 1-1
    II. Organization & Command Structure 1-2
        A. U.S. Navy Administrative Organization 1-2
        B. U.S. Navy Operational Organization 1-3
    III. Composition & Capability of Major Deployable Elements 1-4
        A. Carrier Strike Group (CSG) 1-4
        B. Amphibious Ready Group (ARG)/Marine Expeditionary Unit (MEU) 1-4
        C. Surface Strike Group (SSG) 1-5
        D. Naval Fleet Auxiliary Force (NFAF) 1-5
    IV. Capabilities 1-6
        Future Force 1-6
        A. Mine Warfare 1-6
        B. Aircraft Carriers & Naval Aviation 1-8
        C. Surface Combatants 1-10
        D. Submarine Forces 1-12
        E. Amphibious Warfare 1-14
        F. MSC Support Ships 1-16
    V. Navy Reserve 1-18

II. U.S. Marine Corps 1-19
    I. Mission 1-19
    II. Organization and Structure 1-19
    III. Marine Air/Ground Task Force (MAGTF) 1-21
    IV. Types of MAGTFs (MEF/MEB/MEU/SPMAGTF) 1-22
    V. USMC Concepts and Capabilities 1-25
    VI. Major Marine Corps Equipment 1-26

III. U.S. Coast Guard 1-27
    I. USCG Organization 1-27
    II. Coast Guard Forces 1-28
    III. USCG Roles and Mission 1-29
    IV. Ports, Waterways & Coastal Security (PCWS) 1-30

IV. Naval/Marine Special Warfare Forces 1-33
    U.S. Special Operations Command (USSOCOM) 1-34
    I. Naval Special Warfare (NSW) 1-33
    II. U.S. Marine Corps Special Operations Forces 1-38

V. Strategic Lift/Force Projection 1-39
    I. Force Projection 1-39
    II. U.S. Transportation Command (USTRANSCOM) 1-40
    III. U.S. Maritime Administration (MARAD) 1-42
    IV. Strategic Deployment 1-44
    V. Strategic Lift Port Operations 1-46

Chap 2: Maritime Operations

I. Operational Warfare at Sea..2-1
   I. Navy Doctrine 2-1
   II. History/Background 2-2
   III. The Maritime Domain 2-4
   IV. Joint Operational Areas 2-6
     - Maritime Theater Structure 2-7
   V. Operational Level of Command (OLC).................2-8

II. Range/Spectrum of Maritime Operations............2-9
   I. Levels of War..2-9
     A. Strategic Level of War..2-10
     B. Operational Level..2-10
     - Unified Action...2-11
     C. Tactical Level 2-11
   II. Maritime Operations Across the Range of Military Operations...................2-12
     A. Military Engagement, Security Cooperation, and Deterrence................2-13
     B. Crisis Response or Limited Contingency Operations 2-13
     C. Major Operations or Campaigns Involving Large-Scale Combat............2-13
   IV. Major Naval Operations..2-14
     A. Major Operations: Fleet versus Fleet.................2-15
     B. Major Operations: Fleet versus Shore 2-15
     C. Major Operations versus Enemy Maritime Trade 2-15
     D. Major Naval Operations to Defend/Protect Maritime Trade..............2-15
     E. Destruction/Protection of Seaborne Nuclear Deterrent Forces.........2-15
     F. Major Operations in Support of Ground Forces on the Coast..............2-15
   IV. Aegis Ballistic Missile Defense (BMD)..................2-16
   V. Cyber Warfare 2-16
     - U.S. Cyber Command (USCYBERCOM)............2-16
     - U.S. Fleet Cyber Command (U.S. Tenth Fleet)...2-18
     - Naval Network Warfare Command (NETWARCOM) 2-18
   VI. Maritime Operational Threat Response (MOTR) ..2-17

III. Objectives of Naval Operations 2-19
   Strategic Offensive versus Defensive Posture.........2-19
   Operational Factors in Naval Warfare 2-20
   I. Sea Control...2-19
     - Choke-Point Control 2-22
     - Basing/Deployment Area Control.........................2-22
     - Obtaining Air Superiority.......................................2-22
     - Open Ocean vs. Narrow Seas 2-23
     A. Obtaining Sea Control.......................................2-24
     B. Maintaining Sea Control....................................2-26
     C. Exercising Sea Control 2-26
     - Projecting Power Ashore 2-26
   II. Sea Denial 2-27
     - Sea Denial Dynamics...2-28
   III. Attack on Maritime Trade 2-30
     - Objectives 2-30
     - Tenets..2-30
     - Offensive Mining 2-31
     - Commercial Blockade 2-31

IV. Defense and Protection of Maritime Trade 2-31
- Defense of Maritime Trade 2-32
- Protection of Maritime Trade 2-32

IV. Levels of Maritime Command 2-33
Operational vs. Administrative Control 2-33
Command Authority 2-36
Impact of Level of Authority on Command at the Operational Level 2-36
I. Command Relationships 2-34
   A. U.S. Joint Doctrine Command Relationships 2-34
   B. Allied and Multinational Maritime Command 2-35
II. Joint Force Components 2-37
   A. Unified Command 2-37
   B. Subordinate Unified Command 2-37
   C. Joint Task Force (JTF) 2-38
III. Joint Force Commander 2-38
   - Possible Components in a Joint Force 2-38
   - Availability of Forces for Joint Operations 2-39
   A. Commander, Joint Task Force (CJTF) 2-40
   B. Service Component Commander 2-42
   C. Functional Component Commander 2-44

V. Joint Force Maritime Component Commander (JFMCC) 2-45
I. Establishing a JFMCC 2-45
   A. JFMCC Area of Operations 2-46
   B. Integration with Joint Campaign Planning 2-46
   C. Movement and Maneuver of Maritime Forces 2-52
   D. Intelligence, Surveillance, and Reconnaissance (ISR) 2-53
II. JFMCC Functions 2-47
III. JFMCC Roles and Responsibilities 2-48
IV. Notional Joint Maritime Operations Organization and Processes 2-50
V. Undersea/Antisubmarine Warfare (USW/ASW) 2-54
VI. Maritime Fires Support 2-56
VII. Joint Maritime Operations Targeting 2-58
VIII. Considerations for Employing a JFMCC 2-60

VI. Navy Tactical Headquarters and Task Organization 2-61
I. Notional Relationship of Command Level to Level of War 2-61
II. Navy Task Organization (Task Forces, Groups, Units and Elements) .. 2-62
III. Navy Tactical Headquarters (CSG, ESG, SSG) 2-64
IV. Navy Composite Warfare Commander (CWC) 2-65
   A. Officer in Tactical Command/Composite Warfare Commander
   (OTC/CWC) 2-66
   B. Principle Warfare Commanders (PWC) 2-68
   C. Functional Coordinators 2-70

VII. Multinational Operations 2-71
Alliance 2-71
Coalition 2-71
I. Multinational Chains of Command 2-71
   A. National Command 2-71
   B. Multinational Command 2-72
   Command Structures of Forces in Multinational Operations 2-73
II. Maritime Multinational Forces 2-73
III. Civil-Military Coordination and Dealing with Nonmilitary Agencies 2-74
   (Interagency, Intergovernmental & NGOs)

Table of Contents-3

Chap 3: Maritime Operations Center (MHQ/MOC)

I. MHQ / Maritime Operations Center (MOC) 3-1
   I. The Maritime Headquarters (MHQ) 3-1
      A. Command Structure of the MHQ 3-2
      B. Maritime Operations Structure (MOC) 3-3
         - MOC Centers - Intelligence, Operations, and Logistics 3-6
         - MOC Director 3-6
         - B2C2WG (Boards, Bureaus, Centers, Cells, and Working Groups) 3-6
      C. Fleet Management and Operational Structures 3-4
   II. Battle Rhythm 3-8
   III. Operational-Level Functions 3-10
      A. Command and Control 3-11
      B. Fires 3-11
      C. Intelligence 3-12
      D Movement and Maneuver 3-13
      E. Protection 3-13
      F. Sustainment 3-14

II. Forming & Transitioning the MOC Staff 3-15
   I. Forming the MOC Staff 3-15
   II. Staff Transition From Enduring to Emergent Operations 3-16
   III. Balancing Enduring and Emergent Mission Requirements 3-17
   IV. Seven-Minute Drills 3-17
   V. Commander's Considerations During Staff Formation 3-18

III. Maritime Operational Command 3-19
   I. Inherent Relationships of Command to Level of War 3-19
   II. Exercising Operational Command 3-20
      A. Operational-Level Functions 3-20
      B. Role of the Operational Commander in C2 3-20
      C. Level of Control 3-21
      D. Control Areas 3-24
   II. Maritime Domain Awareness (MDA) 3-22
      - Maritime Domain Awareness Goals 3-23
      - Maritime Domain Awareness Process and Functions 3-23
      - Maritime Domain Awareness Critical Tasks 3-23

IV. Operational & Maritime Law 3-25
   The Relationship Between Law and Policy 3-25
   I. Rules of Engagement (ROE) & Rules For the Use of Force (RUF) ... 3-26
   II. Legal Regimes of Oceans and Airspace 3-27
   III. Law of Armed Conflict Fundamentals 3-29
   IV. Legal Basis to Stop, Board, Search, and Seize Vessels 3-30
   V. Piracy 3-31
   VI. Other Fundamental Legal Issues Relevant to the C/JFMCC 3-32

Chap 4: The Maritime Operations Process

Operations Process Overview..4-1
  I. Operations Process Activities 4-2
  II. Operations Process Supporting Topics....................................4-4

I. Planning 4-5
  - The Navy Planning Process (NPP) 4-5
  - Naval Planning and the Levels of War 4-6

II. Preparation...4-7
  I. Preparation Functions..4-7
  II. Integrating Processes and Continuing Activities During Preparation....4-8
    A. Intelligence Preparation of the Operational Environment (IPOE) 4-8
    B. Targeting 4-9
    C. Intelligence, Surveillance, and Reconnaissance (ISR) Synchronization ..4-9
    D. Security..4-9
    E. Operational Protection 4-9
    F. Battlespace Management..4-10
  III. Preparation Activities 4-9
  IV. Rehearsals 4-10
    - Joint Operations Rehearsals...4-10
    - Navy Rehearsals..4-10

III. Execution 4-11
  I. Execution Fundamentals 4-11
    A. Competing Demands 4-12
    B. Situational Understanding..4-12
    C. Commander's Critical Information Requirements (CCIRs).....4-12
    D. Decisionmaking During Execution 4-12
  II. Commander's Decision Cycle 4-14
    A. Assess..4-15
    B. Plan..4-15
    C. Direct 4-15
    D. Monitor...4-15
    E. Communications...4-15
  III. Risk Assessment 4-16
    - Accepting Risk..4-16
    - Applying Risk Management..4-17
    - Risk Management for a JFMCC/NCC 4-17
  IV. Event Horizons 4-18
    - Near 4-18
    - Mid 4-18
    - Far...4-18

IV. Assessment 4-19
  I. The Assessment Process 4-19
    A. Measure of Performance (MOP).......................................4-19
    B. Measure of Effectiveness (MOE) 4-19
  II. Combat Assessment..4-22
    A. Battle Damage Assessment (BDA) 4-22
    B. Munitions Effectiveness Assessment (MEA).....................4-22
    C. Reattack Recommendation 4-23
  III. Assessment Estimate Development and Integration with the NPP 4-24

Table of Contents-5

Chap 5: Naval Planning

Navy Planning Overview 5-1
    I. The Navy Planning Process (NPP) 5-2
    II. Nesting of the Navy Planning Process Within Other Planning Processes 5-4
        - Navy Planning and the Levels of War 5-4
        - Navy Planning Process and Effects 5-4
        - Joint Operation Planning and Execution System (JOPES) 5-5
    III. Roles in the Navy Planning Process 5-6
    IV. Intelligence Preparation of the Operational Environment (IPOE) 5-8
    V. Operational Art & Design 5-10
    VI. Design in Military Operations 5-14

I. Mission Analysis 5-17
    Inputs 5-18
    Process 5-19
        1. Identify Source(s) of Mission 5-20
        2. Determine Support Relationships 5-20
        3. Analyze the Higher Commander's Mission 5-20
        4. Determine Specified, Implied, and Essential Tasks 5-22
        5. State the Purpose 5-23
        6. Identify Externally Imposed Limitations 5-23
        7. Analyze Available Forces and Assets 5-26
        8. Determine Critical Factors, Centers of Gravity, and Decisive Points ... 5-26
        9. Develop Planning Assumptions 5-27
        10. Conduct Initial Risk Assessment 5-27
        11. Develop Proposed Mission Statement 5-27
        12. Conduct Mission Analysis Briefing 5-28
        13. Develop Initial Commander's Intent 5-28
        14. Develop Commander's Critical Information Requirements (CCIR) ... 5-30
        15. Develop Commander's Planning Guidance 5-32
        16. Develop Warning Order 5-34
    Outputs 5-34
    Key Points 5-34

II. Course of Action Development 5-35
    Inputs 5-36
    Process 5-36
        1. Analyze Relative Combat Power 5-36
        2. Generate Course of Action Options 5-38
        3. Test for Validity 5-39
        4. Recommend Command and Control Relationships 5-39
        5. Prepare Course of Action Sketch and Statement 5-39
        6. Prepare Course of Action Briefing 5-41
        7. Develop Course of Action Analysis and Evaluation Guidance 5-41
    Outputs 5-41
    Key Points 5-41

III. Course of Action Analysis (Wargaming) 5-47
    Inputs 5-48
    Process 5-48
        1. Organize for Wargaming 5-50
        2. List all Friendly Forces 5-50

3. Review Assumptions 5-50
4. List Known Critical Events 5-50
5. Determine the Governing Factors 5-51
6. Select the Wargaming Method 5-51
7. Record and Display Results 5-52
8. Wargame the Combat Actions and Assess the Results 5-52
9. Refine Staff Estimates 5-56
10. Update and Refine Intelligence Preparation of the Operational 5-56
   Environment (IPOE) Products
Outputs 5-58

IV. Course of Action Comparison & Decision 5-59
Inputs 5-59
Process 5-60
1. Perform Course of Action Comparison 5-60
2. Perform Course of Action Evaluation 5-61
3. Make Final Tests for Feasibility and Acceptability 5-61
4. State Commander's Decision 5-61
   - Sample Decision Brief 5-64
5. Prepare Synchronization Matrix 5-64
6. Develop the Concept of Operation (CONOPS) 5-66
7. Refine Intelligence Preparation of the Operational Environment (IPOE) ...5-66
Outputs 5-66
Key Points 5-66

V. Plans and Orders Development 5-67
Inputs 5-68
Process 5-70
1. Prepare Plans and Orders 5-70
   - Operation Plans & Orders (OPLANs/OPORDs) 5-71
2. Reconcile Plans and Orders 5-72
3. Backbrief and Crosswalk Plans and Orders 5-72
4. Commander Approves Plans and Orders 5-72
Outputs 5-72
Key Points 5-72

VI. Transition 5-73
Inputs 5-73
Process 5-74
1. Transition Briefing 5-74
   - Roles of the FPC, FOPS, COPS (and Internal/External Transitions) ...5-75
2. Transition Drills 5-76
3. Confirmation Brief 5-76
Outputs 5-76
Key Points 5-76

Chap 6: Naval Logistics

I. Fundamentals of Naval Logistics 6-1
I. The Mission of Naval Logistics 6-1
II. Levels of Logistics Support 6-2
III. Process Elements 6-3
IV. Principles of Logistics 6-6
V. High Yield Logistics (HYL) 6-8
VI. Focused Logistics 6-10

Table of Contents-7

II. Naval Logistics Planning ... 6-11
- Operational Logistics (OPLOG) ... 6-11
- I. The Logistics Planning Process ... 6-12
- II. Naval Logistics Planning ... 6-12
- III. Logistics Support to the Navy Planning Process (NPP) ... 6-14
- IV. Logistics Planning Considerations ... 6-18

III. Logistics Command and Control Systems ... 6-19
- I. Navy Logistics System Organization ... 6-19
- II. Theater Logistics Command and Control ... 6-20
- III. Naval Theater Logistics Command and Control ... 6-22
- IV. Multinational Theater Logistics Command and Control ... 6-24
- V. Logistics Information Systems ... 6-26

IV. Naval Theater Distribution ... 6-27
- I. The Defense Supply System ... 6-27
- II. The Naval Supply System ... 6-27
- III. The Marine Corps Supply System ... 6-28
- IV. The Naval Transportation System ... 6-28
  - The Defense Transportation System ... 6-29
- V. The Hub and Spoke Concept of Navy Theater Distribution ... 6-30
- VI. The Logistics Pipeline ... 6-32
- VII. Force Projection ... 6-33

Chap 7: Navy Theater Security Cooperation

I. Navy Theater Security Cooperation Overview ... 7-1
- I. Guidance ... 7-1
- II. Security Cooperation and Assistance - Planning and Activities ... 7-2
  - A. Security Cooperation ... 7-2
  - B. Security Assistance ... 7-3
  - C. Theater Security Cooperation (TSC) ... 7-3
- III. Theater Security Cooperation Activities ... 7-4

II. Security Cooperation Activity Planning ... 7-5
- I. Planning ... 7-5
- II. General Security Cooperation Activity Checklist ... 7-6

III. Building Relationships, Capabilities & Capacity ... 7-9
- Mission Commander ... 7-9
- Social Norms and Customs ... 7-9
- I. Building Relationships ... 7-10
- II. Building Capacities ... 7-12
  - A. Exercises ... 7-12
  - B. Training ... 7-13
  - C. Information Sharing and Intelligence Cooperation ... 7-13
- III. Building Capacity ... 7-15

Index

Index ... Index-1 to Index-4

I. U.S. Navy

Ref: NWC 3153K, Joint Military Operations Reference Guide (Jul '09), chap. 1.

The U.S. Navy was founded on 13 October 1775 and the Department of the Navy was established on 30 April 1798. The Department of the Navy has three principal components: the Navy Department, consisting of executive offices mostly in Washington, D.C., the operating forces, including the Marine Corps, the reserve components, and in time of war, the U.S. Coast Guard (in peace it is a component of the Department of Homeland Security), and the shore establishment.

I. Mission and Purpose

The mission of the Navy is to maintain, train and equip combat-ready Naval forces capable of winning wars, deterring aggression and maintaining freedom of the seas. In order to successfully carry out this broad mission the Navy maintains capability to perform the following key functions as directed by DOD Directive 5100.1:

- Conduct of prompt and sustained combat incident to operations at sea, including operations of sea-based aircraft and land-based naval air components – specifically, forces to **seek out and destroy enemy naval forces** and to suppress enemy sea commerce, to gain and maintain general naval supremacy, to control vital sea areas and to protect vital sea lines of communication, to establish and maintain local superiority (including air) in an area of naval operations, to seize and defend advanced naval bases, and to conduct such land, air, and space operations as may be essential to the prosecution of a naval campaign.

- Provide naval forces, including naval close air support and space forces, for the conduct of joint amphibious operations

- Organize, train, equip, and provide forces for strategic nuclear warfare to support strategic deterrence

- Organize, train, equip, and provide forces for reconnaissance, anti-submarine warfare, protection of shipping, aerial refueling and mine laying

- Provide the afloat forces for strategic sealift

- Provide air support essential for naval operations

- Organize, train, equip, and provide forces for appropriate air and missile defense and space operations unique to the Navy

- Interdict enemy land power, air power, space power and communications through operations at sea

- Conduct close air and naval support for land operations

The U.S. Navy is an instrument of national power, employed to prevent conflict and, if necessary, prevail in war. It is organized, trained and equipped primarily to fight at and from the sea and to influence events on land. U.S. Navy forces are uniquely suited to overcoming diplomatic, military, and geographic impediments to access – an increasing challenge in the contemporary environment – while respecting the sovereign territory of nations. Self-sustaining, seabased expeditionary forces provide persistent presence by operating forward to conduct security cooperation, build partnerships, prevent and deter conflict, communicate our Nation's intent, respond to crises and, when necessary, facilitate the introduction of additional naval, joint, or multinational forces, as well as interagency, international, or non-governmental organizations.

See pp. 2-9 to 2-18 for discussion of the range/spectrum of maritime operations and pp. 2-19 to 2-32 for discussion of the objectives of naval operations.

II. Organization & Command Structure

Ref: NWC 3153K, Joint Military Operations Reference Guide (Jul '09), p. 2-4.

Navy operating forces have a dual chain of command. **Administratively**, they report to the Chief of Naval Operations for the execution of their Title 10 responsibilities (man, train, equip, maintain). All Navy operating units have an administrative chain of command which runs through the appropriate Type Commander (TYCOM): (Surface Forces (SURFOR), Air Forces (AIRFOR), Submarine Forces (SUBFOR), Special Warfare (NAVSPECWARCOM), Expeditionary Forces (Navy Expeditionary Combat Command (NECC)), Auxiliary and sealift forces (Military Sealift Command)). Type Commanders have responsibility for the training and readiness of their "type" forces to include maintenance, manning, and equipping as well as the training and assignment of personnel.

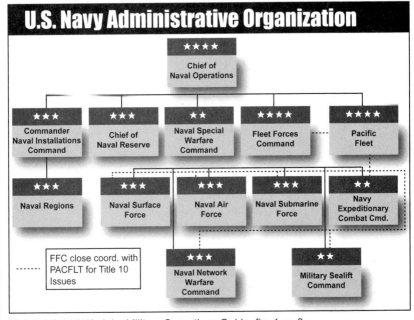

Ref: NWC 3153K, Joint Military Operations Guide, fig. 1, p. 3.

Operationally, Navy forces report to the appropriate Unified Combatant Commanders through the assigned Navy component commander. As Navy units enter the area of responsibility for a Geographic Combatant Commander, they fall under operational control of the appropriate numbered fleet commander (via the Navy component commander).

Naval Shore Forces

The shore establishment provides support to the operating forces (known as "the fleet") in the form of: facilities for the repair of machinery and electronics; communications centers; training areas and simulators; ship and aircraft repair; intelligence and meteorological support; storage areas for repair parts, fuel, and munitions, and medical / dental facilities.

Operational Organization (Numbered Fleets)

The operational level of command for Navy forces is the Numbered Fleet commander. All operational units operate under the Operational Control (OPCON) of the numbered fleet commander within whose area of operations they are located. The Forces for Memorandum contained within the GFMIG specifies the NCC and numbered fleet (if any) assignments to each CCDR.

Maritime Forces

U.S. Navy Operational Organization

Ref: NWC 3153K, Joint Military Operations Guide, fig. 2, p. 4.

Principal Headquarters

Navy principal headquarters are defined as those Navy commands, other than Navy component commands and numbered fleet commands, that have operational responsibilities and report directly to a CCDR. This includes Naval Network Warfare Command (NAVNETWARCOM) and some submarine task force commands. Principal headquarters operate at the operational level supporting the CCDR with functional area expertise.

Task Forces

Below the numbered fleet staff level, the operational chain of command is task oriented. Naval fleets are organized into task forces. Each task force is responsible to the Fleet Commander for certain functions related to the assigned units. The numbering system for Task Forces subordinated to numbered fleets is derivative of the number of the fleet (e.g. designations of 7th fleet Task Forces take the form CTF-7x.) Forces are further organized below the task force level. An individual Carrier Strike Group (CSG) or Amphibious Ready Group (ARG) within a given fleet constitutes a Task Group (e.g. USS RONALD REAGAN CSG operating in the 5th Fleet could be designated CTG 50.1). REAGAN CSG units operating together for a specific task, perhaps the air defense units within the CSG, would receive a separate Task Unit designation (e.g. REAGAN CSG air defense units could be designated CTU 50.1.2). Individual units within the CTU are designated as Task Elements (e.g. the commanding officer of the Guided Missile Destroyer MAHAN operating in the Arabian Gulf could be designated CTE 50.1.2.3). This organizational scheme is scalable and tailorable to meet any operational situations.

See pp. 2-62 to 2-63 for discussion of Navy task organization (task forces, task groups, task units, and task elements). See p. 2-64 for discussion of Navy tactical headquarters to include Navy strike groups (CSG/ESG/SSG) and expeditionary strike forces. See pp. 2-65 to 2-70 for discussion of the Officer in Tactical Command/Composite Warfare Commander.

(Maritime Forces) I. U.S. Navy 1-3

III. Composition & Capability of Major Deployable Elements

Ref: NWC 3153K, Joint Military Operations Reference Guide (Jul '09), p. 11-14.

The basic fighting units within the Navy are ships, submarines and aircraft squadrons which are predominantly O-5 level commands. Navy O-6 level "Major Commands" include larger ships (cruisers (CG), amphibious assault ships (LHD/LHA), destroyer, amphibious or submarine squadrons (DESRON/PHIBRON/SUBRON), air wings (Carrier (CVW) or Type) and aircraft carriers (CVN). Similar type ships (i.e.: destroyers or amphibious ships) are organized under squadrons (DESRONs or PHIBRONs). Aircraft squadrons are organized under wings.

See p. 3-28 for discussion of Navy tactical headquarters to include Navy strike groups (CSG/ESG/SSG) and expeditionary strike forces.

Carrier Strike Group (CSG)

The CSG is a flexible, heavy strike group that can operate in any threat environment, in the littorals or open ocean. CSG capabilities support initial crisis response missions and may be undertaken in non-permissive environments characterized by multiple threats including, but not limited to: anti-ship missiles, ballistic missiles, fighter/attack aircraft, electromagnetic jammers, cruise missile equipped surface combatants, submarines (nuclear and diesel) and terrorist threats. Typically a carrier strike group will have:

- 1 CSG command staff (one or two star flag led staff)
- 1 Destroyer Squadron (DESRON) staff (O-6 led staff)
- 1 Aircraft Carrier (CVN)
- 1 Carrier Air Wing (CVW)
- 5 surface combatant ships (CG/DDG/FFG/LCS)
- Minimum 3 cruise missile land attack (TLAM) capable ships
- Minimum 4 air/missile defense capable ships
- Minimum 2 helo capable ships with embarked helo detachments
- 1 cruise missile land attack (TLAM) / ASW submarine (SSN)
- 1 multi-product logistic support ship (T-AOE, T-AKE or T-AO and T-AE)
- 1 logistics helo detachment

Amphibious Ready Group (ARG)/Marine Expeditionary Unit (MEU)

The ARG/MEU is the routine rotational amphibious force package employed by the Navy Marine Corps team. The baseline ARG/MEU consists of 3 amphibious ships with naval support elements and an embarked MEU without an embarked Flag or General led staff. It may or may not deploy with surface combatants and a submarine depending upon the mission requirements. In the event a requirement exists for an ARG/MEU to be led by a Flag or General officer the amphibious force package will be referred to as an Expeditionary Strike Group. Centered on the flexibility and readiness of a combined Marine Expeditionary Unit and an Amphibious Ready Group, the total ARG/MEU provides operational freedom and expanded warfare capabilities, not only by land with embarked Marines, but at sea as well. *See also pp. 1-22 to 1-24.*

An ARG/MEU consists of the following:

- 1 Amphibious Squadron (PHIBRON) Staff (O-6 led staff)
- 3 Amphibious ships
- 1 Amphibious Assault Ship (LHA or LHD)
- 1 Amphibious Landing Ship Transport Dock (LPD)

- 1 Amphibious Landing Ship Dock (LSD)
- Naval Support Elements (NSE)
- 1 Assault Craft Unit (ACU) detachment with 1 – 4 displacement landing craft (LCU)
- 1 Assault Craft Unit (ACU) detachment with 3 – 6 non-displacement landing craft (LCAC)
- 1 Beachmaster Unit (BMU) detachment
- 1 SAR/logistics helo detachment
- 1 Tactical Air Control squadron (TACRON) detachment
- 1 Fleet Surgical Team (Level II medical capability)
- Marine Expeditionary Unit (MEU) (*See also pp. 1-22 to 1-24.*)
  - Command Element (O-6 led MEU staff)
  - Ground Combat Element (reinforced battalion landing team)
  - Air Combat Element (Composite squadron of fixed and rotary wing aircraft)
  - Logistics Combat Element (Combat Logistics Battalion)
- 3 surface combatant ships (optional depending on resources/mission)
- 2 air/missile defense surface combatant ships (CG/DDG)
- 1 multi-mission surface combatant ship (CG/DDG/FFG/LCS)
- 2 SUW/ASW capable helo detachments (3-4 helos)

Surface Strike Group (SSG)

The SSG is a surface group that can operate independently or in conjunction with other maritime forces. SSG capabilities support crisis response missions or sustained missions and may be employed in limited non-permissive environments characterized by multiple threats. TLAM / standard missile equipped SSGs provide deterrence and immediate contingency response, while maintaining the ability to conduct maritime security operations and other tasks. SSGs are primarily designed to be an independent, sea-based, mobile group that will provide sea control and strike power to support joint and allied forces afloat and ashore. SSG capabilities include passive surveillance and tracking, passive defense and early warning, strike operations, sea control, as well as the multi-platform capabilities inherent within the SSG. When so equipped, SSGs capabilities include maritime ballistic missile defense (BMD). An SSG composition is tailored to its assigned mission, but a nominal SSG is as follows:

- 3 surface ships
- 2 strike / cruise missile land attack (TLAM) capable surface combatants (CG/DDG)
- 1 surface combatant ship or amphibious ship (CG/DDG/LCS/FFG/LHA/LHD/LPD/LSD)

Naval Fleet Auxiliary Force (NFAF)

The 40 ships of Military Sealift Command's Naval Fleet Auxiliary Force are the supply lines to U.S. Navy ships at sea. These ships provide virtually everything that Navy ships need, including fuel, food, ordnance, spare parts, mail and other supplies. NFAF ships enable the Navy fleet to remain at sea, on station and combat ready for extended periods of time. NFAF ships also conduct towing, rescue and salvage operations or serve as floating medical facilities. All NFAF ships are operated by civilian mariners (CIVMARS) under Military Sealift Fleet Support Command (MSFSC). They provide all of the Navy's combat logistics services to the fleet. MSFSC maintains exercises command and control through regional sealift logistics commands located in Norfolk (SEALOGLANT), San Diego (SEALOGPAC), Naples (SEALOGEUR), Bahrain (SEALOGCENT) and Far East (Singapore SEALOGFE).

IV. Capabilities

Naval Ship Designations

When dealing with naval forces, one encounters a series of acronyms designating ship types. These letter designations for warships, adopted by the U.S. Navy around the turn of the century, have since been used worldwide as universal shorthand for warship types. A T-Designation such as T-AE, T-AFS, T-AO, etc., denotes Naval Fleet Auxiliary Force (NFAF) vessels, owned by the USG and operated by the Military Sealift Command (MSC) with civil service merchant marine crews and embarked naval detachments. These vessels are formally known as United States Naval Ships or USNS and can be identified by the blue and gold striping on their stacks.

Naval Ship Designations

AS	Submarine Tender	LHD/LHA	Amphibious Assault Ship
CVN	Carrier (Nuclear Power)	LPD	Amphibious Transport Dock
CG	Guided Missile Cruiser	LSD	Landing Ship, Dock
DDG	Missile (Anti-air) Destroyer	MCM	Mine Countermeasures Ship
FFG	Frigate, Guided Missile	T-AE	Ammunition Ship
PC	Patrol Craft	T-AFS	Combat Stores Ship
SSN	Submarine, Nuclear Attack	T-AH	Hospital Ship
SSBN	Ballistic Missile Submarine	T-AKE	Dry Cargo/Ammunition Ship
SSGN	Guided Missile Submarine	T-AO	Fleet Oiler
LCC	Amphibious Command Ship	T-AOE	Fast Combat Support Ship
LCS	Littoral Combat Ship	T-AGOS	Ocean Surveillance Ship
		T-ATF	Fleet Ocean Tug
		T-ARS	Salvage and Rescue ship

Ref: NWC 3153K, Joint Military Operations Guide, fig. 9, p. 14.

A. Mine Warfare

Naval mines are cheap, reliable and easy to obtain. The "weapons that wait" can pose a significant threat to any military operation where the transportation and the sustainability of forces in theater is accomplished by sea. A mine countermeasures operation is slow and labor-intensive. There are three types of mine countermeasures operations:

- **Mine hunting** - methods to determine where (and just as important where not) the mines are located. It is usually conducted by SONAR or visual means.

- **Minesweeping** - active measure to counter mines. Mines may be: contact, acoustic, magnetic, seismic, pressure or a combination thereof.

- **Mine neutralization** - active destruction of known mine(s). Accomplished by the AN/SLQ-48 submersible vehicle or Navy Mine Countermeasures EOD teams.

Mine Warfare

Ship Type	Class	No.	Warfare Missions	Equipment
MCM Mine Countermeasures Ship	AVENGER 1050 tons 224 ft 39 ft beam 13 kts 84 crew	14	MIW	Two .50 cal mg, AN/SLQ-48 vehicle, AN/SQQ-30 sonar, AN/SQQ-32. Mechanical, acoustic and influence sweep gear.

Ref: NWC 3153K, Joint Military Operations Guide, fig. 15, p. 24.

1-6 (Maritime Forces) I. U.S. Navy

Future Force
Ref: NWC 3153K, Joint Military Operations Reference Guide (Jul '09), p. 13.

While the 2006 Quadrennial Defense Review did not include a specific numerical goal for Navy ships, then CNO (and current CJCS) ADM Michael Mullin, set forth a plan which advocated a 313 ship Navy. This number remains the "target" although as of June 2009 we stand at 283 ships. Regardless of the numerical target, the challenges with maintaining a modern and capable fleet (ships, submarines and aircraft) in the current budget environment and with the persistent difficulties on achieving budget targets during project development and construction are immense. QDR 2006 called for a larger fleet centered on 11 carrier strike groups. Selected Navy procurement programs include:

CVN 78 Program
CVN 78 (future USS GERALD R. FORD) represents the first major investment in CVN design since the 1960s. Improvements over CVN 77 include reduced manning with associated reduced total ownership costs and an increased sortie generation of 25 percent.

DDG 51 Guided Missile Destroyer
Fifty six DDG 51 guided missile destroyers have been delivered with six in various stages of construction.

DDG 1000 Multi-Mission Destroyer
The Navy's new multi-mission destroyer designed to provide precision strike and sustained volume fires to support Joint forces inland and conduct independent attacks against land targets. The DDG 1000 program emphasizes "sensor-to-shooter" connectivity in order to provide a naval or Joint Task Force commander the multi-mission flexibility to engage a wide variety of land targets while simultaneously defeating maritime threats. The ultimate number of DDG 1000s is unknown and could be as low as 2 -3.

Littoral Combat Ship (LCS)
The littoral combat ship will be optimized for war fighting in the littoral environment. It will be a theater-based asset designed to counter enemy access-denial weapons such as diesel-electric-powered submarines, mines, and fast patrol boats. LCS will include modular mission payloads that provide operational flexibility to match the threat. LCS units also will be attached to strike groups as required, to give them enhanced protection when operating near shore. The Navy plans to acquire 55 littoral combat ships. Of this number, one ship is in commission and one is under construction. If approved, the FY'10 budget will purchase three additional ships.

F-35 Joint Strike Fighter (JSF)
The JSF will replace the Navy's aging inventory of F/A-18s and will be capable of meeting all air-to-ground and air-to-air combat requirements. In a bigger context, the JST will replace at least 13 different aircraft for 11 nations.

Virginia Class Submarine
With 5 Virginia class submarines currently commissioned, 5 under construction and 8 more budgeted the Navy is on track for 18 total.

B. Aircraft Carriers & Naval Aviation
Ref: NWC 3153K, Joint Military Operations Reference Guide (Jul '09), pp. 24-25.

A critical piece to U.S. Navy power and capability, U.S. Navy aviation sets the U.S. apart from the rest of the world's navies. The Carrier Air Wing, with its strike fighters (F/A-18 variants Hornets/Super Hornets), electronic warfare (EA-6B Prowler), airborne early warning (E-2C Hawkeye), Logistics (C-2 Greyhound) and rotary wing (SH/MH-60 variants) provide the U.S. with unmatched sea control and power projection capabilities all launched from sovereign U.S. territory in international waters.

Land based naval aviation aircraft fill critical maritime capability needs in the areas of intelligence, surveillance and reconnaissance (ISR), command and control (C2) and logistics. Maritime patrol aircraft (P-3C/AIP) provide long dwell surveillance with an onboard sensor and weapons suites designed to detect and destroy enemy ships and submarines. The P-3 AIP Orion, with its full motion video and real-time downlink to troops on the ground make it a high demand asset in support of land forces. EP-3 Aires provides multi-intelligence collection capability. The P-3 airframe is reaching the end of its serviceable lifespan and will soon be replaced by the multi-mission maritime aircraft (MMA) designated P-8A Poseidon which is built on a commercial 737 airframe. Logistics needs are filled by a variety of aircraft, but predominantly the C-40A (737 airframe). Almost all Navy air logistics capability is resident within the Navy reserve.

Aircraft Carrier (CVN)
The aircraft carrier is a multipurpose platform. It has the flexibility to base various types of aircraft in order to conduct anti-air, strike, anti-surface, and anti-submarine warfare missions simultaneously. The carriers are capable of over 30 kts and have substantial endurance (16 days of 24hr/day aviation fuel). The embarked air wing helps provide protection to both the carrier and the escort ships. The carrier has a limited ability to provide underway replenishment and/or vertical replenishment to support ships in company.

Carrier Air Wing (CVW)
Typical wing composition on a carrier includes:
- VFA / VMFA*(Fighter Attack) - 2 Squadrons of 10 FA-18A+/C Hornets 20
- VFA (Fighter Attack) - 2 Squadrons of 12 FA-18E/F Super Hornets 24
- VAW (Early Warning) - 1 Squadron of 4 E-2C Hawkeyes 4
- VAQ (Electronic Warfare) - 1 Squadron of 4 EA-6B Prowlers 4
- VRC (Onboard delivery) - 1 Detachment of 2 C-2A COD 2
- HSC (SUW/NSW/CSAR/MCM) - 1 Squadron of 8 MH-60Ss 8
- HSM (SUW/ASW) - 1 Squadron of 11 MH-60Rs 11
- TOTAL = 73

Flight Deck Operations
The Carrier Air Wing Commander (CAG) performs major command functions in directing and administering the employment of embarked aviation squadrons. There are two common methods of organizing aircraft launches and recoveries. First, Cyclic Operations, which consists of several scheduled launch/recovery cycles per flight day. A cycle is normally 1.25 to 1.5 hours long, enabling 7-8 cycles in a 12-hour flying day—producing upwards of 120 sorties. Cyclic operations provide predictability for the flight deck, but are inflexible. Aircraft cannot be easily launched or recovered outside of prescribed times due to fueling, rearming, and deck spotting (various aircraft locations on the flight deck) evolutions for the next cycle. Flexible Deck/Battle Flexible Deck Operations mean that aircraft can land anytime, not just once a cycle. For warfare commanders, "flex deck" operations mean greater flexibility to "get an aircraft now."

Naval Aviation

Aircraft Type	Warfare Missions/Armament & Equipment
FA-18A+/C Hornet FA-18 E/F Super Hornet	**Missions** STK, MIW, AAW, ASuW, Air Refueling Tanker (E/F only), SEAD **Arms** Sparrow, Sidewinder, Harpoon, HARM, AMRAAM, SLAM ER, JDAM, JSOW, LGB, MK-80 series bombs, 20mm cannon, mines, cluster munitions ,ATFLIR, SHARP (E/F only), AESA radar (E/F only)
EA-18G Growler	**Missions** Suppression of Enemy Air Defense (SEAD) **Arms/Equip** AMRAAM, HARM, ALQ -99 / ALQ -218 jamming /targeting system, AESA radar.
EA-6B Prowler	**Missions** SEAD **Arms/Equip** HARM, ALQ -99 jamming transmitter and receiver system
E-2C Hawkeye	**Missions** AEW, CCC **Equip** AN/APS 145 radar, Cooperative Engagement Capability (CEC)
C-2A Greyhound	**Mission** Carrier onboard delivery to/from carrier **Arms** None
SH-60B Seahawk LAMPS III	**Missions** ASW, SAR, VERTREP **Arms/Equip** MK -46/50 torpedo, Hellfire Missile, sonobuoys, and door gun, APS 124 search radar, FLIR, ALQ-142 ESM system
SH-60F Seahawk	**Missions** ASW, SAR, VERTREP, **Arms/Equip** MK -46/50 torpedo, Hellfir e, sonobuoys, dipping sonar
HH-60H Seahawk	**Missions** CSAR, VERTREP, Naval Special Warfare (NSW) **Arms/Equip** Door gun, Hellfire missile
MH-60R Seahawk	**Missions** ASW, ASuW, CSAR, VERTREP, NSW, LOG **Arms/Equip** Hellfire, Mk -50 torpedo, doorgun, surface search radar, FLIR, dipping sonar, sonobouys. ALQ-210 ESM system
MH-60S Knighthawk	**Missions** VERTREP, LOG,SAR **Arms/Equip** Hellfire, 2.75" rockets, 2x 7.62 mm doorguns
MH-53 Sea Dragon	**Missions** Airborne mine countermeasure (AMCM), VERTREP, assault support **Arms/Equip** Mk 105 magnetic minesweeping sled, side scan sonar, mechanical minesweeping system
P-3C Orion	**Missions** ASW, ASuW, MIW, C2W, CCC **Arms** MK -80 series bombs, Mk-54 torpedo, mines, Harpoon, Maverick, SLAM ER, sonobuoys, cluster munitions
EP-3 Aries III	**Missions** SIGINT **Equip** AN/APX -134 BIG LOOK radar, COMM / IFF / ESM / IR suites

Ref: NWC 3153K, Joint Military Operations Guide, fig. 16, p. 25.

C. Surface Combatants

Ref: NWC 3153K, Joint Military Operations Reference Guide (Jul '09), p. 15-17.

Surface combatants are multi-mission platforms that can operate independently, in company with a carrier, amphibious forces, or in convoy as escorts. Additional missions include surface fire support, blockade, screening, search and rescue, tracking, ELINT collection, tactical deception, surveillance, evacuation, harassment and landing force. Types of surface combatants include:

Surface Combatants

Ship Type	Class	No.	Warfare Missions	Equipment
CG-47 Guided Missile Cruiser (CG 47 – 51 decommissioned)	TICONDEROGA 9,600 tons 567 ft 55 ft beam 30+ knots 388 crew	22 of 27 built remain in service	AAW, EW, SUW, CCC, USW, STW	2 x 5 in (127 mm) 54 caliber Mark 45 dual purpose guns; 2 x 20 mm Phalanx CIWS Mark 15 guns; CG-52 on replace these with 2 x 61-cell Mark 41 VLS each armed with a mix of ASROC, Tomahawk, SM-2 and ESSM, LAMPS (2)
DDG-51 Destroyer Guided Missile	ARLEIGH BURKE 8300 tons 466 ft 59 ft beam 30+ kts 323 crew	55 of 62 built	AAW, SUW, USW, STW, EW, CCC	VLS for Tomahawk, ASROC, standard missiles, Harpoon (canisters), 5"/54 cal gun, CIWS, torpedo tubes, ESM, LAMP (2) (DDG - 72 and later)
FFG Frigate	OLIVER HAZARD PERRY 4100 tons 445 ft 45 ft beam 29 kts, 200 crew	30 of 40 built remain in service	AAW, SUW, USW, EW, CCC	76 mm gun, CIWS, LAMPS (2)
LCS Littoral Combat Ship	FREEDOM 3079 tons 378 ft, 57 ft beam 45 kts 50 crew	1 (55 planned)	AMW	RAM, 1x57mm gun, 2 MH60 R/S helicopters or 1 MH60 R/S and 3 Firescout VTUAV's.

Ref: NWC 3153K, Joint Military Operations Guide, fig. 11, p. 17.

The Ticonderoga Class Cruiser (CG)//
The 22-ship Ticonderoga-class guided missile cruiser provides the muscle of the surface combatant fleet. The Aegis Weapon System provides unprecedented defensive capability against high performance aircraft and cruise missiles. The SPY-1 phased array radar enables it to control all friendly aircraft units operating in its area and has the capability for surveillance, detection, and tracking of enemy aircraft and missiles. Recent AEGIS system upgrades provide coupled with the SM-3 missile provide proven IAMD capability as demonstrated by USS LAKE ERIE in shooting down a tumbling US satellite. Towed array sonar and LAMPS MK III helos provide ASW capability. Tomahawk vertical launch systems provide land attack capability. Ticonderoga class has hangars for two LAMPS helos and is capable of 30+ knots. Endurance depends on speed (2500 NM at 30 kts to 8000 NM at 14 kts).

The Arleigh Burke Class Destroyer (DDG) (62 ship class)
Like the larger Ticonderoga-class cruisers, DDG 51's combat systems center around the Aegis combat system and the SPY-ID, multi-function phased array radar. The combination of Aegis, the Vertical Launching System, an advanced anti-submarine warfare system, advanced anti-aircraft missiles and Tomahawk, make the Burke class formidable ships. DDG51 class ships, if in receipt of system upgrades, possess IAMD capabilities equivalent to those on the CG-47 class.

The Oliver Hazard Perry Class Frigate (FFG) (40 ship class-21 remain in commission)
Still in the active fleet serving as escorts as well as conducting drug-interdiction or maritime interception operations. They were built to escort amphibious readiness groups, underway replenishment groups and convoys with particular emphasis on Air Defense and ASW. This class of ship has had its MK 13 guided missile launcher removed and thus no longer has the ability to launch either the Standard Missile or the Harpoon Missile. In spite of this, FFGs provide valuable forward presence for security cooperation, maritime security and many other tasks in support of the Joint Force commander.

The Freedom Littoral Combat Ship (LCS) (planned 55 ship class-1 in commission)
LCS is a fast, agile, focused-mission platform designed for operation in near-shore environments yet capable of open-ocean operation. It is designed to defeat asymmetric "anti-access" threats such as mines, quiet diesel submarines and fast surface craft. The LCS 1 Freedom class consists of two different hull-forms: a semi-planing mono-hull, and an aluminum trimaran – designed and built by two industry teams, respectively led by Lockheed Martin and General Dynamics. These seaframes will be outfitted with reconfigurable payloads, called Mission Packages, which can be changed out quickly. Mission packages are supported by special detachments that will deploy manned and unmanned vehicles and sensors in support of mine, undersea and surface warfare missions.

D. Submarine Forces

Ref: NWC 3153K, Joint Military Operations Reference Guide (Jul '09), pp. 17-18.

Attack Submarines (SSN)

Attack submarines are designed to seek and destroy enemy submarines and surface ships; project power ashore with Tomahawk cruise missiles and Special Operation Forces; carry out Intelligence, Surveillance, and Reconnaissance (ISR) missions; support Carrier Strike Groups; and engage in mine warfare. There are three classes of SSNs now in service. Los Angeles (SSN 688) class submarines are the backbone of the submarine force with 45 now in commission. Thirty-one Los Angeles class are equipped with 12 Vertical Launch System tubes for firing Tomahawk cruise missiles. The Navy also has three Seawolf class submarines. USS Seawolf (SSN 21) is exceptionally quiet, fast, well armed, and equipped with advanced sensors. Though lacking Vertical Launch Systems, the Seawolf class has eight torpedo tubes, which can also fire Tomahawks, and can hold up to 50 weapons in its torpedo room. The third ship of the class, USS Jimmy Carter (SSN 23), has a 100-foot hull extension called the multi-mission platform. This hull section provides for additional payload to accommodate advanced technology used to carry out classified research and development and for enhanced warfighting capabilities. The Navy is now building the next-generation SSN, the Virginia (SSN 774) class. The Virginia class is tailored to excel in a wide range of warfighting missions. These include anti-submarine and surface ship warfare; special operation forces; strike; intelligence, surveillance, and reconnaissance; carrier and expeditionary strike group support; and mine warfare. The Virginia class has several innovations that significantly enhance their warfighting capabilities with an emphasis on littoral operations.

Ballistic Missile Submarines (SSBN)

The Navy's fleet ballistic missile submarines, often referred to as "Boomers," serve as an undetectable launch platform for intercontinental missiles. They are designed specifically for stealth and the precision delivery of nuclear warheads. The 14 Ohio class SSBNs have the capability to carry up to 24 submarine-launched ballistic missiles (SLBMs) with multiple independently-targeted warheads. The SSBN's primary weapon is the Trident II D-5 missile. SSBNs are specifically designed for extended deterrent patrols. To increase their at-sea time, the Ohio class have three large-diameter logistics hatches that allow sailors to rapidly transfer supply pallets, equipment replacement modules and machinery components, significantly reducing the time required for in-port replenishment and maintenance. The Ohio class design allows the submarines to operate for 15 or more years between major overhauls. On average, the submarines spend 77 days at sea followed by 35 days in-port for maintenance. Each SSBN has two crews, Blue and Gold, which alternate manning the submarines while on patrol. This maximizes the SSBN's strategic availability while maintaining the crew's training readiness and morale at high levels.

Guided Missile Submarine (SSGN)

Ohio class guided-missile submarines (SSGN) provide the Navy with an unprecedented combination of strike and special operation mission capability within a stealthy, clandestine platform. Armed with tactical missiles and equipped with superior communications capabilities, SSGNs are capable of directly support dozens of Special Operation Forces (SOF) in America's global war on terrorism. The SSGN Program Office converted four SSBNs into SSGNs in a little more than five years at a significantly lower cost than building a new platform and in a similar time span. Each SSGN is capable of carrying up to 154 Tomahawk or Tactical Tomahawk land-attack cruise missiles. The missiles are loaded in seven-shot Multiple-All-Up-Round Canisters (MACs) in 22 of 24 missile tubes. These missile tubes can also accommodate additional stowage canisters for SOF equipment, food, and other consumables, extending the amount of forward-deployed time for on board SOF forces. The missile tubes also promise additional capability to host future payloads such as new types of missiles, unmanned aerial vehicles, and unmanned undersea vehicles. Each submarine has the capacity to host up to 66 SOF personnel at a time. Additional berthing was installed in the missile compartment to accommodate the added personnel and other measures have been taken to extend the amount of time that the SOF forces can spend deployed aboard the SSGNs. Two lock-out chambers (permanently fixed in the first two missile tubes) allow clandestine insertion and retrieval of SOF personnel. Both the Dry Deck Shelter (DDS) and the Advanced SEAL Delivery System (ASDS) can mount atop the lockout chambers, greatly enhancing the SSGNs' SOF capabilities. During conversion, each SSGN received the Common Submarine Radio Room and two High-Data-Rate antennas for significantly enhanced communication capabilities. These additions allow each SSGN to serve as a forward-deployed, clandestine Small Combatant Joint Command Center—a new concept that will be fully tested in the first few SSGN deployments.

Submarines

Ship Type	Class	No.	Warfare Missions	Equipment
SSBN ballistic missile	OHIO 18,700 tons 560 ft 42 ft beam 20+ kts 155 crew	14	USW, SUW, STW, MIW	24 tubes Trident missiles, 4 torpedo tubes
SSGN guided missile	OHIO 18,700 tons 560 ft 42 ft be am 20+ kts 155 crew	4	USW, SUW, STW, MIW	154 Vertical Launch tubes for Tomahawk four torpedo tubes
SSN-688 attack	LOS ANGELES 6,900 tons 360 ft 33 ft beam 20+ kts 133 crew	45 of 55 built remain in service	USW, SUW, STW, MIW	Tomahawk, MK48 torpedoes MK 37 mines
SSN-774 attack	VIRGINIA 7,800 tons 377 ft 34 ft beam 25+ kts 113 crew	5	USW, SUW, STW, MIW	Tomahawk, MK48 torpedoes, mines, unmanned undersea vehicles
SSN-21	SEAWOLF 9,150 tons 353 ft 40 ft beam 25+ kts 133 crew	3	USW, SUW, STW, MIW	Tomahawk, MK48 torpedoes, mines

Ref: NWC 3153K, Joint Military Operations Guide, fig. 12, p. 19.

(Maritime Forces) I. U.S. Navy 1-13

E. Amphibious Warfare

Ref: NWC 3153K, Joint Military Operations Reference Guide (Jul '09), pp. 19-21.

Amphibious warships provide flexible and multi-function support to embarked Marines and support the Marine Corps tenets of Operational Maneuver From the Sea (OMFTS) and Ship to Objective Maneuver (STOM). Much more than just troop transports, they must be able to sail in harm's way and provide a rapid buildup of combat power ashore – via both air and surface – in the face of opposition. This requirement necessitates inherent survivability and self-defense capabilities as well as the ability to seamlessly conduct Task Force operations. Because of their inherent capabilities, these ships have been and will continue to be called upon to also support humanitarian and other contingency missions on short notice.

Tarawa-class LHAs and Wasp-class LHDs

The Tarawa-class LHAs and Wasp-class LHDs provide the Marine Corps with a means of ship-to-shore movement by helicopter in addition to movement by landing craft. LHDs/LHAs have extensive storage capacity and can accommodate Landing Craft Utility (LCU) and Landing Craft, Air Cushion (LCAC) boats. They embark, maintain and operate various helos (CH-46, CH-53, AH-1, UH-1, MH-60S), AV-8s and MV-22s from their aviation space with a typical combat embarkation being 31 aircraft. They carry large numbers of troops (1000+) and robust C2 spaces for the embarked Navy and Marine command elements. They also have large hospital capability.

Austin-class LPDs

The versatile Austin-class LPDs provide substantial amphibious lift for Marine troops and their vehicles and cargo. Additionally, they serve as the secondary aviation platform for Expeditionary Strike Groups. The oldest of the class turned 42 this year. As the new San Antonio-class LPDs enter service, Austin-class LPDs will be decommissioned. The ships of the LPD 17 class are a key element of the Navy's seabase transformation. Collectively, these ships functionally replace over 41 ships (LPD 4, LSD 36, LKA 113, and LST 1179 classes of amphibious ships) providing the Navy and Marine Corps with modern, seabased platforms that are networked, survivable, and built to operate with 21st century transformational platforms, such as the MV-22 Osprey, the Expeditionary Fighting Vehicle (EFV), and future means by which Marines are delivered ashore.

Dock Landing Ships

Dock Landing Ships support amphibious operations including landings via Landing Craft Air Cushion (LCAC), conventional landing craft and helicopters, onto hostile shores. These ships transport and launch amphibious craft and vehicles with their crews and embarked personnel in amphibious assault operations. LSD 41 was designed specifically to operate LCAC vessels. It has the largest capacity for these landing craft (four) of any U.S. Navy amphibious platform. It will also provide docking and repair services for LCACs and for conventional landing craft. LSD 49 – 52 were modified as "Cargo Variant" in order to increase cargo capacity at the expense of landing craft space. The ships differ from LSD 41 – 48 by reducing its well deck size and associated LCAC embarkation capacity (from 4 to 2) in favor of additional cargo capacity.

Amphibious Warfare

Ship Type	Class	No.	Warfare Missions	Equipment
LHA Amphibious Assault Ship	TARAWA 39,300 tons 820 ft 106 ft beam 24 kts 950 crew 1900 troops	2	AMW	6' 25mm MG, CIWS, RAM, NSSMS, can take LCU or LCAC. 2 helos, 6 AV-8A. Good medical capability.
LHD Amphibious Assault Ship	WASP 40,500 tons 844 ft 106 ft beam 22+ kts 1015 crew 1875 troops	8	AMW	Same as Tarawa, can take AV-8B, three LCAC, Outstanding C5I for AMW. 8/50 cal gun.
LSD Dock Landing Ship (2 varients)	WHIDBEY ISLAND (8) & HARPERS FERRY (4) 15,800 tons 609 ft 84 ft beam 20+ kts 340 crew 340 troops	12	AMW	CIWS, helo capable, 4 LCAC capable, LCU also. HARPERS FERRY class carriers only 2 LCAC
LPD-4 Amphibious Transport Dock	AUSTIN 17,000 tons 570 ft 84 ft beam 21 kts 388 crew 900 troops	5	AMW	CIWS, 3"/50 cal guns, large flight deck, large troop capacity. All vessels 35+ years old.
LPD-17	SAN ANTONIO 24,900 tons 684 ft 105 ft beam 22+ kts 495 crew	4	AMW	RAM, VLS, CIWS, 50 cal mg, 2 LCAC, 2 surgical operating rooms, large flight deck.
LCC Amphibious Command Ship	BLUE RIDGE 18,874 tons 634 ft 108 ft beam 23 kts 842 crew	2	AMW, C3	CIWS, command and control ship for amphib ops, fleet flagships (6th and 7th fleets), helo capable except for CH-53.

Note: LCAC - 200 tons, 88 ft, 47 ft beam, 70 ton max payload (1 x M1A1 tank); range 200 miles @ 40 kts (sea state and payload dependent)

Ref: NWC 3153K, Joint Military Operations Guide, fig. 13, p. 20-21.

F. MSC Support Ships
Ref: NWC 3153K, Joint Military Operations Reference Guide (Jul '09), pp. 21-23.

Military Sealift Command provides critical support to the Fleet with government owned or contracted vessels manned by contracted or civil service mariners. Most ships have embarked military detachments. Government owned MSC ships are designated USNS while contracted vessels are referred to by the appropriate maritime designator (MV-motor vessel; SS-Steamship; etc). Regardless of their ownership status, all MSC ships fall under the OPCON of the appropriate military commander and they enjoy sovereign immunity under international law. MSC ships are organized in 4 major areas:

Naval Fleet Auxiliary Force (NFAF)
Supply lines to U.S. Navy ships at sea. NFAF ships provide underway replenishment (UNREP) services for fuel, food, ammunition, spare parts and other supplies to keep the U.S. Navy fleet at sea, on station and operating at the highest possible tempo. They also have specialized fleet ocean tugs, rescue-salvage ships and hospital ships to support fleet requirements.

Special Mission Ships (SMS)
Military Sealift Command's Special Mission Program has 25 ships that provide operating platforms and services for a wide variety of U.S. military and other U.S. government missions. Oceanographic and hydrographic surveys, underwater surveillance, missile tracking, acoustic surveys, command and control, and submarine and special warfare support are among the missions these ships carry out.

Prepositioning Ships
Military Sealift Command's Prepositioning Program is an essential element in the U.S. military's readiness strategy. Afloat prepositioning strategically places military equipment and supplies board ships located in key ocean areas to ensure rapid availability during a major theater war, a humanitarian operation or other contingency. MSC's 32 prepositioning ships support the Army, Navy, Air Force, Marine Corps and Defense Logistics Agency. More information on prepositioning ships is in the Strategic Lift chapter of this handbook.

Sealift Program
Military Sealift Command's Sealift Program provides high-quality, efficient and cost-effective ocean transportation for the Department of Defense and other federal agencies during peacetime and war. More than 90 percent of U.S. war fighters' equipment and supplies travel by sea. The program manages a mix of government-owned and long-term-chartered dry cargo ships and tankers, as well as additional short-term or voyage-chartered ships. More information on the Sealift program is in the Strategic Lift chapter of this handbook.

Military Sealift Command

Ship Type	Class	No.	Warfare Missions	Equipment
T-AOE Fast Combat Support Ship	SUPPLY 48,000 tons 752 ft. 107 ft. beam 29kts 670 crew	4	Combat Logistics (ammo, supply, & petroleum products)	(2) MH - 60 helos, 150,000 barrels oil, 1800 tons ammo, 400 tons dry stores 250 tons frozen stores
T-AE Ammunition Ship	KILUAUEA 18,088 tons 564 ft. 81 ft beam 20 kts 383 crew	4	Ammunition Replenishment	(2) MH- 60 helos Deliver ammo via helo or ship- ship cable
T-AO Fleet Oiler	HENRY J. KAISER 42,000 tons 677 ft. 97 ft beam 20kts 80 crew	14	Petroleum Product Replenishment	180,000 barrels fuel, 600 tons cargo ammo
T-AFS Combat Stores Ship	MARS & SIRIUIS 48,000 tons 581 ft 79 ft beam 20 kts 110-130 crew	3	Combat Logistics (food, stores, & repair parts)	(2) MH- 60helos 596,000 cubic feet of store space
T-AKE Dry Cargo & Ammunition Ship	LEWIS AND CLARK 41,000 tons 689 ft 105 ft beam 20kts 130 crew	6	Ammunition and stores replenishment Also Sea Basing	(2) MH - 60 helos, Delivery via helo or ship-ship cable 783,000 cubic feet of store space plus 18,000 bbls of fuel.
T-AH Hospital Ship	MERCY 69,000 tons 894 ft 105 ft beam 17 kts 63 civil service mariners 956 Naval medical staff 258 Naval support staff	2 Normally kept in reduced operating status		12 full-equipped operating rooms, 1000 bed hospital facility, digital radiological services, medical laboratory pharmacy, optometry lab, CAT-scan, two oxygen producing plants. Helo deck capable of landing large military helicopters. Side ports to embark patients at sea.
T-ATF Fleet Ocean Tugs	POWHATAN 2,300 tons 240 ft 42 ft beam 15 kts 20 crew	4	Towing, salvage, and recovery	Each vessel is equipped with a ten-ton capacity crane and a bollard pull of at least 54 tons. A deck grid is fitted aft which contains one-inch bolt receptacles spaced 24 inches apart. This allows for the bolting down of a wide variety of portable equipment. There are two fire pumps supplying three fire monitors with up to 2,200 gallons of foam per minute. A deep submergence module can be embarked to support naval salvage teams.
T-AGOS	VICTORIOUS & IMPECCABLE	5	Collect, process and transmit accoustic data in support of undersea surveillance requirements	2 surveillance towed-array sensor system (SURTASS)
AS Submarine Tender	L.Y. SPEAR 23,000 tons 645 ft 85 ft beam 20 kts 1325 crew	2	Repair	Virtually no defensive systems, outstanding at sea repair capability. Can repair surface ships as well.
ARS* Rescue and Salvage Ship	SAFEGUARD 3282 tons 255 ft 51 ft beam 14 kts 100 crew		Rescue & Salvage	2-25mm guns, 7.5 ton and 40 ton booms, hauling force of 150 tons, conducts firefighting, diving, salvage, and towing operations

*One ship of the class has been transferred to MSC - USNS GRASP (T-ARS-51).

Ref: NWC 3153K, Joint Military Operations Guide, fig. 14, p. 22.

Maritime Forces

(Maritime Forces) I. U.S. Navy 1-17

V. Navy Reserve

Ref: NWC 3153K, Joint Military Operations Reference Guide (Jul '09), pp. 27-28.

The United States Navy Reserve is organized into two general types of units:

1. Commissioned Units

Reserve units, with organic equipment such as aviation squadrons, Naval Reserve Force (NRF) ships, cargo handling battalions, mobile inshore undersea warfare units, and mobile construction battalions. These units are tasked to deliver a complete operational entity to the operating force and are commanded by either Active or Reserve component officers and staffed primarily by Selected Reserve Personnel. Naval Reserve Force ships are under operational control of the Commanders-in-Chief, Atlantic or Pacific Fleet, while those designated as Reserve Frigate Training ships come under the operational control of Commander, Surface Group Six, who is assigned to Commander, Naval Reserve Force. Thirty two percent of all Selected Reserve personnel are assigned to commissioned units.

2. Augmentation Units

Augmentation units are units that augment active component units with trained personnel. Such units are tailored to augment designated ships, special warfare commands, intelligence staffs, etc. Their function is to allow for peak operations for an indefinite period of time. They also provide surge capability, and then sustain the high level of activity to support deployed forces.

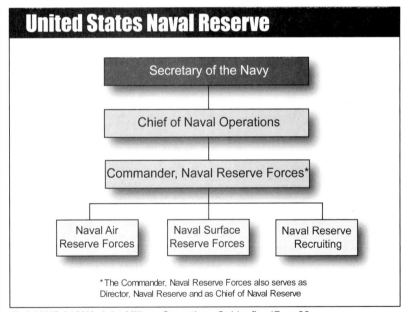

Ref: NWC 3153K, Joint Military Operations Guide, fig. 17, p. 28.

II. U.S. Marine Corps

Ref: NWC 3153K, Joint Military Operations Reference Guide (Jul '09), chap. 3 and www.usmc.mil.

The Marine Corps is organized as a general purpose "force in readiness" to support national needs. Deploying for combat as a combined-arms Marine Air/Ground Task Force (MAGTF), the Marine Corps provides the Nation with a responsive force that can conduct operations across the spectrum of conflict. The Marine Corps' most important responsibility is to win the nation's battles.

I. Mission of the U.S. Marine Corps

The primary mission, as stated in the National Security Act of 1947, "...is to provide Fleet Marine Forces of combined arms together with supporting air components, for service with the fleet..." This act also states that the Marine Corps minimum peacetime structure shall consist of "...not less than three combat divisions and three aircraft wings, and such other land combat, aviation and other services as may be organic therein..." In addition, the Marine Corps maintains a fourth Marine division and aircraft wing in reserve.

II. Organization and Structure

The Marine Corps is divided into four broad categories:

A. Headquarters, U.S. Marine Corps

Headquarters, U.S. Marine Corps (HQMC) consists of the Commandant of the Marine Corps and those staff agencies that advise and assist him in discharging his responsibilities prescribed by law and higher authority. The Commandant is directly responsible to the Secretary of the Navy for the total performance of the Marine Corps. This includes the administration, discipline, internal organization, training, requirements, efficiency, and readiness of the service.

Also, as the Commandant is a member of the Joint Chiefs of Staff, HQMC supports him in his interaction with the Joint Staff. The Commandant also is responsible for the operation of the Marine Corps material support system.

B. Operating Forces

Operating forces — the heart of the Marine Corps — comprise the forward presence, crisis response, and combat power that the Corps makes available to U.S. unified combatant commanders. The Marine Corps has established three permanent combatant-level service components in support of unified commands with significant Marine forces assigned: U.S. Marine Corps Forces Command (MARFORCOM), U.S. Marine Corps Forces Pacific (MARFORPAC), and U.S. Marine Corps Forces, Special Operations Command (MARSOC). The Commander, MARFORCOM is assigned to the Commander, U.S. Joint Forces Command (JFCOM). He provides the 2d Marine Expeditionary Force (II MEF) and other unique capabilities to JFCOM. Likewise, the Commander, MARFORPAC is assigned to the Commander, U.S. Pacific Command. He provides I and III MEFs to PACOM. The Commander, MARSOC is assigned to the Commander, Special Operations Command (SOCOM). He provides assigned forces to SOCOM. These assignments reflect the peacetime disposition of Marine Corps forces.

Marine forces are apportioned to the remaining geographic combatant commands — the U.S. Southern Command (SOUTHCOM); U.S. Northern Command (NORTHCOM); U.S. European Command (EUCOM); U.S. Central Command (CENTCOM); U.S. Africa Command (AFRICOM); and U.S. Forces Korea (USFK) for contingency planning, and are provided to these commands when directed by the Secretary of Defense. Listed below are the Marine Corps service component headquarters.

- **Marine Forces Command (MARFORCOM).** Located in Norfolk, VA and commanded by a three-star general. Commander, MARFORCOM provides the II Marine Expeditionary Force (MEF) and activated Marine Forces Reserve units to the Commander, USJFCOM.
- **Marine Forces Pacific (MARFORPAC).** Located at Camp H.M. Smith, HI and commanded by a three-star general. Commander, MARFORPAC provides I and III MEFs to Commander, USPACOM.
- **Marine Forces, Special Operations Command (MARSOC).** Located at Camp Lejeune, NC and currently commanded by a two-star general. Commander, MARSOC provides assigned forces to Commander, USSOCOM.
- Other Service Component Commands. The Marine Corps maintains a service component headquarters with U.S. Central Command, U.S. European Command, U.S. Southern Command, U.S. Africa Command, and U.S. Northern Command. Marine forces are apportioned, but not assigned to these unified geographic commands.
- **Marine Forces Central Command (MARCENT).** A three-star headquarters located in Tampa, FL. Commander, MARCENT also serves as Commanding General, I MEF, located at Camp Pendleton, CA.
- **Marine Forces European Command (MARFOREUR).** A one or two-star headquarters located in Stuttgart, Germany.
- **Marine Forces Southern Command (MARFORSOUTH).** A one or two-star headquarters located in Miami, FL.
- **Marine Forces Africa Command (MARFORAF).** A one or two-star headquarters currently located in Stuttgart, Germany. Commander, MARFORAF also commands MARFOREUR.
- **Marines Forces Northern Command (MARFORNORTH).** A three-star headquarters located in New Orleans, LA. Commander, MARFORNORTH also commands Marine Forces Reserve.

C. Marine Corps Reserve

The United States Marine Corps Reserve (MARFORRES) is responsible for providing trained units and qualified individuals to be mobilized for active duty in time of war, national emergency, or contingency operations, and provide personnel and operational tempo relief for active component forces in peacetime. MARFORRES, like the active forces, consists of a combined arms force with balanced ground, aviation, and combat service support units. MARFORRES is organized under the Commander, MARFORRES. Their headquarters is located in New Orleans, LA.

D. Supporting Establishment

Marine Corps bases and stations — often referred to as the "fifth element" of the MAGTF — comprise the personnel, bases, and activities that support the Marine Corps' operating forces. This infrastructure consists primarily of 15 major bases and stations in the United States and Japan, as well as the personnel, equipment, and facilities required to operate them. These bases and stations fall under several regional commands to include Marine Corps Installations-East (MCIEast), MCI-West, and MCI-Pacific.

The supporting establishment also includes the Marine Corps Logistics Command (MCLC) and Training and Education Command (TECOM). Additionally, the supporting establishment includes civilian activities and agencies that support Marine forces.

III. Marine Air/Ground Task Force (MAGTF)

Ref: NWC 3153K, Joint Military Operations Reference Guide (Jul '09), p. 77-78.

The MAGTF is a balanced, air-ground combined arms task organization of Marine Corps forces under a single commander, structured to accomplish a specific mission or a number of missions across the range of military operations (ROMO).

MAGTFs are flexible, task-organized forces that are capable of responding rapidly to a broad range of combat, crisis, and conflict situations. MAGTFs vary in size and capability according to the mission, threat, and operating environment. The MAGTF is primarily organized and equipped to conduct amphibious operations as part of naval expeditionary forces. MAGTFs are also capable of sustained combat or peace operations ashore.

Each MAGTF, regardless of size or mission, has the same basic structure. A MAGTF consists of four core elements: Command, Aviation, Ground, and Logistics Combat Element. As the Ground Combat Element grows in size, the Aviation, Logistics, and Command elements typically become larger.

There are four basic MAGTF organizations: Marine Expeditionary Force (MEF), Marine Expeditionary Brigade (MEB), Marine Expeditionary Unit (MEU), and Special Purpose MAGTFs (SPMAGTF).

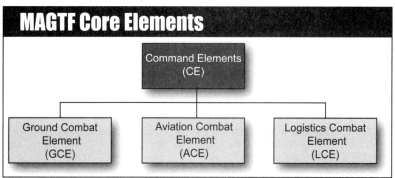

Ref: NWC 3153K, Joint Military Operations Guide, fig. 18, p. 77.

1. Command Element (CE)
The CE contains the MAGTF headquarters and other units that provide intelligence, communication, and administrative support. The CE is scalable and task organized to provide the command, control, communications, computers, intelligence, and joint interoperability necessary for effective planning and execution of operations.

2. Ground Combat Element (GCE)
The GCE is task organized to conduct ground operations to support the MAGTF mission. This element includes infantry, artillery, reconnaissance, armor, light armor, assault amphibian, engineer, and other forces, as needed. The GCE can vary in size and composition.

3. Aviation Combat Element (ACE)
The ACE conducts offensive and defensive air operations and is task organized to perform those functions of Marine aviation required to support the MAGTF mission.

4. Logistics Combat Element (LCE)
The LCE is task organized to provide the full range of combat service support functions and capabilities necessary to maintain the continued readiness and sustainability of the MAGTF as a whole. The LCE may vary in size and composition.

IV. Types of MAGTFs (MEF, MEB, MEU, SPMAGTF)

Ref: NWC 3153K, Joint Military Operations Reference Guide (Jul '09), pp. 78-82.

There are four basic MAGTF organizations: Marine Expeditionary Force (MEF), Marine Expeditionary Brigade (MEB), Marine Expeditionary Unit (MEU), and Special Purpose MAGTFs (SPMAGTF).

MAGTF SIZE (Largest to Smallest)	ELEMENT		
	GCE	ACE	LCE
Marine Expeditionary Force (MEF)	Marine Division (MARDIV)	Marine Aircraft Wing (MAW)	Marine Logistics Group (MLG)
Marine Expeditionary Brigade (MEB)	Marine Regiment (RLT or RCT)	Marine Aircraft Group (MAG)	Combat Logistics Regiment (CLR)
Marine Expeditionary Unit (MEU)	Battalion Landing Team (BLT)	Reinforced Helicopter/Fixed Wing Squadron	Combat Logistics Battalion (CLB)
Special Purpose MAGTF (SPMAGTF)	Elements of a MARDIV	Elements of a MAW	Elements of a MLG

A. Marine Expeditionary Force (MEF)

The MEF is the largest standing MAGTF and the principal Marine Corps war fighting organization. It is capable of missions across the range of military operations through amphibious and sustained operations ashore in any environment. Each MEF is comprised of a Command Element (CE), Marine Division (GCE), Marine Aircraft Wing (ACE), and a Marine Logistics Group (LCE). The three standing MEFs provide a reservoir of capabilities and combat power from which all smaller MAGTFs are formed. There are three standing MEFs: I MEF, II MEF, and III MEF.

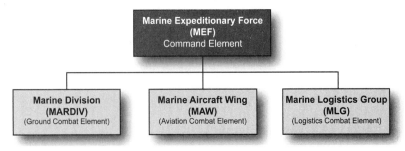

Each MEF is commanded by either a Lieutenant General or Major General and consists of anywhere from 20,000 to 90,000 personnel. A MEF generally deploys on U.S. Navy amphibious ships with support from Military Sealift Command (MSC) and Maritime Pre-positioned Force (MPF) vessels, as well as Air Mobility Command (AMC). A MEF deploys with 60 days of supplies for sustained operations ashore.

I Marine Expeditionary Force (I MEF)
Headquartered at Camp Pendleton, CA; units located in California and Arizona:
- 1st Marine Division (1st MARDIV) - Camp Pendleton, CA
- 3rd Marine Aircraft Wing (3rd MAW) - Miramar San Diego, CA
- 1st Marine Logistics Group (1st MLG) – Camp Pendleton, CA

II Marine Expeditionary Force (II MEF)
Headquartered at Camp Lejeune, NC; units located in North and South Carolina:
- 2nd Marine Division (2nd MARDIV) - Camp Lejeune, NC
- 2nd Marine Aircraft Wing (2nd MAW) - Cherry Point, NC
- 2nd Marine Logistics Group (2nd MLG) - Camp Lejeune, NC

III Marine Expeditionary Force (III MEF)
Headquartered in Okinawa, Japan; units located in Hawaii and Japan:
- 3rd Marine Division (3rd MARDIV) - Okinawa, Japan
- 1st Marine Aircraft Wing (1st MAW) - Okinawa, Japan
- 3rd Marine Logistics Group (3rd MLG) - Okinawa, Japan

B. Marine Expeditionary Brigade (MEB)

The MEB is a medium sized non-standing MAGTF that is task organized to respond to a full range of crises, from forcible entry to humanitarian assistance. MEBs are not standing organizations that are formed only in times of need. An example is post 9/11; the 4th MEB and 2nd MEB were formed to respond to combat and peacekeeping contingencies in Afghanistan and Iraq.

A MEB is commanded by a Brigadier General or Major General and consists of anywhere from 3,000 to 20,000 personnel. It also generally deploys on U.S. Navy amphibious ships with support from MSC and MPF vessels. It deploys with 30 days of supplies for sustained operations ashore.

Marine Expeditionary Brigade elements consist of:
- Command Element (CE)
- Ground Combat Element (GCE) = Marine Regiment (RLT or RCT)
- Aviation Combat Element (ACE) = Marine Aircraft Group (MAG)
- Logistics Combat Element (LCE) = Combat Logistics Regiment (CLR)

Types of MAGTFs (Continued)

Ref: NWC 3153K, Joint Military Operations Reference Guide (Jul '09), pp. 78-82.

C. Marine Expeditionary Unit (MEU)

The standard forward deployed Marine expeditionary organization. A MEU is task organized to be a forward deployed presence and designed to be the "first on the scene" force. A MEU is capable of a wide range of small scale contingencies, to include:

- Amphibious raids/limited objective attacks
- Noncombatant evacuation operations (NEO)
- Security operations /Counter-Intelligence operations
- Tactical recovery of aircraft and/or personnel (TRAP)
- Humanitarian/civic action operations

Prior to deployment, a MEU undergoes an intensive six-month training program, focusing on its conventional and maritime operations missions. The training culminates with a thorough evaluation and certification. In addition to possessing conventional capabilities, a MEU, when augmented with a Marine Special Operations Company (MSOC) provided by MARSOC, may be designated as a MEU (Special Operations Capable) or MEU(SOC). A MEU is commanded by a Colonel and consists of anywhere from 1,500 to 3,000 personnel. MEUs typically deploy for six-month deployments aboard U.S. Navy amphibious ships. They deploy with it 15 days of supplies for sustained operations ashore.

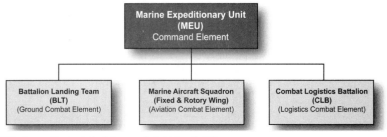

MEU elements consist of (see p. 1-4 and p. 1-21):

- Command Element (CE)
- Ground Combat Element (GCE) = Battalion Landing Team (BLT)
- Aviation Combat Element (ACE) = Composite Marine Squadron (Rotary wing with a complement of fixed wing aircraft, depending on mission)
- Logistics Combat Element (LCE) = Combat Logistics Battalion (CLB)

D. Special Marine Air/Ground Task Force (SPMAGTF)

The SPMAGTF is a non-standing MAGTF temporarily formed to conduct a specific mission. It is normally formed when a standing MAGTF is unavailable or inappropriate. Their designation derives from the mission they are assigned, the location in which they will operate, or the name of the operation in which they will participate. (i.e., SPMAGTF Somalia, SPMAGTF Katrina etc.). These MAGTFs vary in size and composition based on the individual mission. As with the MEU, the SPMAGTF may be the forward element of a larger MAGTF. Regardless of size, all MAGTFs are "expeditionary" forces. An expeditionary force is a capability, vice a structure. Any size MAGTF could be referred to as a Marine "expeditionary" capability.

V. USMC Concepts and Capabilities

Expeditionary in nature, with special emphasis in conducting a wide range of operations from the sea, and providing a combined arms team and a national swing force, the United States Marine Corps is capable of conducting worldwide stability operations; limited objective operations; amphibious operations and sustained operations ashore. USMC doctrine is based on maneuver warfare. Maneuver seeks to shatter enemy cohesion through a series of rapid, violent, and unexpected actions. Operational mobility, surprise, speed, and flexibility allow MAGTFs to pit their strengths against enemy vulnerabilities. Emphasis is on Operational Maneuver from the Sea (OMFTS).

The MAGTFs provide a continuum of capabilities to support naval, unified combatant commanders, and national requirements. These MAGTFs are joined by other unique Marine capabilities to help the Corps deal with a full range of conventional and unconventional threats and assignments. Listed below are a few of these other unique capabilities.

1. Global Response Forces (GRF)

The Marine Corps GRFs are standing contingency forces that can respond rapidly to emerging crises anywhere in the world. COMMARFORPAC and COMMARFORCOM maintain GRFs in continuous states of readiness, enabling the Corps to provide combatant commanders with the appropriate GRF as soon as the SecDef directs. Marine GRFs provide great versatility: they can be immediately employed from U.S. Navy amphibious ships, fly into a crisis area and marry-up with equipment from the Maritime Prepositioning Force or conduct security and enabling functions as the lead element of a MAGTF. Additionally, the Chemical and Biological Incident Response Force (CBIRF) – a unique Marine Corps capability – maintains a high state of readiness to respond.

2. Maritime Prepositioning Force (MPF)

The MPF is a strategic power-projection capability that combines the lift capacity, flexibility, and responsiveness of surface ships with the speed of strategic airlift. Strategically positioned around the globe, the Maritime Prepositioning Ships (MPS) of the MPF provide the Geographic Combatant Commanders (GCC) with persistent forward presence and rapid crisis response. The MPF is organized into three Maritime Prepositioning Ships Squadrons (MPSRON): MPSRON-1, based in the Mediterranean; MPSRON-2, based in the Indian Ocean; and MPSRON-3, based in the Western Pacific. Each interoperable MPSRON is designed to couple with a Fly-In Echelon (FIE) to support the rapid closure of a Marine Expeditionary Brigade (MEB). In addition to force closure, each MPSRON can sustain a MEB-size force for 30 days.

3. Response

When needed, an MPSRON moves to the crisis region and offloads either in port or in-stream off-shore. Offloaded equipment and supplies are then married up with Marines and sailors arriving at nearby airfields. The end result is a combat-ready MAGTF rapidly established ashore, using minimal reception facilities. The MAGTF combat capability provided by MPF supports GCC military operations that defeat adversaries and win wars, but has also supported regional crises that require rapid and effective humanitarian assistance and disaster relief.

4. Marine Special Operations Command (MARSOC)

Although the notion of a Marine "special forces" contribution to the U.S. Special Operations Command (USSOCOM) was considered as early as the founding of USSOCOM in the 1980s, it was resisted by the Marine Corps. After a three-year development period, the Corps agreed in 2006 to supply a 2,600-man unit to USSOCOM.

5. Logistics

MARSOC, as the U.S. Marine Corps service component of USSOCOM, trains, organizes, equips, and when directed by CDRUSSOCOM, deploys task organized, scalable, and responsive U.S. Marine Corps special operations forces worldwide in support of combatant commanders and other agencies. MARSOC Headquarters is located at Marine Corps Base, Camp Lejeune, N.C.

VI. Major Marine Corps Equipment

Ref: NWC 3153K, Joint Military Operations Reference Guide (Jul '09), p. 85.

USMC Ground Equipment

Combat Vehicles	Description
AAV Amphibious Assault Vehicle	Troop carrier: 18 troops, 3 crew or 10k cargo. Comes in C^2 variant and a recovery vehicle variant. Water 8+ MPH, land 45+ MPH, Range (land) 300 miles.
EFV Expeditionary Fighting Vehicle	Troop carrier scheduled to replace the AAV. Fielding programmed for FY 2010. 17 troops, 3 crew. C^2 variant, recovery vehicle variant, and other variants planned or under development. Water 25 knots, Land 45 MPH.
LAV Light Armored Vehicle	Serves as assault and recon vehicle. Provides tactical mobility. Amphibious (for river crossings), 6 MPH water, 62 MPH land. Crew of three, 4 troops. May come as anyone of three variants; C^2, logistics, and recovery. Equipped with 25 mm cannon. TOW, mortar (81 mm), air defense and logistics variants.
M1A1 Abrams Tank	A stabilized 120mm main gun, powerful 1,500 hp turbine engine, and special armor, make the M1A1 particularly suitable for attacking or defending against armor forces. Equipped with 50 cal and 7.62 MGs; Speed - 41.5 mph; Weight - 67.6 tons; Crew - 4
M198 Howitzer M777 Howitzer	The M198 is a 155mm medium-towed artillery piece. Sustained rate of fire - 1 rd/minute, max rate - 4 rds/minute; Range – 18,150 meters (30K with rocket assist); Crew: 11-15 men. By 2010, the M777 155mm howitzer will replace the M198. The M777 is 5,000 lbs lighter, and thus more expeditionary. Firing ranges are similar; however, the M777's more advanced fire direction system will improve firing precision, speed, and flexibility.

USMC Aircraft

Aircraft	Warfare Missions
AV-8B Harrier	630 MPH, Ferry range 2100 NM. V/STOL aircraft, short or vertical launch capability. Ordnance load 16,500 lbs. Night operating capability.
F/A-18	Speed - supersonic; Ferry Range - 2,000 NM (2300 miles), Armament – carries an assortment of air-to-air and air-to-ground weapons, including Sparrows, Sidewinders, AMRAAMs, Harpoons, and Maverick missiles; GBU / CBU bombs; Night operating capability.
EA-6B	Max Speed - .99 Mach; Cruise Speed - .72 Mach; Range – 850 NM (978 miles), unlimited with aerial refueling; Armament – ALQ 99 Tactical Jamming System; USQ-113 Communications Jammer, High Speed Anti-Radiation Missile (HARM); Mission - Airborne Electronic Warfare (EW)
CH-53E	Speed – 150 Kts (173 MPH); Range – 540 NM (621 miles) w/o refueling, unlimited with aerial refueling; Crew – 4; Payload – 55 troops, or 70k lb cargo; Mission – cargo movement primary, troop assault secondary
CH-46E	Speed - 145 Kts (167 MPH); Range - 132 NM (152 miles); Crew – 4; Payload -- 9-16 Troops, 15 litters, or 2k-4k lb cargo; Mission – troop assault primary, cargo movement secondary. CH-46 is being replaced by V-22 Osprey.
V-22 Osprey	275 MPH cruise, 300 MPH dash, 24 Troops or 12 litters, cargo capacity: 10k internal, 15k external. Has potential SOF application. 1st Osprey SQDN became operational in FY 05.
AH-1W Sea Cobra	190 kts, range 256 NM, crew of two. 20 mm nose gun turret, 2.75" and 5.0" rockets, Hellfire and TOW missiles, Sidewinder and Sidearm missiles.
UH-1N Huey	121 kts, range 172 NM. 8-10 troops or 6 litters may be armed if required.

III. U.S. Coast Guard

Ref: NWC 3153K, Joint Military Operations Reference Guide (Jul '09), chap. 5 and www.uscg.mil.

I. USCG Organization

The United States Coast Guard (USCG) is a military branch of the United States involved in maritime law, mariner assistance, and search and rescue, among other duties of coast guards elsewhere. One of the seven uniformed services of the United States, and the smallest armed service of the United States, its stated mission is to protect the public, the environment, and the United States economic and security interests in any maritime region in which those interests may be at risk, including international waters and America's coasts, ports, and inland waterways.

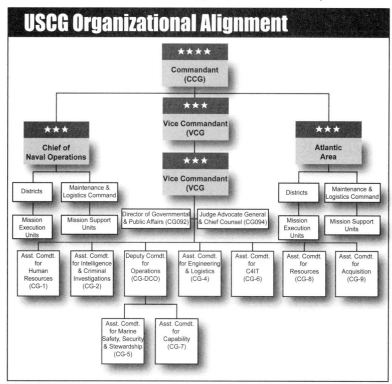

Ref: NWC 3153K, Joint Military Operations Guide, fig. 27, p. 121.

Today's U.S. Coast Guard, with nearly 42,000 men and women on active duty, is a unique force that carries out an array of civil and military responsibilities touching almost every facet of the U.S. maritime environment.

Its core roles are to protect the public, the environment, and U.S. economic and security interests in any maritime region in which those interests may be at risk, including international waters and America's coasts, ports, and inland waterways.

(Maritime Forces) III. U.S. Coast Guard 1-27

A Military Service

The legal basis for the Coast Guard is Title 14 of the United States Code, which states: "The Coast Guard as established January 28, 1915, shall be a military service and a branch of the armed forces of the United States at all times." Upon the declaration of war or when the President directs, the Coast Guard operates under the authority of the Department of the Navy. As members of a military service, Guardians on active duty and in the Reserve are subject to the Uniform Code of Military Justice and receive the same pay and allowances as members of the same pay grades in the other four armed services.

II. Coast Guard Forces

A. Workforce

Mission success is made possible by the combined activities of Coast Guard operational and support personnel. This teamwork is key to ensuring Coast Guard readiness, agility, and operational excellence.

The Coast Guard's full-time workforce is made up of approximately 40,000 active duty military personnel and over 7,000 civilian employees. They are augmented when necessary by small numbers of civilians working under contract. This entire workforce could fit into an average size major league baseball stadium.

The Coast Guard Reserve offers citizens the opportunity to serve in the military part-time while maintaining a separate civilian career. The Reserve provides the Coast Guard highly trained and well qualified personnel for active duty in time of war and national emergency, and for augmentation of Coast Guard forces during natural or man-made disasters or accidents. The Coast Guard Reserve, numbering over 10,000 members, provides the Coast Guard surge capacity and flexibility to respond to all threats and all hazards.

Nearly 30,000 strong, the men and women of the uniformed all-volunteer U.S. Coast Guard Auxiliary spend thousands of hours each year, often on their personal vessels and aircraft, helping to carry out Coast Guard missions. On some waterways, Auxiliarists are the principal Coast Guard personnel serving the public. They are probably best known for their boating safety classes and courtesy vessel safety checks. However, since 1997 they have supported all Coast Guard missions except those involving military operations or law enforcement. The Coast Guard Auxiliary is the only all-volunteer component within the Department of Homeland Security. All together, this small service with a very big job numbers only about 87,000 personnel. By comparison, the next smallest U.S. armed force is the Marine Corps with over 198,000 active duty members alone.

B. Operational Force Structure

Coast Guard field operational units can be grouped according to three types of forces. These are multi-mission shore-based forces.

1. Sector Commands

Coast Guard sector commands focus service delivery on major port regions within the U.S. and its territories. Sector commands are a consolidation of Coast Guard shore-based field operational units. These include boat stations, aids to navigation teams, and prevention and response forces such as vessel inspectors, port operations forces, communications centers, and mission controllers. Sector Commanders possess specific legal authorities for statutorily defined areas:

- **Captain of the Port (COTP)**, with authority over maritime commerce
- **Federal Maritime Security Coordinator (FMSC)**, with authority over maritime security
- **Officer in Charge of Marine Inspection (OCMI)**, with authority over vessel standards compliance

III. USCG Roles and Mission

Ref: NWC 3153K, Joint Military Operations Reference Guide (Jul '09), pp. 122-123.

Since 1915, when the Coast Guard was established by law as an armed force, we have been a military, multi-mission, maritime force offering a unique blend of military, law enforcement, humanitarian, regulatory, and diplomatic capabilities. These capabilities underpin our three broad roles: **maritime safety, maritime security, and maritime stewardship.**

Ref: NWC 3153K, Joint Military Operations Guide, fig. 29, p. 123.

By law, the Coast Guard has 11 missions (listed in order of percentage of operating expenses):

- Ports, waterways, and coastal security
- Drug interdiction
- Aids to navigation
- Search and rescue
- Living marine resources
- Marine safety
- Defense readiness
- Migrant interdiction
- Marine environmental protection
- Ice operations
- Other law enforcement

IV. Ports, Waterways & Coastal Security (PWCS)
Ref: www.uscg.mil.

The Homeland Security Act of 2002 divided the Coast Guard's eleven statutory missions between homeland security and non-homeland security. Reflecting the Coast Guard's historical role in defending our nation, the Act delineated Ports, Waterways and Coastal Security (PWCS) as the first homeland security mission. The Commandant of the Coast Guard designated PWCS as the service's primary focus alongside search and rescue.

The PWCS mission entails the protection of the U.S. Maritime Domain and the U.S. Marine Transportation System (MTS) and those who live, work or recreate near them; the prevention and disruption of terrorist attacks, sabotage, espionage, or subversive acts; and response to and recovery from those that do occur. Conducting PWCS deters terrorists from using or exploiting the MTS as a means for attacks on U.S. territory, population centers, vessels, critical infrastructure, and key resources. PWCS includes the employment of awareness activities; counterterrorism, antiterrorism, preparedness and response operations; and the establishment and oversight of a maritime security regime. PWCS also includes the national defense role of protecting military outload operations.

PWCS is a new name for the Coast Guard's mission previously called Port and Environmental Security (PES). PES included port security, container inspection, and marine firefighting.

The Coast Guard's systematic, maritime governance model for PWCS employs a triad consisting of domain awareness, maritime security regimes, and maritime security and response operations carried out in a unified effort by international, governmental, and private stakeholders.

Maritime Domain Awareness (MDA)
Maritime domain awareness means the effective understanding of anything associated with the maritime domain that could impact the security, safety, economy, or environment of the U.S. Attaining and sustaining an effective understanding and awareness of the maritime domain requires the collection, fusion, analysis, and dissemination of prioritized categories of data, information, and intelligence. These are collected during the conduct of all Coast Guard missions. Awareness inputs come from Field Intelligence Support Teams, Maritime Intelligence Fusion Centers, Nationwide Automatic Identification System and other vessel tracking systems, and public reporting of suspicious incidents through America's Waterway Watch. *See p. 3-22 for discussion of MDA.*

- **America's Waterway Watch (AWW).** AWW is a combined effort of the Coast Guard and its Reserve and Auxiliary components, continues to grow, enlisting the active participation of those who live, work or play around America's waterfront areas. Coast Guard Reserve personnel concentrate on connecting with businesses and government agencies, while Auxiliarists focus on building AWW awareness among the recreational boating public.

Maritime Security Regimes
Maritime security regimes comprise a system of rules that shape acceptable activities in the maritime domain. Regimes include domestic and international protocols and/or frameworks that coordinate partnerships, establish maritime security standards, collectively engage shared maritime security interests, and facilitate the sharing of information. Domestically, the Coast Guard-led Area Maritime Security Committees carry out much of the maritime security regimes effort. Abroad, the Coast Guard works with individual countries and through the International Maritime Organization, a specialized agency of the United Nations. Together, regimes and domain awareness inform decision makers and allow them to identify trends, anomalies, and activities that threaten or endanger U.S. interests.

Defeating Terrorism
Defeating terrorism requires integrated, comprehensive operations that maximize effectiveness without duplicating efforts. Security and response operations consist of counterterrorism and antiterrorism activities.

- **Counterterrorism.** Counterterrorism activities are offensive in nature. The Maritime Security Response Team (MSRT) is a highly specialized resource with advanced counterterrorism skills and tactics. The MSRT is trained to be a first responder to potential terrorist situations; deny terrorist acts; perform security actions against non-compliant actors; perform tactical facility entry and enforcement; participate in port level counterterrorism exercises; and educate other forces on Coast Guard counterterrorism procedures.

- **Antiterrorism.** Antiterrorism activities are defensive in nature. As a maritime security agency, the Coast Guard uses its unique authorities, competencies, capacities, operational capabilities and partnerships to board suspect vessels, escort ships deemed to present or be at significant risk, enforce fixed security zones at maritime critical infrastructure and key resources, and patrol the maritime approaches, coasts, ports, and rivers of America. Coast Guard cutters, boats, helicopters, and shoreside patrols are appropriately armed and trained.

Small Vessel Security Strategy (SVSS)
The intent of the Small Vessel Security Strategy (SVSS) is to reduce potential security and safety risks from small vessels through the adoption and implementation of a coherent system of regimes, awareness, and security operations that strike the proper balance between fundamental freedoms, adequate security, and continued economic stability. Additionally, the strategy is intended to muster the help of the small vessel community in reducing risks in the maritime domain.

The SVSS is designed to guide efforts to mitigate the potential security risks arising from small vessels operating in the maritime domain. While guiding DHS efforts, this strategy acknowledges that to effectively reduce risk, all maritime security partners—Federal, state, local, and Tribal partners and the private sector as well as international partners—must work together to develop, implement, and undertake cooperative actions to reduce both security and safety risks from misuse of small vessels.

Much of the recent U.S. maritime security efforts have focused on regulating cargo containers and large vessels at official Ports of Entry (POE). Examples of such regulations include the 96-hour Advance Notice of Arrival, cargo manifest/crew list transmittal within 24 hours of departure, and the carriage requirement for the Automatic Identification System (AIS). This strategy broadens the focus of federal interest, taking into consideration small vessels regardless of type. The small vessel community includes a wide-range of vessels, from small commercial vessels, such as uninspected towing vessels and passenger vessels, to commercial fishing vessels and recreational boats, whether personal watercraft or large power and sail boats.

Maritime Safety and Security Teams (MSSTs)
Twelve Maritime Safety and Security Teams (MSSTs) enforce security zones, conduct port state control boardings, protect military outloads, ensure maritime security during major marine events, augment shoreside security at waterfront facilities, detect Weapons of Mass Destruction, and participate in port level antiterrorism exercises in their homeports and other ports to which elements of an MSST may be assigned for operations.

Viewing maritime initiatives and policies as part of a larger system enables a better understanding of their relationships and effectiveness. A well designed system of regimes, awareness, and operational capabilities creates overlapping domestic and international safety nets, layers of security, and effective stewardship making it that much harder for terrorists to succeed.

- **Search and Rescue Mission Coordinator (SMC)**, with authority over rescue operations
- **Federal On-Scene Coordinator (FOSC)**, with authority over oil and hazardous material spill response and preparedness

Coast Guard Sector commands are the principal enforcers of ports, waterways, and coastal laws and regulations. As such, they are the Coast Guard's key operational link to federal, state, local, tribal, and private sector partners.

2. Maritime Patrol and Interdiction Forces (MPFs)

Coast Guard cutters, aircraft, and their crews make up the second type of forces. These multi-mission platforms are assigned operations domestically or globally, and enable maritime presence, patrol, response, and interdiction throughout the maritime domain. With their military command, control, and communications networks, they allow the Coast Guard to deter criminal activity and respond to threats and natural or man-made emergencies.

The Coast Guard can also provide these uniquely capable forces to the Department of Defense for national security contingencies. Our newest cutters and aircraft are highly adaptable and capable of meeting current and future homeland and national security needs around the world. Networked and mobile, our cutters and aircraft provide domain awareness and coordinate multi-mission, interagency operations. Although maritime patrol and interdiction forces work principally in the offshore and international environments, they can also operate near shore or within ports. This is critical following a disaster or major disruption to local command, control, and communications capabilities. As the Nation's only provider of Polar icebreaking capabilities, the Coast Guard enables unique access and capabilities in the Polar Regions.

3. Deployable Specialized Forces (DSF) and Deployable Operations Group (DOG)

Deployable Specialized Forces (DSFs) are rapidly transportable elements with specialized skills in law enforcement, military port security, hazardous spill response, and other such missions. These specialized teams provide the Coast Guard with surge capability and flexibility. The **Deployable Operations Group (DOG)** oversees, coordinates, and integrates Coast Guard DSFs, which include some reserve-based units. The DOG also works with other DHS components and government agencies to develop integrated, multi-agency, force packages to address maritime threats and hazards.

Forces within the DOG include:

- **Maritime Safety and Security Teams (MSSTs)**, which include security and boat forces
- **Maritime Security Response Team (MSRT)**, which has specialized capabilities for law enforcement
- **Tactical Law Enforcement Teams (TACLETs) and Law Enforcement Detachments (LEDETs)**, which deploy wherever needed for law enforcement missions
- **Port Security Units (PSUs)**, which provide expeditionary port security
- **National Strike Force (NSF)**, which provides high-end pollution and hazardous material response

IV. Naval/Marine Special Warfare Forces

Ref: NWC 3153K, Joint Military Operations Reference Guide (Jul '09), chap. 6 and NSW Press Kit/Fact File (2010).

Special Operations Forces must be highly trained, properly equipped and deployed to the right place at the right time for the right missions. As key members of Joint, Interagency, and Internationally teams, SOF will employ all assigned authorities and apply all available elements of power to accomplish assigned missions.

I. Naval Special Warfare (NSW)

Naval Special Warfare (NSW) forces conduct maritime special operations in support of joint and naval operations. Principal core tasks are SR, DA, FID and CT. NSW forces are deployed under the OPCON of either a naval component or joint force commander. ADCON is retained by the parent command.

A. NSW Organization

Naval Special Warfare (NSW) is comprised of approximately 8,300 total personnel, including more than 2,300 active-duty Special Warfare Operators, known as SEALs, 600 Special Warfare Boat Operators, also known as Special Warfare Combatant-craft Crewmen (SWCC), 900 reserve personnel, 3,650 support personnel and more than 880 civilians.

Naval Special Warfare Command

Naval Special Warfare Command in San Diego, Calif., leads the Navy's special operations force and the maritime component of United States Special Operations Command (USSOCOM), headquartered at MacDill Air Force Base, Tampa, Fla.

See p. 1-37.

Naval Special Warfare Groups

Naval Special Warfare Groups are major commands that train, equip and deploy components of NSW squadrons to meet the exercise, contingency and wartime requirements of geographic combatant commanders, Theater Special Operations Commands, and numbered fleets located around the world.

NSWTG and NSWTU are task organized force packages deployed to joint and fleet warfighting commanders to plan, coordinate, command and conduct NSW operations.

- A **NSWTG** is task organized to provide command and control of one or more NSWTU.
- A **NSWTU** is composed of a command and control element, a support element, and a combination of one or more SEAL or SDV platoons, and/or special boat detachments.

NSW Squadrons

NSW Squadrons are built around deployed SEAL Teams and include senior leadership, SEAL Delivery Vehicle Teams, Special Boat Teams, and support technicians such as mobile communications teams, tactical cryptologic support and explosive ordnance disposal specialists. Naval Special Warfare Squadrons are among the most responsive, versatile and effective force packages fighting the global war on terrorism today.

U.S. Special Operations Command (USSOCOM)
www.socom.mil

The Department of Defense (DOD) activated U.S. Special Operations Command (US-SOCOM) April 16, 1987, at MacDill Air Force Base, Fla. DOD created the new unified command in response to congressional action in the Goldwater-Nichols Defense Reorganization Act of 1986 and the Nunn-Cohen Amendment to the National Defense Authorization Act of 1987. Congress mandated a new four-star command be activated to prepare Special Operations Forces (SOF) to carry out assigned missions and, if directed by the president or secretary of defense (SECDEF), to plan for and conduct special operations. To enable USSOCOM to carry out its mission, Congress gave the new command specific authorities and responsibilities called Title 10.

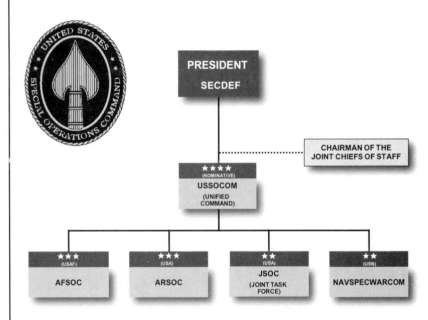

CDRUSSOCOM is a functional CCDR who exercises COCOM of all Active and RC SOF minus US Army Reserve civil affairs and PSYOP forces stationed in CONUS. When directed, CDRUSSOCOM provides US based SOF to a geographic CCDR who exercises COCOM of assigned and OPCON of attached SOF through a CDR of a theater SO command or a joint SO task force in a specific operational area or to prosecute SO in support of a theater campaign or other operations. When directed, CDRUSSOCOM can establish and employ a joint SO task force as the supported CDR. In addition to functions specified in Title 10, USC, Section 167, USSOCOM:
- Serves as the SOF joint force provider
- Integrates and coordinates DOD PSYOP capabilities to enhance interoperability and support USSTRATCOM's IO responsibilities and other CCDRs' PSYOP planning and execution
- Serves as the lead CCDR for planning, synchronizing, and as directed, executing global operations against terrorist networks in coordination with other CCDRs
- Exercises C2 of selected SO missions as directed

USSOCOM Title 10 Authorities and Responsibilities
In addition to the service-like authorities of developing training and monitoring readiness, some of the authorities Congress gave USSOCOM are unique responsibilities for a unified command. USSOCOM is not dependent on the Army, Navy, Marine Corps or Air Force for its budget or to develop and buy new equipment, supplies or services for the command. USSOCOM has its own budgetary authorities and responsibilities through a specific Major Force Program (MFP-11) in DOD's budget. Additionally, USSOCOM has its own acquisition authorities, so it can develop and buy special operations-peculiar equipment, supplies or services.

USSOCOM Mission
United States Special Operations Command (USSOCOM) has a two-fold mission: (1) Provide fully capable special operations forces to defend the United States and its interests; and (2) Synchronize planning of global operations against terrorist networks.

USSOCOM Priorities
- Deter, Disrupt & Defeat Terrorist Threats
- Plan & conduct special operations
- Emphasis persistent, culturally attuned engagement
- Foster interagency cooperation
- Develop & Support our People and Families
- Focus on quality
- Care for our people and families
- Train & educate the joint warrior/diplomat
- Sustain and Modernize the Force
- Equip the operator
- Upgrade SOF mobility
- Obtain persistent intelligence, surveillance & reconnaissance systems

Before the Sept. 11, 2001, terrorist attacks on the United States, USSOCOM's primary focus was on its supporting command mission of organizing, training and equipping SOF and providing those forces to support the geographic combatant commanders and U.S. ambassadors and their country teams. The president further expanded USSOCOM's responsibilities in the 2004 Unified Command Plan. The Unified Command Plan assigned USSOCOM responsibility for synchronizing Department of Defense plans against global terrorist networks and, as directed, conducting global operations. USSOCOM receives reviews, coordinates and prioritizes all DoD plans that support the global campaign against terror, and then makes recommendations to the Joint Staff regarding force and resource allocations to meet global requirements.

USSOCOM Components
USSOCOM's components are U.S. Army Special Operations Command (USASOC), Naval Special Warfare Command (NAVSPECWARCOM), Air Force Special Operations Command (AFSOC), Marine Corps Forces Special Operations Command (MARSOC), the Joint Military Information Support Command (JMISC) and the Joint Special Operations University. USSOCOM also has a sub-unified command, Joint Special Operations Command (JSOC). USSOCOM has approximately 56,000 active duty, Reserve and National Guard Soldiers, Sailors, Airmen, Marines and Department of Defense (DOD) civilians assigned to the headquarters, its components, and subordinate unified command.

B. NSW Forces

Because SEALS are experts in special reconnaissance and direct action missions — the primary skill sets needed to combat terrorism — NSW is postured to fight a globally-dispersed enemy, whether ashore or afloat, before they can act. NSW forces can operate in small groups and have a continuous presence overseas with their ability to quickly deploy from Navy ships, submarines and aircraft, overseas bases and forward-based units. The proven ability of NSW forces to operate across the spectrum of conflict and in operations other than war, and provide real-time, first-hand intelligence offer decision makers immediate and multiple options in the face of rapidly changing crises around the world.

Sea-Air-Land (SEAL) teams

Sea-Air-Land (SEAL) teams are CONUS-based commands established to train, equip, deploy and support SEAL platoons to conduct NSW in support of joint and fleet commanders. Each team consists of six 16-man SEAL platoons composed of two officers and 14 enlisted SEAL operators and requisite support personnel. SEALs conduct clandestine missions infiltrating their objective areas by fixed- and rotary-wing aircraft, Navy surface ships, combatant craft, submarines and ground mobility vehicles. When directed, a SEAL team deploys as a Naval Special Warfare Squadron (NSWSQRN) disbursing its forces in smaller Task Units or Task Elements to plan, coordinate, command, and conduct special operations.

Special Boat Teams (SBT)

Special Boat Teams (SBT) are manned by Special Warfare Combatant-craft Crewmen (SWCC) who operate and maintain state-of-the-art surface craft to conduct coastal patrol and interdiction and support special operations missions. Focusing on infiltration and exfiltration of SEALs and other SOF, SWCCs provide dedicated rapid mobility in shallow water areas where larger ships cannot operate. They also bring to the table a unique SOF capability: Maritime Combatant Craft Aerial Delivery System — the ability to deliver combat craft via parachute drop.

SEAL Delivery Vehicle (SDV)

The SEAL Delivery Vehicle (SDV) team comprises specially trained SEALs and support personnel who conduct undersea operations from SDVs, Dry Deck Shelters (DDS), and the Advanced SEAL Delivery System (ASDS). DDS deliver SDVs and specially trained forces from modified submarines. When teamed with their host submarines, the ASDS and SDV platforms provide the most clandestine maritime delivery capability in the world.

Naval Special Warfare Combat Service Support Teams

Naval Special Warfare Combat Service Support Teams provide full-spectrum logistics support to SEAL (sea, air, land) Teams, Special Boat Teams, NSW Task Groups/Task Units. Tasking for each CSST includes crisis-action and logistics planning and coordination; in-theater contracting, small purchase and leasing actions; and comprehensive forward operating base support.

Naval Special Warfare Development Group (NAVSPECWARDEVGRU)

NAVSPECWARDEVGRU provides centralized management for the test, evaluation, and development of current and emerging technologies applicable to NSW. It also develops maritime, ground, and airborne tactics for NSW.

Naval Special Warfare Center (NAVSPECWARCEN)

NAVSPECWARCEN serves as the schoolhouse for NSW selection, training, tactics and doctrine development. In addition to conducting and managing the basic SEAL and Special Warfare Combatant Craft Crewman training programs, the Center also provides instruction and training in NSW for selected allied military personnel.

Naval Special Warfare Command
Ref: NSW Press Kit/Fact File (2010).

Naval Special Warfare Command in San Diego, Calif., leads the Navy's special operations force and the maritime component of United States Special Operations Command (USSOCOM), headquartered at MacDill Air Force Base, Tampa, Fla.

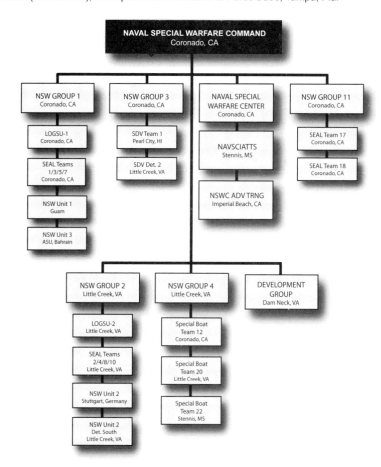

Naval Special Warfare (NSW) is comprised of approximately 8,300 total personnel, including more than 2,300 active-duty Special Warfare Operators, known as SEALs, 600 Special Warfare Boat Operators, also known as Special Warfare Combatant-craft Crewmen (SWCC), 900 reserve personnel, 3,650 support personnel and more than 880 civilians.

II. U.S. Marine Corps Special Operation Forces

In October 2005, the Secretary of Defense directed the formation of a Marine component of U.S. Special Operations Command. On Feb. 24, 2006, Marine Corps Forces Special Operations Command (MARSOC) activated at Camp Lejeune, N.C. as a major command. MARSOC is under the operational control of USSOCOM and reports directly to the Commandant of the Marine Corps for service matters.

A. Mission

As the Marine component of USSOCOM, MARSOC is tasked by the commander of USSOCOM to train, organize, equip and when directed by commander of USSOCOM, deploy task organized, scaleable and responsive U.S. Marine Corps Special Operations Forces worldwide in support of combatant commanders and other agencies. MARSOC has been directed to conduct foreign internal defense, direct action and special reconnaissance. Commander, USSOCOM assigns MARSOC missions based on USSOCOM priorities. MARSOC units then deploy under USSOCOM deployment orders.

MARSOC core tasks are Foreign Internal Defense (FID), Special Reconnaissance (SR), Information Operations (IO), Unconventional Warfare (UW), Direct Action (DA), and Counter terrorism (CT).

B. Organization

MARSOC Forces are organized as follows:

Marine Special Operations Battalions (MSOB)

Marine Special Operations Battalions (MSOB) are organized, trained and equipped to deploy for worldwide missions as directed by MARSOC. The 1st MSOB activated on Oct. 26, 2006 and is headquartered at Camp Pendleton, Calif. The 2d MSOB was activated May 15, 2006 and is headquartered at Camp Lejeune, N.C. Each battalion consists of four Marine Special Operations Companies (MSOC) and is task-organized with personnel uniquely skilled in special equipment support, intelligence and fire-support. Each MSOC is commanded by a Marine major and capable of deploying task-organized expeditionary Special Operations Forces for special reconnaissance and direct-action missions in support of the geographic combatant commanders.

The Marine Special Operations Advisor Group (MSOAG)

The Marine Special Operations Advisor Group (MSOAG), which consists of a Headquarters Company and 3rd and 4th Marine Special Operations Battalions, provides tailored military combat-skills training and advisor support for identified foreign forces in order to enhance their tactical capabilities and to prepare the environment as directed by USSOCOM. Marines and Sailors of the MSOAG train, advise and assist friendly host-nation forces—including naval and maritime military and paramilitary forces—to enable them to support their governments' internal security and stability, to counter subversion and to reduce the risk of violence from internal and external threats. MSOAG deployments are coordinated by MARSOC, through USSOCOM, in accordance with engagement priorities within the Global War on Terrorism.

Marine Special Operations Support Group (MSOSG)

The Marine Special Operations Support Group (MSOSG) is located at Camp Lejeune and provides combat support and combat service support to MARSOC Units, to include: logistics, communication and intelligence.

Marine Special Operations School (MSOS)

The Marine Special Operations School (MSOS), also located at Camp LeJeune, screens, assesses, selects, trains and certifies Marine personnel as special operations forces and has responsibility for doctrine development.

V. Strategic Lift/ Force Projection

Chap 1

Maritime Forces

Ref: NWC 3153K, Joint Military Operations Reference Guide (Jul '09), chap. 7.

The ability of the U.S. military to successfully carry out its assigned tasks per our National Security Strategy and National Military Strategy depends greatly on its capability to deploy forces, equipment, and sustainment to a theater of operations within a given period of time. While logistics includes all those supporting activities required to sustain a deployed force, strategic mobility defines that part of the logistics process which transports people, equipment, supplies, and other commodities by land, sea, and air, to enable military force projection. In fact, the operational commander must have a clear understanding of the capabilities and limitations of the strategic mobility process if he or she is going to successfully execute a major operation or campaign. Force selection, phasing of operations, and risk assessment are directly tied to the ability to project forces and support from the United States to the area of responsibility, area of operation, or theater of war.

I. Force Projection

Force projection is the military element of national power that systemically and rapidly moves military forces in response to requirements across the spectrum of conflict. It is a demonstrated ability to alert, mobilize, rapidly deploy, and operate effectively anywhere in the world. Force projection encompasses a range of processes including mobilization, deployment, employment, sustainment, and redeployment. These processes have overlapping timelines, are continuous, and can repeat throughout an operation. Force projection operations are inherently joint and require detailed planning and synchronization.

- **Mobilization.** Mobilization is the process of assembling and organizing resources to support national objectives in time of war and other emergencies. Mobilization includes bringing all or part of the industrial base and the Armed Forces of the United States to the necessary state of readiness to meet the requirements of the contingency.

- **Deployment.** Deployment is the movement of forces to an operational area in response to an order.

- **Employment.** Employment prescribes how to apply force and/or forces to attain specified national strategic objectives. Employment concepts are developed by the combatant commands (COCOM) and their component commands during the planning process. Employment encompasses a wide array of operations—including but not limited to—entry operations, decisive operations, and post-conflict operations.

- **Sustainment.** Sustainment is the provision of personnel, logistics, and other support necessary to maintain and prolong operations or combat until successful accomplishment or revision of the mission or national objective.

- **Redeployment.** Redeployment involves the return of forces to home station or demobilization station.

Refer to The Sustainment & Multifunctional Logistician's SMARTbook (Warfighter's Guide to Logistics, Personnel Services, & Health Services Support) for complete discussion of force projection, deployment and redeployment, and RSO&I operations.

(Maritime Forces) V. Strategic Lift 1-39

II. U.S. Transportation Command
www.transcom.mil

USTRANSCOM mission is to develop and direct the Joint Deployment and Distribution Enterprise to globally project strategic national security capabilities; accurately sense the operating environment; provide end-to-end distribution process visibility; and responsive support of joint, U.S. government and Secretary of Defense-approved multinational and non-governmental logistical requirements.

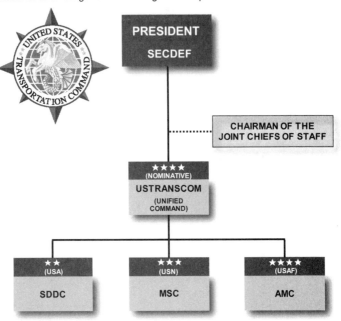

The Commander, US Transportation Command (CDRUSTRANSCOM) is a functional CCDR who is responsible to:
- Provide common-user and commercial air, land, and sea transportation, terminal management and aerial refueling to support global deployment, employment, sustainment, and redeployment of US forces
- Serve as the mobility joint force provider
- Provide DOD global patient movement, in coordination with geographic CCDRs, through the Defense Transportation Network
- Serve as the distribution process owner

At every moment of every day, around the globe, USTRANSCOM's superb force of soldiers, sailors, airmen, Marines, Coast Guardsmen, DOD civilians and commercial partners accomplishes a wide array of joint mobility missions. With its people, trucks, trains, aircraft, ships, information systems and infrastructure, USTRANSCOM provides the U.S. with the most responsive strategic mobility capability the world has ever seen.

During an average week, USTRANSCOM conducts more than 1,900 air missions, with 25 ships underway and 10,000 ground shipments operating in 75 percent of the world's countries. As of October 2004, the command has moved more than 1.9 million passengers; 1,108,987 tons by air; 3.7 million tons by sea; and delivered more than 53.7 billion barrels of fuel by ship.

USTRANSCOM oversees the strategic mobility process in both peace and war. USTRANSCOM's charter is to maintain and operate a deployment system for orchestrating the transportation aspects of worldwide mobility planning, integrate deployment-related Information Management systems, and provide centralized wartime traffic management. Actual movement is executed by USTRANSCOM component commands: Surface Deployment and Distribution Command (SDDC-Army), Military Sealift Command (MSC-Navy), and Air Mobility Command (AMC-Air Force). The Department of Transportation's Maritime Administration (MARAD) bridges MSC, U.S. flag commercial companies, and U.S. unions for sealift procurement and operations. Each element of the strategic mobility triad (airlift, sealift, and prepositioning) has distinct advantages and disadvantages in terms of response time, availability of carrying assets, carrying capacity and throughput, and vulnerability.

A. Military Surface Deployment and Distribution (SDDC)

SDDC is a major command of the U.S. Army. As a transportation component of the USTRANSCOM, SDDC is the CONUS transportation manager and provides worldwide common-use ocean terminal services and traffic management services to deploy, employ, sustain and redeploy US forces on a global basis.

B. Military Sealift Command (MSC)

As a component command of USTRANSCOM, MSC provides common-user sealift across the range of military operations. MSC adjusts and controls the total number of ships under its COCOM to meet demand. Under normal peacetime conditions, the MSC force consists of government-owned ships as well as privately-owned ships under charter to MSC. When demand increases, MSC can expand its fleet by acquiring additional sealift from a variety of resources and through a number of different acquisition programs. MSC resources available to the DTS beyond MSC's active peacetime fleet include:

- **Fast Sealift Ships (FSS)**. The FSSs are former containerships, purchased by the Navy and converted to RO/RO configuration with on-board cranes and self-contained ramps that enable the ships to off-load onto lighterage while anchored at sea or in ports where sore facilities for unloading equipment are unavailable. The vessels are specially suited to transport heavy or bulky unit equipment such as tanks, large wheeled vehicles and helicopters. The present eight ships have a joint, one-time lift capability of approximately 1.3 million square feet and also have a container capability. These ships are capable of carrying 150,000 square feet of Army, combat, combat support, or combat service support equipment at a speed of 27 knots. On Oct. 1 2007, The U.S. Maritime Administration began operating all eight FSS. The FSS will transfer to the RRF on Oct. 1, 2008 and will lose their USNS designation.

- **Large Medium Speed Roll-On/Roll–off Ships (LMSR)**. An LMSR is similar to any other RO/RO ship in that it is specially designed to carry wheeled and tracked vehicles as all or most of its cargo. A LMSR differs from most other RO/RO ships in that it is faster, larger and has cranes and hatches to support LO/LO operations. Ships carry two Army heavy brigades pre-positioned afloat, and 11 LMSR ships will be lay berthed in CONUS to deploy Army equipment. These ships can maintain a speed of 24 knots.

- **Pre-positioned Ships**. MSC has a large fleet of pre-positioned ships that can be used for common-user sealift once they discharge their cargo.

C. Air Mobility Command (AMC)

AMC is a U.S. Air Force major command headquartered at Scott Air Force Base, Illinois. As the Air Force component command of USTRANSCOM, AMC provides air lift, air refueling, and aeromedical evacuation services for deploying, employing, sustaining and redeploying U.S. forces wherever they are needed worldwide.

III. U.S. Maritime Administration (MARAD)
Ref: NWC 3153K, Joint Military Operations Reference Guide (Jul '09), pp. 159-160.

MARAD has primary federal responsibility for ensuring the availability of efficient water transportation service to American shippers and consumers. MARAD seeks to ensure that the United States enjoys adequate shipbuilding and repair service, efficient CONUS ports, effective intermodal water and land transportation systems, and reserve shipping capacity in time of national emergency. MARAD administers federal laws and programs designed to support and maintain a US merchant marine capable of meeting the Nation's shipping needs for both domestic and foreign commerce and national security. MARAD advances the capabilities of the maritime industry to provide total logistic support (port, intermodal, ocean shipping, and training) to the military Services during war or national emergencies through the following:

- In accordance with DOD readiness criteria, maintaining an active Ready Reserve Force (RRF) fleet of strategic sealift, which is a component of the inactive National Defense Reserve Fleet (NDRF), to support emergency and national security sealift needs
- Administer funding for the maintenance of the RRF and NDRF
- Administering the Maritime Security Program and priorities and allocations of the VISA
- Acquiring US flag, US-owned, and other militarily useful merchant ships in accordance with appropriate authorities from the Merchant Marine Act of 1936 and the emergency Foreign Vessels Acquisition Act of 1954
- Ensuring readiness preparation and coordination of commercial strategic ports for mobilization through the National Port Readiness Network
- Administering the Vessel War Risk Insurance Program (Title 12, Merchant Marine Act of 1936)
- Sponsoring merchant mariner training programs for both licensed and unlicensed seamen and ensuring reemployment rights for merchant marines who crew sealift vessels during a sealift crisis

Ready Reserve Force (RRF)
The RRF is the most significant source of government –owned early deployment shipping in terms of both the number of ships and overall cargo –carrying capability. It includes 50 ships kept in reserve by MARAD to meet surge shipping requirements for DOD. MARAD maintains these vessels in 4-, 5-, 10- or 20-day readiness status. Most are berthed on the three CONUS seacoasts, with one port in Tsuneshi, Japan as well. They consist of commercial or former military vessels of high military utility including RO/RO, sea barge, lighter aboard ship (LASH), container, tanker, crane, and breakbulk ships. Some of these vessels have had their military capabilities enhanced with the addition of systems such as the modular cargo delivery system and the offshore petroleum discharge system (OPDS).

The National Defense Reserve Fleet (NDRF)
The National Defense Reserve Fleet provides an additional reserve of ships for national defense and national emergency purposes. The NDRF consists of dry cargo vessels, tankers, military auxiliaries, and other ship types. In addition to maintaining ships for US-TRANSCOM logistics, the Missile Defense Agency sponsors 2 ships for missile tracking. Vessels are either owned by the Maritime Administration or held for other Government agencies on a reimbursable basis. As of 1 Apr 2009, the NRDF consisted of 180 vessels.

The US Flag Fleet
Ships from the US flag fleet are routinely chartered by MSC to meet government shipping demands. Shipping contracts are also negotiated for government cargo that does

not have to move on dedicated shipping. When an expansion of government requirements occurs such that voluntary US and foreign flag charters no longer meet requirements, it is the US flag fleet that is expected to respond to meet the requirements. There are three acquisition processes, not counting voluntary chartering, available for DOD acquisition of additional US flag shipping. They are the Voluntary Intermodal Sealift Agreement (VISA), the Voluntary Tanker Agreement (VTA), and requisitioning.

- **Commercial Charter.** MSC frequently charters US and foreign flag ships during peacetime to provide additional sealift capacity. Charter is a routine commercial transaction that can be accomplished in a little as two days. However, all chartered ships may not be immediately available in time of crisis. Depending on ship location, the time required to arrive at the designated loading port may be as much as 30 days.
- **The VISA.** VISA is the primary sealift mobilization program. It is an intermodal capacity-oriented program vice a ship-by-ship oriented program. All major US flag carriers are enrolled in VISA. This constitutes more than 90 percent of the US flag dry cargo fleet. The worldwide intermodal system provided by these carriers provides extensive and flexible capabilities to the Department of Defense. The types of ships enrolled in the VISA program includes containerships, RO/RO ships, LASH vessels, combination RO/RO and containerships, heavylift ships, breakbulk ships, and tugs and barges. VISA is activated upon approval of the Secretary of Defense. A **joint planning advisory group (JPAG)** is central to the successful implementation of VISA and is comprised of representatives from USTRANSCOM, SDDC, MSC, DLA, MARAD, and intermodal industrial transportation representatives.
- **The VTA.** The VTA is a method of acquiring additional petroleum product carriers once the commercial market is no longer responsive. It is a cooperative effort by industry and government to meet military requirements for product tankers. It is activated by MARAD at the request of the Secretary of Defense.
- **Liner Agreements.** SDDC, a component of USTRANSCOM, arranges for common user ocean services by either establishing new contracts or utilizing existing contracts with commercial carriers offering liner service on scheduled trade routes. The liner service established by these contracts may be for container or break bulk service responding to either unit or sustainment requirements.
- **Requisitioning.** SECTRANS is authorized to requisition any vessel which is majority owed by US citizens, whether registered under the US or Foreign flag, whenever the POTUS proclaims that the security of the nation makes it advisable, or during any national emergency declared by the proclamation of the POTUS (and/or concurrent resolution of the Congress) under the authority of Section 902 of the Merchant Marine Act of 1936 (46 US Code (USC) 1242).

Foreign Flag Ships

When US flag ships are unavailable, foreign flag ships can be acquired for DOD use through three different methods: voluntary charter, allied shipping agreements, and requisitioning of effective US control shipping.

- **Voluntary Charter.** During peacetime, MSC will charter foreign flag ships whenever US flag ships are unavailable. This ability allows MSC to enter the foreign charter market and quickly expand its fleet whenever the need arises.
- **Allied Shipping Agreements.** Allied shipping agreements, arranging for vessels received through allied nations, can either be pre-negotiated and in existence or they can be drawn up on an emergency basis as the need arises.
- **Effective United States-Controlled Ships (EUSCS).** EUSCS are ships owned by US citizens or companies that are registered in countries that have no prohibition on requisitioning of these vessels by the United States. These ships may be requisitioned by the United States under authority of Section 902, Merchant Marine Act of 1936 (title 46, USC, section 1242).

IV. Strategic Deployment

Ref: NWC 3153K, Joint Military Operations Reference Guide (Jul '09), pp. 142-143.

The deployment process is an essential enabler that allows the U.S. Armed Forces to project force to accomplish the will of our national leadership. Given its key role, great attention must be given to thorough planning as it is difficult, if not impossible, to recover from mistakes made in the deployment phase. Both joint, and service, planners are faced with a plethora of issues that must be successfully addressed in order to ensure the commander's intent is met. These issues tend to focus on the advantages/disadvantages of each leg of the strategic mobility triad such as response time, availability of transportation assets, logistics throughput, and asset vulnerability.

More specifically, joint and service planners need to provide for the following considerations: transportation facilities, transportation facility support forces and equipment, operation of APODs/SPODs and their associated command relationships (includes POL, MHE/Cargo handling equipment), on and off-load operations, base defense/force protection, joint airspace and sea control (air and sea lines of communications), intelligence, weather, the threat, countermeasures to the threat, air and sea refueling, and the communications requirements of the deploying force. The Joint Operation Planning and Execution System (JOPES) and the joint operation planning process (JOPP) provide the processes, formats, and systems which link planning for joint force projection to the execution of joint operations.

- **Joint Deployment Distribution Operations Center (JDDOC)**. The USTRANSCOM Deployment Distribution Operations Center (DDOC) serves as the focal point to orchestrate and optimize DTS operations in support of the Combatant Commanders and other customers.

- **Theater Opening (TO)**. Theater opening (TO) is the ability to rapidly establish and initially operate ports of debarkation (air, sea, and rail), to establish the distribution system and sustainment bases, and to facilitate port throughput for the reception, staging, and onward movement of forces within a theater of operations. It is a complex joint process involving the GCC and strategic and joint partners such as USTRANSCOM, its components, and DLA.

- **Joint Task Force Port Opening (JTF-PO)**. JTF-PO provides a rapid and worldwide deployable entity designed to support geographic combatant commanders throughout the range of military operations as they conduct their mission of supporting our nation's interests abroad. JTF-PO is designed to reverse the historic shortcomings associated with the rapid opening of ports worldwide, including ad hoc C2 and lack of continuous visibility of cargo moving from the ports of debarkation through the theater of operations. Consistent and deliberate joint training, a robust C2 suite, to include ITV; and dedicated surface movement control units enable JTF-PO to effectively and efficiently address previous deficiencies of global transportation movement. The JTF-PO APOD combines fielded Air Force and Army units to open an airport and prepare it for logistics operations in as little as 72 hours. Similar to the APOD, USTRANSCOM has developed, and is in the process of fielding, the SPOD capability which will provide the rapid opening of a seaport in support of logistics operations.

* Civil Reserve Air Fleet (CRAF)

A unique and significant part of the nation's mobility resources is the Civil Reserve Air Fleet. Selected aircraft from the U.S. airlines, contractually committed to CRAF, support Department of Defense airlift requirements in emergencies when the need for airlift exceeds the capability of military aircraft.

The Joint Deployment Process

The joint deployment process is divided into four phases: planning, pre-deployment activities, movement, and Joint Reception, Staging, Onward Movement, and Integration (JRSOI). They are iterative and often occur simultaneously throughout an operation, and are sometimes depicted in a "backwards" planning sequence:

1. Deployment Planning

Deployment planning occurs during both JOPES contingency planning (if sufficient planning time is available prior to mission execution) and crisis action planning (CAP). It is conducted at all command levels and by both the supported and supporting commanders. Deployment planning activities include all action required to plan for a deployment and employment of forces. Deployment planning activities must be coordinated among the supported combatant command responsible for accomplishment of the assigned mission, the Services, and the supporting combatant commands providing forces for the joint force mission. Normally, supported CCDRs, their subordinate commanders, and their Service components are responsible for providing mission statements, theater support parameters, inter-theater lift requirements, applicable host nation (HN) environmental standards, and pre-positioned equipment planning guidance during pre-deployment activities.

2. Pre-deployment Activities

Pre-deployment activities are all actions taken by the joint planning and execution community (JPEC) before actual movement to prepare to execute a deployment operation. It includes continued refinement of OPLANs, from the strategic to the tactical level at the supported and supporting commands. It includes sourcing forces, completion of operation specific training, and mission rehearsals. As early as possible in pre-deployment, movement requirements are identified and lift support requirements are reported and scheduled. Personnel and equipment movement preparation is competed and verified. Accompanying sustainment is identified and prepared for movement.

3. Movement

Movement includes the movement of self-deploying units and those that require lift support. It includes movements within CONUS, deployments within an AOR, and end-to-end origin to destination strategic moves.

4. Joint Reception, Staging, Onward Movement, & Integration (JRSOI)

JRSOI, is the critical link between deployment and employment of the joint forces in the OA. It integrates the deploying forces into the joint operation and is the responsibility of the supported CCDR. Deployment is not complete until the deploying unit is a functioning part of the joint force. The time between the initial arrival of the deploying unit and its operational employment is potentially the period of its greatest vulnerability. During this transition period, the deploying unit may not fully sustain itself, defend itself, or contribute to mission accomplishment because some of its elements have not attained required mission capability. JRSOI planning is focused on the rapid integration of deploying forces and capabilities to quickly make them functioning and contributing elements of the joint force.

5. Redeployment Operations

Similar to deployment, redeployment operations encompass four phases. These are: redeployment planning, pre-redeployment activities, movement, and JRSOI. These phases describe the major activities inherent in moving deployed forces and materiel from their current deployed location through integration into another theater or to the home and/or demobilization station. Redeployment operations are dependent on the supported combatant commander's defined end state, concept for redeployment, or requirement to support another JFC's concept of operations.

V. Strategic Lift Port Operations

Critical components of the DTS are military and commercial ports supporting the air and maritime movement of unit and non-unit personnel, equipment, and cargo. These ports could be owned and operated by SDDC, AMC, a Service, geographic combatant commanders, or commercial or HN authorities. They may be either sophisticated fixed locations or heavily dependent on deployable mission support forces or joint logistics over-the-shore (JLOTS) assets to accomplish the mission. The significant surface and air cargo handling capabilities that exist in the Services should be used jointly rather than in isolation to maximize the throughput capability of these essential transportation modes.

The extensive use of containers and 463L pallets makes container handling equipment (CHE) and MHE essential elements of the DTS. Ensuring that these assets are available early allows for the efficient loading and unloading of ships and aircraft and increases the rate at which a port can be cleared. Without these assets, the DTS may come to a halt.

Single Port Manager (SPM)

The SPM performs those functions necessary to support the strategic flow of deploying and redeploying forces, unit equipment, and sustainment supply in the SPOEs and APOEs and hand-off to the geographic combatant commander in the SPODs and APODs. The Department of Defense uses the SPM approach for all worldwide common-use aerial and seaport operations. As outlined in the Unified Command Plan, USTRANSCOM has the mission to provide worldwide common-user aerial and seaport terminal management and may provide terminal services by contract. Thus USTRANSCOM, through AMC and SDDC, will manage common-use aerial ports and seaports for the geographic combatant commander. In areas not served by a permanent USTRANSCOM presence, USTRANSCOM will deploy an AMC air mobility squadron and/or aerial port mobile flight and tanker air mobility control element and an SDDC port management cell to manage the ports in concert with a designated port operator.

- **SDDC.** As USTRANSCOM's surface TCC, SDDC performs SPM functions necessary to support the strategic flow of the deploying forces' equipment and sustainment supply in the SPOE and hand-off to the geographic combatant commander in the SPOD. SDDC has port management responsibility through all phases of the theater port operations continuum, from a bare beach (e.g., JLOTS) deployment to a commercial contract fixed-port support deployment. When necessary, in areas where SDDC does not maintain a manned presence, a deployment support team will be established to direct water terminal operations, including supervising movement operations, contracts, cargo documentation, CONUS security operations, arrange for support, and the overall flow of information. As the single seaport manger, SDDC is also responsible for providing strategic deployment status information to the combatant commander and to manage the workload of the SPOD port operator based on the combatant commander's priorities and guidance. SDDC transportation groups and other SDDC units operate ports that use contracted labor. If Army stevedores are used, transportation groups assigned to the combatant commander operate the port.

- **AMC.** As USTRANSCOM's air TCC, AMC performs SPM functions necessary to support the strategic flow of the deploying forces' equipment and sustainment supply in the APOE and hand-off to the geographic combatant commander in the APOD. AMC has port management responsibility through all phases of the theater aerial port operations continuum, from a bare base deployment to a commercial contract fixed-port support deployment. AMC is the single aerial port manager and, where designated, operator of common-user APOEs and/or APODs.

I. Operational Warfare at Sea

Ref: NWP 3-32, Maritime Operations at the Operational Level of War (Oct '08), chap. 1.

I. Navy Doctrine

Navy doctrine is a statement of officially sanctioned beliefs, war-fighting principles, and terminology that describes and guides the proper use of the Navy in maritime operations. The Army, Air Force, and Marine Corps focus and build doctrine for the execution of missions on or above land. With its focus on land operations joint doctrine tacitly reflects the fact that the job of gaining and maintaining maritime superiority or supremacy — of engaging and winning battles in the maritime domain — falls almost exclusively to the Navy. Joint doctrine is authoritative guidance and takes precedence over individual Service doctrine, which must be consistent with joint doctrine. As a body of best practices or norms, Navy doctrine "is authoritative but requires judgment in application." Doctrine is not an impediment to a commander's exercise of imagination. Rather, doctrine is a framework of fundamental principles, practices, techniques, procedures, and terms that guides a commander in employing his force to accomplish the mission.

Maritime Operations

Maritime Superiority — That degree of dominance of one force over another that permits the conduct of maritime operations by the former and its related land, maritime, and air forces at a given time and place without prohibitive interference by the opposing force. (JP 1-02. Source: JP 3-32).

Maritime Supremacy — That degree of maritime superiority wherein the opposing force is incapable of effective interference. (JP 1-02)

Ref: NWP 3-32, Maritime Operations at the Operational Level of War, p. 1-1.

The principles discussed within doctrine are enduring, yet they evolve based on policy and strategy, in light of new technology or organizations, from lessons gained from experience, and insights derived from operational analysis. Navy doctrine standardizes terminology, training, relationships, responsibilities, and processes. Its focus is on how to think about operations, not what to think about operations. Doctrine provides a basis for analysis of the mission and its objectives and tasks, and developing the commander's intent and associated planning guidance. It provides a foundation for training and education. Doctrine is distinct from concepts in that it describes operations with extant capabilities and is subject to policy, treaty, and legal constraints, while concepts, whether near-term or futuristic in nature, can explore new methods, structures, and systems employment without the same restrictions.

NWP 3-32, Maritime Operations at the Operational Level of War outlines the Navy operational-level fundamentals, command, control, and organization. It is also a bridge between the theory of operational art and the practical specific guidance that Navy commanders and staffs require to accomplish their mission. It is prepared to complement existing joint and Navy doctrine and provides a general guide to the application of command at the operational level of war and the staff organization and functionality required to support the operational commander.

II. History/Background

Naval forces (and more specifically, navies) have a rich culture of operational freedom dating back to the days of sail, when simple mission-type orders, such as "go forth and do the king's work," focused on the Crown's intent. Ships "enjoyed" independence with long voyages and no further communications with their senior headquarters until their return months and years later from deployment. The success of the mission relied heavily upon the tactical commander's ability to interpret the intentions of the senior headquarters and then translate it into tactical actions. The senior headquarters exercised decentralized command, providing broad guidance to the tactical commanders and then relying upon these commanders' initiative to take advantage of opportunities for mission success. This decentralized command was not a choice but rather an acknowledgment of the inability to communicate with ships once they had gone out of sight.

Soon after the Navy's **transition from sail to steam** came the advent of over-the-horizon communications. These communications were neither reliable nor had the ability to pass large amounts of data. Accordingly, the decentralized command style from the era of sail continued for maritime forces throughout World Wars I and II. Communications have since evolved to produce high data rate, long range, and reliable networks. No longer is the senior headquarters unable to communicate with the tactical commander. These networks link sensors and systems to provide senior headquarters and tactical commanders greater understanding of the operational environment. While impressive, these advances in information systems for the Navy are constrained by the laws of physics, which limit the amount and types of information that can be passed, given the bandwidth limitations of afloat commands. Therefore, senior maritime headquarters (MHQ) still exercise decentralized command, albeit using a much enhanced awareness of the operational environment.

Society's **move from the industrial age to the information age** not only signaled a change to the maritime commander's understanding of the operational environment, but also resulted in Navy organizational change.

Prior to the **Civil War**, the Navy rarely had ships operating together as a single organized group. The introduction of steam propulsion, armored hulls, and increased fire power during the Civil War brought about Navy organizational changes. By 1907 the Navy was assigning vessels to the Pacific and Atlantic fleets. In 1913 the Navy recognized that in order to maintain span of control/command, the fleets needed to be sub-divided. The Navy elected to subdivide the fleet into forces, which were defined as "all the vessels of the fleet that are of the same type or class or that are assigned to the same duty." It followed that a vessel was assigned an administrative force commander and an operational force commander. This arrangement of dual chains of command remains in effect today.

During **World War I**, the Navy organization evolved into the three dimensions of surface, subsurface, and the air. In World War II, maritime operations within the operational environment expanded to include not only dominance on the high seas, but also the establishment of sea control in the littorals and amphibious landings of Army and Marine forces. Today's maritime operations have further expanded to include a spectrum of activities from peacetime operations to high-intensity conflict, commonly referred to as the range of military operations.

Success in the modern maritime operational environment requires working with elements of the joint force, multinational partners, and maritime commanders at the tactical, operational, and strategic levels. The Navy has a rich tradition of inter-service cooperation with an aim toward accomplishing joint objectives at the strategic and operational levels of war. Sea control has been a central element in many of our nation's conflicts. The dispatch of maritime forces to confront the Barbary pirates at the beginning of the nineteenth century is illustrative of the role of maritime power in securing strategic lines of communications (LOCs). In the War of 1812, British maritime forces demonstrated the strategic value of sea control as American trade atrophied and British land forces were offered operational flexibility with little fear of losing vital sea lines of communications (SLOCs). American victories in the Great Lakes in 1813 provided an early glimpse of U.S. maritime power's role in operational sea denial. The Mexican War and the American Civil War cemented the requirement for close relationships between ground and maritime forces operations. Whether the operational objective was in support of the introduction of ground forces deep into an enemy's heartland or the strangling of an enemy's economic arteries, the joint campaigns contained major maritime and land operations that were inexorably linked to a common objective and operational design.

During **World War II** commanders used the factors of time, space, and forces to exercise control of Navy, Marine, Army, and Army Air Forces in the maritime domain. In the Pacific, Admiral Nimitz was a master of helping set the conditions for his subordinate commanders and sequencing operations to provide his commanders the opportunity to execute their tasking. He assigned forces to his subordinates and organized his commands (i.e., assigned or attached forces to subordinate commanders) to best support his assigned missions. Then, through his skillful issuance of mission-type orders and clear commander's intent, he allowed his subordinates to combine their initiative with their tactical skills to execute the mission. By managing the risk to forces and the mission, he gave his commanders the freedom and flexibility they required to adapt their planning, thereby taking advantage of opportunities presented to them in the fog of war.

The **Cold War** naval planning focused on the blue-water campaign to defeat the Soviet threats by attrition. Following the Cold War, the focus of maritime operations within the operational environment evolved. Today's threat may not be clearly defined and may very easily be asymmetric in nature. Cyberspace has created another dimension to the operational environment. Although the traditional military-on-military force threat remains, today's commanders must deal with these potentially complex missions with a more holistic approach. Operating in today's maritime operations requires a command and control (C2) system and processes that support planning and execution from the strategic, through operational, to tactical levels. The tactical level accomplishes missions in support of tasks that produce tactical and operational impacts toward operational and strategic objectives in support of strategic and national goals.

There has been little opportunity for the Navy to plan and execute a major operation or campaign (sequence of major operations) since World War II. As a result, the operational art and C2 capabilities associated with command at the maritime operational level have not had the opportunities to evolve and adapt to the modern operational environment. The maritime headquarters with maritime operations center (MHQ with MOC) is focused on defining and developing operational-level headquarters around the globe with some degree of baseline commonality.

The **MHQ with MOC** provides a framework from which Navy commanders at the operational level exercise C2. C2 entails processes (planning, directing, coordinating, and controlling forces and operations) and systems (personnel, equipment, communications, facilities, and procedures employed by commander) as they relate to the exercise of authority and direction over assigned or attached forces and organizations. MHQ with MOC organization does not mimic the World War II headquarters of ADM Nimitz. Similarly, the C2 model from World War II cannot be replicated for today's environment of multi-mission naval platforms operating in a very technologically advanced environment.

See chap. 3 for discussion of the maritime headquarters (MHQ) and the maritime operations center (MOC).

III. The Maritime Domain

Ref: NWP 3-32, Maritime Operations at the Operational Level of War (Oct '08), p. 1-4 to 1-6. See pp. 3-22 to 3-23 for discussion of maritime domain awareness (MDA).

Unlike the other components of the joint force, the maritime component routinely conducts operations across all of the domains, described as air, land, maritime, space, and the information environment. The maritime domain is defined as "the oceans, seas, bays, estuaries, islands, coastal areas, and the airspace above these, including the littorals." This joint definition has fundamental implications for the Navy's role in joint operations. The Navy is the principal warfighting organization that conducts operations over, on, under, and adjacent to the seas: overlying airspaces, surfaces, sub-surfaces, and the ocean bottom, as well as the shoreline infrastructures that affect maritime operations.

The maritime domain also contains social, economic, political, military, and legal components. About 70 percent of the world's surface is covered by the oceans and seas. Naval forces operate from the deep waters of the open ocean to the generally shallower waters fronting the coastlines of the continental land mass and large offshore islands. In the event of regional conflict, small coastal navies operating in the proximity of these straits can pose serious problems for the operations of larger navies. There are several thousand straits in the world's ocean, but only about 200 have some international importance. These straits are the hubs and the most vulnerable segments of sea communications linking "narrow seas" with other seas or open ocean areas. They can also be used to effectively block the exit or entry of hostile naval forces or the transit of an enemy's merchant ships.

The seas/oceans are extremely important to the economic prosperity of many countries. Maritime trade is the principal means of transporting raw materials and manufactured goods. About 96 percent of the entire world's trade by weight is still carried by ships. The sea remains the primary, and by far the most cost-effective, means for the movement of international trade. The importance of the world's oceans and seas to the economic well being and security of all nations has perhaps never been greater than it is today. Approximately 80 percent of all countries border the sea, and nearly 95 percent of the world's population lives within 600 miles of it. About 60 percent of the politically significant urban areas around the world are located within some 60 miles of the coast, and 70 percent are within 300 miles. Approximately 40 percent of all the world's cities with populations of 500,000 or more are located on a coast. By 2025, it is projected 60 percent of the world's population will live in cities, most of which will be in littoral areas. The littorals are economically significant because all seaborne trade originates and ends there.

About 50,000 large ships carry approximately 80 percent of the world's trade. Each year 1.9 billion tons of petroleum, or some 60 percent of all oil produced, are shipped by sea. Some 75 percent of the world's maritime trade and 50 percent of its daily oil consumption pass through a handful of international straits. There are some 4,000 ports involved in maritime trade, including 30 so-called mega-ports.

Diplomatic and political concerns related to the maritime domain have also increased. Many maritime nations, but especially the smaller ones, have tried to extend their claims over offshore resources in order to obtain additional economic benefits. These claims have led in turn to numerous disputes over the exact extent of maritime borders and EEZ's. This is highlighted in diplomatic and legal tension over some archipelagic waters and international straits, since a country's naval forces face certain constraints and restrictions when operating in internal waters, territorial seas, contiguous zones, EEZs, and continental shelves claimed by coastal states. International law provides free and legal access for ships up to the territorial seas, and right of innocent passage for either transiting territorial waters without entering internal waters or proceeding in either direction between the high seas and internal waters.

The United States currently recognizes approximately 150 navies in the world, which are categorized by size. Small navies are mainly used for policing duties in their nation states' territorial waters and EEZs. Medium-size navies have significant capabilities for challenging a much stronger opponent even beyond the EEZ. Larger navies have the capability to project power in many littoral areas of the world's ocean. The great majority of the world's navies are small and capable of operating only in their respective littoral waters. Only a few navies, such as the U.S. Navy, are capable of sustained employment far from their country's shore. In addition, most maritime nations also maintain air forces capable of conducting operations over the adjacent sea/ocean areas.

The U.S. Navy is manned, trained, and equipped to execute maritime operations. At its core, maritime operations are focused on the application of sea power to achieve sea control. However, sea power is not exclusively synonymous with naval warfare. It is a much broader concept that entails at least four elements: the control of international trade and commerce; the usage and control of ocean resources; the operations of navies in war; and the use of navies and maritime economic power as instruments of diplomacy, deterrence, and political influence in time of peace.

The U.S. Navy's culture has evolved to meet the uniqueness of operations in the maritime domain. With mission-tailored and multi-mission platforms, the U.S. Navy is capable of attaining maritime superiority or supremacy. By necessity, modern naval platforms are multi-mission, each with a wide range of capabilities specifically designed to counter threats in the maritime domain and to project power throughout all domains. Navy platforms operate in a very dynamic environment that includes ships and aircraft from potential adversaries and neutral parties. These ships and aircraft are constantly in motion, thereby presenting the operational commander with an added challenge of gaining and maintaining situational awareness (SA).

Employing these uniquely adapted platforms within the highly fluid, multidimensional maritime domain (consisting of undersea, surface, air, land, space, and the information environment) is the purview of the maritime commander. Forward-deployed Navy forces (NAVFOR) shape the operational environment as a matter of routine. Through port visits and multinational exercises they assist in shaping perceptions and influencing the behavior of adversaries and allies, developing allied and friendly military capabilities for self-defense and coalition operations, improving information exchange and intelligence sharing, and providing U.S. forces with peacetime and contingency access.

Navy forces demonstrate the nation's resolve and intentions from international waters. Forward-deployed Navy forces are often times the nation's first responders to events requiring military response outside the homeland. When adversarial action occurs beyond our borders, as a key element of the joint force that is readily capable of employing credible combat power forward, Navy forces are often times the first on-scene force to exercise the deterrence phase of a joint operation or campaign.

Air and land forces will normally arrive in the JOA using a combination of air and sea lift. The primary means to transport supplies to sustain these forces is normally sea lift; a successful sea-lift operation requires the Navy to achieve and maintain maritime superiority within the JOA's maritime domain. The Navy also achieves and maintains sea control over the SLOCs to and from the maritime domain of the JOA.

The Navy is not limited to the maritime domain. It has capabilities that can project fires overland through employment of naval surface fires support (NSFS), Tomahawk land attack missiles (TLAMs), expeditionary forces, and strike/fighter aircraft in support of other components in the joint force. Support of another component in the joint force requires the joint force commander (JFC) to establish a support or other command relationship between the components. Typically, the maritime component supports other joint components when prosecuting targets deep inland, but it may be in a supported role for certain operations (e.g., amphibious operations/non-combat evacuation) in the littorals.

(Maritime Operations) I. Operational Warfare at Sea

IV. Joint Operational Areas

Ref: JP 3-0 (Change 1), Joint Operations (Feb '08), pp. II-15 to II-17.

Operational area is an overarching term encompassing more descriptive terms for geographic areas in which military operations are conducted. Operational areas include, but are not limited to, such descriptors as AOR, theater of war, theater of operations, JOA, amphibious objective area (AOA), joint special operations area (JSOA), and area of operations (AO). Except for AOR, which is normally assigned in the UCP, the GCCs and other JFCs designate smaller operational areas on a temporary basis. Operational areas have physical dimensions comprised of some combination of air, land, and maritime domains. JFCs define these areas with geographical boundaries, which facilitate the coordination, integration, and deconfliction of joint operations among joint force components and supporting commands. The size of these operational areas and the types of forces employed within them depend on the scope and nature of the crisis and the projected duration of operations.

Operational Areas Within a Theater

This example depicts a combatant commander's area of responsibility (AOR), also known as a theater. Within the AOR, the combatant commander has designated a theater of war. Within the theater of war are two theaters of operations and a joint special operations area (JSOA). To handle a situation outside the theater of war, the combatant commander has established a theater of operations and joint operations area (JOA), within which a joint task force (JTF) will operate. JOAs could also be established within the theater of war or theaters of operations.

Ref: JP 3-0, Joint Operations, fig. II-3, p. II-16.

Ref: NWP 3-32, Maritime Operations at the Operational Level of War (Oct '08), p. 2-2 to 2-3.

1. Maritime Control Area
An area generally similar to a defensive sea area in purpose except that it may be established anyplace on the high seas. Maritime control areas are normally established only in time of war.

2. Area of Operations (AO)
An operational area defined by the joint force commander for land and maritime forces. Areas of operation do not typically encompass the entire operational area of the joint force commander but should be large enough for component commanders to accomplish their missions and protect their forces. Also called AO.

3. Operational Area
An overarching term encompassing more descriptive terms for geographic areas in which military operations are conducted. Operational areas include, but are not limited to, such descriptors as area of responsibility, theater of war, theater of operations, joint operations area, amphibious objective area, joint special operations area, and area of operations.

4. Joint Operations Area (JOA)
An area of land, sea, and airspace, defined by a geographic combatant commander or subordinate unified commander, in which a joint force commander (normally a joint task force commander) conducts military operations to accomplish a specific mission

5. Theater of Operations
An operational area defined by the geographic combatant commander for the conduct or support of specific military operations. Multiple theaters of operations normally will be geographically separate and focused on different missions. Theaters of operations are usually of significant size, allowing for operations in depth and over extended periods of time.

Maritime Theater Structure
Ref: NWC 4020A, On Naval Warfare, (Milan Vego, Sept '08), pp. 27 to 29.

A **maritime theater** is usually divided into smaller parts to ensure the most efficient use of subordinate forces in both peacetime and war. This can be done either formally or informally. The smaller the part of the theater the less the need to formally delineate its boundaries. Optimally, the division of the available space should be based on the scale and number of the military objectives to be accomplished by subordinate levels of command.

A maritime theater can comprise one or more **maritime theaters of operations**; and each of these in turn can encompass several **maritime areas of operations**. The latter in turn usually encompass several undeclared **naval (or maritime) combat zones (sectors)**.

Each militarily organized space encompasses an arbitrarily determined **area of interest (AOI)**. For a maritime theater commander, such an area encompasses adjacent geographic areas where political, military, economic, or other events and actions have an effect within his respective theater.

A **joint maritime operations area (JOA)** can be declared when two or more services act jointly to accomplish an operational, or sometimes even a strategic, objective in a certain part of the theater in a situation short of war. A joint area of operations encompasses land and adjacent sea and ocean areas and the airspace above them.

Maritime battlefield refers to a part of the theater in which naval battles, engagements, and other tactical actions are planned or conducted. **Battlespace** pertains to a three-dimensional physical space plus a corresponding part of cyberspace in which a tactical commander sets the term of a battle or engagement. A battlespace is usually larger than an **area of operations**, and it can overlap the battlespace of other friendly commanders.

V. Operational Level of Command (OLC)

Defining operational level of command (OLC) is best done through comparison with the tactical and strategic levels of command. One of the primary differences between the three levels is the objectives. Objectives at the operational level of command are more encompassing than those at the tactical level, yet more focused than those at the strategic level.

The OLC is focused on translating strategic objectives into subordinate tasks/missions by specifying the "what, when, where, who, and why" and leaving the "how" to the subordinates. It links the various operations together into a campaign plan and coordinates the six operational functions of fires, C2, intelligence, movement and maneuver, protection, and sustainment. For the Navy, it is also the level at which administrative and operational command authorities are frequently dual-hatted.

The OLC is also the level at which maritime capabilities are integrated with other component capabilities. Working with the other component commanders through the command relationships designated by the JFC, Navy forces are optimized to achieve operational success. Navy component commanders (NCCs) and numbered fleet commanders and their staffs operate principally at the operational level.

Commands/staffs below the numbered fleet level (e.g., task force commanders and their staffs) are involved in operational-level planning and coordination, but their principal focus is command at the tactical level. OLC's focus on learning about an unfamiliar problem(s) and exploit that understanding to create a broad approach to problem solving. This OLC focus creates a design from which the tactical-level commander can follow established planning procedures to create a detailed plan of action.

See also pp. 3-19 to 3-24.

Navy Command

Navy command is unique and reflects its operational environment, traditions, and culture. Despite the change in today's environment, the Navy has retained unique characteristics in the capabilities it provides, as well as the way it functions compared to the other services/components. Navy forces enjoy a high degree of mobility and flexibility. Being essentially self-deploying, they are able to operate in support of strategic objectives without impacting another nation's sovereignty and do not necessarily require host-nation permission for their presence. As such, they provide a unique characteristic of persistent military capabilities to the combatant commander (CCDR) that is essentially immediately available. Navy tactical commanders are expected to take initiative using the operational-level commander's guidance, which defines what needs to be done but not how to do it.

Unique Aspects of Naval Operational Command

Today, the Navy provides Navy operational commanders unique, multi-mission platforms that have numerous capabilities within a single platform. As such, the tactical forces of the Navy operational commander are multi-mission platforms, which provide a unique challenge to the Navy operational commander when apportioning forces to subordinate commanders. The Navy operational commander exercises those operational authorities delegated by higher authority. With the command authority of operational control (OPCON) the commander can organize forces using the task force (TF) construct. The Navy operational commander has a hierarchical chain of command that extends from the commander to his TF subordinate commanders to their subordinate task groups (TGs), further to task units (TU's), and ultimately to task elements (TEs). As such, the commander can form his task organization to suit the mission assigned. This TF command architecture gives the Navy the ability to assemble its forces as required to support various missions in a seamless manner. Consequently, multi-mission platforms are relatively easily reassigned, or re-tasked, to provide their capabilities to subordinate commanders.

See pp. 2-62 to 2-63 for discussion of Navy task organization, to include task forces, task groups, task units, and task elements.

Chap 2
II. Range/Spectrum of Maritime Operations

Ref: NWP 3-32, Maritime Operations at the Operational Level of War (Oct '08), chap. 2.

I. Levels of War

The levels of war are doctrinal perspectives to clarify the links between strategic objectives and tactical actions. The three levels are strategic, operational, and tactical. Understanding the interdependent relationship of all three helps commanders visualize a logical flow of operations, allocate resources, and assign tasks. Actions within the three levels are not associated with a particular command level, unit size, equipment type, or force or component type. Instead, actions are defined as strategic, operational, or tactical based on their impact or contribution to achieving strategic, operational, or tactical objectives.

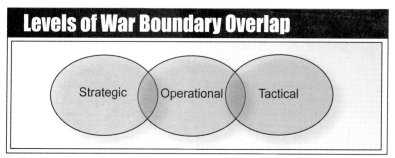

Ref: NWP 3-32, Maritime Opns at the Operational Level of War, fig. 2-1, p. 2-1.

There are no finite limits or boundaries among the three levels. National assets such as intelligence and communications satellites, previously considered principally in a strategic context, today are also significant resources for tactical operations. Commanders at every level must be aware that in a world of constant, immediate communications, any single action may have consequences at all levels.

The operational level is the level of war at which campaigns and major operations are planned, conducted, and sustained to achieve strategic objectives within theaters or other operational areas (OA's). Activities at the operational level:

- Link tactics and strategy by establishing operational objectives needed to achieve the strategic objectives
- Sequence events to achieve the operational objectives
- Initiate actions and apply resources to bring about and sustain these events

The operational level lies between the tactical and strategic levels of war. The boundaries between the tactical, operational, and strategic levels of war overlap and are displayed as three circles with sides overlapping each other. The overlap represents shared activities. Depending on the mission, these shared activities may be many or few.

See also p. 2-62, Notional Relationship of Command Level to Level of War, and p. 3-19, Inherent Relationships to the Levels of War.

(Maritime Operations) II. Range/Spectrum of Maritime Operations 2-9

A. Strategic Level Of War

The strategic level is that level of war at which a nation, often as a member of a group of nations, determines national or multinational (alliance or coalition) strategic objectives and guidance and develops and uses national instruments of power to achieve these objectives. The President establishes policy, which the Secretary of State (SECSTATE) and Secretary of Defense (SecDef) translate into national strategic objectives that facilitate theater-strategic planning. Combatant commanders (CCDR's) usually participate in strategic discussions with the President and SecDef through the Chairman of the Joint Chiefs of Staff (CJCS) and with allies and coalition partners. Thus, the CCDR strategy is an element that relates to U.S. national strategy and operational activities within the theater. Derived from national strategy and policy and shaped by doctrine, military strategy provides a framework for conducting operations. Strategic military objectives define the role of military forces in the larger context of national strategic objectives.

For specific situations that require the employment of military capabilities (particularly for anticipated major combat operations), the President and SecDef typically establish a set of national strategic objectives. The supported CCDR often will have a role in achieving more than one national objective. Some national objectives will be the primary responsibility of the CCDR, while others will require a more balanced use of all instruments of national power with the CCDR in support of other government agencies. Achievement of these objectives should result in attainment of the national strategic end state — the broadly expressed conditions that should exist at the end of a campaign or operation. Once established, the national strategic objectives enable the supported commander to develop the military end state, recommended termination criteria, and supporting military strategic objectives.

Commanders at the strategic level define the military end state using military strategic objectives and the conditions that can support achievement of each objective. The strategic commander defines the time and space along with his military strategic objectives and conditions in his guidance to the operational commander.

B. Operational Level

The operational level links the tactical employment of forces to national and military strategic objectives. The focus at this level is on the design and conduct of operations using operational art, which is defined in the DOD Dictionary of Military and Associated Terms, as "the application of creative imagination by commanders and staffs — supported by their skill, knowledge, and experience — to design strategies, campaigns, and major operations and organize and employ military forces. Operational art integrates ends, ways, and means across the levels of war."

JFC's and their component commanders use operational art to determine when, where, and for what purpose major forces will be employed and to influence the adversary disposition before combat. Operational art governs the deployment of those forces, and their commitment to or withdrawal from battle. Operational art also governs the planning of battles and major operations to achieve operational and strategic objectives.

Commanders at the operational level build campaign/major OPLAN's to achieve the military strategic/operational objectives. A campaign plan is defined as a joint OPLAN for a series of related major operations aimed at achieving strategic or operational objectives within a given time and space. When building and executing the military campaign/major operation plan, commanders at the operational level must ensure military actions are synchronized with those of other government and nongovernmental agencies and organizations, together with international partners, in order to achieve national strategic objectives.

At the heart of the campaign/major OPLAN are operational objectives and associated conditions and tasks for each tactical action contained in the plan. The campaign/major OPLAN evolves from the assessment and planning phases of the commander's decision process and provides the operational-level commander's direction to the tactical commanders. Once the campaign/major OPLAN is provided to tactical-level commanders for execution, the operational-level commander monitors its execution and controls force actions as required.

Unified Action

Operational success often depends on unified action. The construct of unified action highlights the integrated and synchronized activities of military forces and nonmilitary organizations, agencies, and the private sector to achieve common objectives, though in common parlance joint operations increasingly has this connotation. Unified actions are planned and conducted by joint force commanders in accordance with guidance and direction received from the President, Secretary of Defense, and combatant commanders.

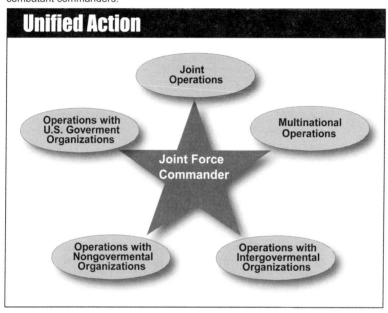

Ref: JP 1, Doctrine for the Armed Forces of the United States, fig. II-1, p. II-3.

C. Tactical Level

Tactics is the employment and ordered arrangement of forces in relation to each other. The tactical level focuses on planning and executing battles, engagements, and activities to achieve military objectives assigned to tactical units or Navy task forces (TF's), task groups (TG's), task units (TU's), and task elements (TE's). An engagement normally is of short duration and includes a wide variety of actions between opposing forces. A battle consists of a set of related engagements, which typically last longer than engagements involving larger forces, such as fleets, armies, and air forces; and normally affect the course of a campaign. Forces at this level generally employ various tactics to achieve their military objectives.

Commanders at the tactical level derive tasking from orders guided by the campaign/major OPLAN. This tasking will include tactical objectives. Tactical objectives often are associated with the specific "target" of an action and are associated with the enemy's centers of gravity (COG's), critical vulnerabilities, and decisive points (DP's). In this context, an objective could be a terrain feature, such as attainment of maritime superiority in a decisive area, the seizing or defending of which is essential to the commander's plan. The objective also could be an enemy force or capability, the destruction of which creates a vulnerability for the adversary.

(Maritime Operations) II. Range/Spectrum of Maritime Operations 2-11

II. Maritime Operations Across the Range of Military Operations

Ref: NWP 3-32, Maritime Operations at the Operational Level of War (Oct '08), p. 2-7 to 2-8.

The Navy is not limited to the maritime domain and conducts activities across the range of military operations.

U.S. Navy Activities - Range Military Ops

Military Engagement, Security Cooperation and Deterrence	Crisis Response and Limited Contingency Operations	Major Operations and Campaigns
• Homeland Defense • Protection of U.S. Economic Interests • Enforcement of Maritime Treaties • Support of Foreign Policy • Support of Military (Theater) Strategy • Support of Peace Operations	• Support of Insurgency • Support of Counterinsurgency • Support of Counterterrorism • Foreign Humanitarian Assistance • Civil Support	• Regional War • Global War

Ref: NWP 3-32, Maritime Opns at the Operational Level of War, table 1-1, p. 1-3.

Expanding webs of political, military, economic, social, informational, and infrastructure systems provides opportunities for regional powers to compete on a broader scale and emerge on the global landscape with considerable influence. Littoral and urban environments and other complex terrain will increasingly characterize areas of operation that may include humanitarian crisis conditions and combat operations. Adaptive adversaries continually will seek new capabilities and new employment methods to counter the United States and its allies. As new capabilities, or new methods of employing capabilities, are developed and become more accessible to more players, the conduct of warfare and crisis resolution will change. The nature of war will remain a violent clash of wills between states or armed groups pursuing advantageous political ends. The conduct of warfare will include combinations of conventional and unconventional, kinetic and non-kinetic, and military and nonmilitary actions and operations, all of which add to the increasing complexity of the operational environment.

Future adversaries may lack the ability or choose not to oppose the United States through traditional military action. These adversaries will challenge the United States and its multinational partners by adopting and employing asymmetric methods across selected air, land, maritime, and space domains as well as the information environment against areas of perceived U.S. vulnerability. Many will act and operate without regard for the customary law of war.

Navy and joint forces must maintain an unsurpassed ability to fight and win our nation's wars. Sea power with its concurrent military and geo-economic focus supports national security goals through operations that do not necessarily include either adversaries or combat. Examples include peacekeeping, humanitarian relief operations, and support to civil authorities, foreign and domestic. These operations can contribute to preventing conflict and may require different types of capabilities or different methods of employing those capabilities than traditionally used to fight wars.

Regardless of the type of operation, the Navy and joint forces will require capabilities and processes to respond in the most efficient manner and to minimize the use of military force to that necessary to achieve the overarching strategic objective. This includes the need for engagement before and after conflict/crisis response, the need for integrated involvement with interagency and multinational partners, and the need for multipurpose capabilities that can be applied across the range of military operations.

A common thread throughout the range of military operations is the involvement of a large number of agencies and organizations — many with indispensable practical competencies and significant legal responsibilities and authorities — that interact with the Navy and our multinational partners. The Navy commander at the operational level develops campaign/major OPLAN's that integrate and synchronize maneuver and direct tactical force commanders' actions in conjunction with the actions of other government and nongovernmental agencies to achieve unified action.

A. Military Engagement, Security Cooperation, and Deterrence

Military engagement, security cooperation, and deterrence shape the operational environment and help keep the day-to-day tensions between nations or groups below the threshold of armed conflict while maintaining U.S. global influence. These activities improve relations with potential allies and coalition partners. These ongoing and specialized activities establish, shape, maintain, and refine relations with other nations and domestic civil authorities (e.g., state governors or local law enforcement). The general strategic and operational objective is to protect U.S. interests at home and abroad.

By virtue of their being forward deployed, Navy forces (NAVFOR) have the ability to support accomplishment of national strategic objectives without impacting a country's sovereignty, and routinely are used to conduct military engagement, security cooperation, and deterrence activities that help "shape" the operational environment.

Operational command of these activities is complicated by their continuous nature, fluid strategic priorities in response to world events, and complex identification of measures (goals) from which to access achievement of strategic and operational objectives. Naval commanders at the operational level develop a type of campaign/major OPLAN called a theater security cooperation plan (TSCP) to ensure Navy force actions are correctly prioritized, sequenced, and timed to support achievement of CCDR engagement, security cooperation, and deterrence strategic objectives. These campaign/major OPLAN's, unlike ones developed for other types of military operations, are being executed continuously. Force availability, changes to geographic combatant commander (GCC) priorities, and world events all impact these campaign/major OPLAN's. Achievement of strategic objectives may take decades. Conditions (measures for assessment of strategic objective achievement) will evolve as the plan progresses.

B. Crisis Response or Limited Contingency Operations

Crisis response or limited contingency operations can be a single small-scale, limited-duration operation or a significant part of a major operation of extended duration involving combat. The associated general strategic and operational objectives are to protect U.S. interests and prevent surprise attack or further conflict. The level of complexity, duration, and resources depends on the circumstances. Many of these operations involve a combination of military forces and capabilities in close cooperation with other government agencies (OGA's), IGO's, and NGO's. A crisis may prompt the conduct of foreign humanitarian assistance (FHA), civil support (CS), noncombatant evacuation operations (NEO's), peace operations (PO), strikes, raids, or recovery operations.

C. Major Operations or Campaigns Involving Large-Scale Combat

Major operations or campaigns involving large-scale combat place the United States in a wartime state. In such cases, the general goal is to prevail against the enemy as quickly as possible, conclude hostilities, and establish conditions favorable to the United States and its multinational partners.

III. Major Naval Operations

Ref: The Newport Papers: Major Naval Operations (Milan Vego, 2008), chaps. 1 & 2.

Major naval operations are the principal methods by which naval forces achieve operational objectives in a conflict at sea. In generic terms, a major naval operation consists of a series of related major and minor naval tactical actions conducted by diverse naval forces and combat arms of other services, in terms of time and place, to accomplish an operational (and sometimes strategic) objective in a given maritime theater of operations. Major naval operations are planned and conducted in accordance with an operational idea (scheme) and common plan. They are normally an integral part of a maritime or land campaign, but they can sometimes be conducted outside of the framework of a campaign.

A **maritime campaign** is predominantly fought on the open ocean and in sea areas adjacent to a continental landmass. One's naval forces would play the most important role in such a campaign. A land campaign in the littoral area would also involve participation of naval forces. In general, the closer to the continental landmass or large archipelagoes one's naval forces have to operate, the more the success of their actions will depend on close cooperation with other services or a high degree of jointness.

Tactical actions in a major naval operation can be fought on the surface, beneath the surface, in the air, and in some cases on the coast. Naval tactical actions can range from actions in which weapons are not used, such as patrols and surveillance, to attacks, strikes, raids, engagements, and naval battles. As the term implies, they are aimed at accomplishing tactical objectives in a given part of a maritime theater.

- In the past, a **naval battle** was the main method of accomplishing a major tactical objective as a part of a major naval operation. It consisted of a series of related attacks, counterattacks, strikes, and counterstrikes coordinated in time and place.

- In the past, a **naval engagement** consisted of a series of related strikes/counterstrikes and attacks/counterattacks aimed to accomplish the most important tactical objective in a naval battle.

- With the advent of missiles and other long-range, highly precise, and lethal weapons it became possible to destroy the enemy force at sea or on the coast at much longer range than by using guns or torpedoes. Hence, a new method of combat force employment called "**strike**" gradually emerged as the principal method of accomplishing major tactical and sometimes operational objectives in war at sea and in the air. Depending on the target to be destroyed or neutralized, one may differentiate tactical, operational, and strategic strikes. A strike can be conducted by a small number of platforms of a single type of force—for example, missile craft, submarines, or attack aircraft (helicopters).

- A broader form of strike is a **naval raid**—conducted by a single or several naval combat arms to accomplish a tactical objective as a part of a major offensive or defensive naval operation. The aim is usually to deny temporarily some position or to capture or destroy an enemy force, coastal installation, or facility. Temporary or local control of the sea is not a prerequisite for the success of these actions. The stronger fleet can also conduct raids to divert the enemy's attention or to force the enemy to react in a secondary sector. A naval raid is usually conducted against an objective that the enemy considers so valuable that its loss or serious degradation cannot be ignored. A larger purpose of a naval raid is to accomplish some temporary advantage, with the threat of future repetition.

- The most frequently conducted tactical action using weapons is a **naval attack**, a combination of tactical maneuver and weapons used to accomplish a minor tactical objective. A naval attack can be conducted independently or as part of a strike or raid. A naval attack can be conducted by a single or several types of platforms.

Types of Major Naval Operations

In generic terms, the main purposes of a major naval operation today in the case of a high-intensity conflict at sea can be fleet versus fleet (destroy the enemy fleet at sea or in its bases); fleet versus shore (conduct an amphibious landing on the opposed shore and destroy enemy coastal installations/facilities); attack against an enemy's maritime trade; defense and protection of friendly maritime trade; destruction of an enemy's (or protection of friendly) sea-based strategic nuclear forces; and support of ground forces operating in the littoral.

A. Major Operations: Fleet versus Fleet

Destruction or neutralization of the main enemy forces at sea and in their bases is the main prerequisite for obtaining and then maintaining sea control in a given part of a maritime theater. This is especially important at the start of the hostilities at sea. In the past, this objective was usually accomplished by a clash of major parts of the opposing fleets in a so-called decisive battle or "general fleet action." Major naval operations to destroy a neutralize a major part of the enemy fleet can be conducted in a distant ocean area or in a narrow sea.

B. Major Operations: Fleet versus Shore

The stronger side at sea will occasionally conduct major naval operations to accomplish operational objectives on the enemy coast, amphibious landings to destroy or annihilate enemy coastal installations/facilities or important military-economic centers deep in the enemy's territory.

C. Major Operations versus Enemy Maritime Trade

Attack on maritime trade in general has always been an important feature of any war at sea. It constitutes a form of pressure on a country dependent on overseas trade for the necessities of life. These actions are meant to destroy or neutralize not only the enemy shipping at sea and in ports but also such other elements of maritime trade as shipyards/ship-repair facilities, port installations/facilities, and railroad/road traffic in the littorals.

D. Major Naval Operations to Defend/Protect Maritime Trade

Both the stronger and the weaker sides at sea would commit large forces to ensuring an uninterrupted flow of maritime traffic throughout the conflict. Obviously, this operational task is more critical, and also usually more difficult, for the weaker side.

E. Destruction/Protection of Seaborne Nuclear Deterrent Forces

Today only large navies, specifically those of the United States and Russia, have the capability to mount major operations aimed at destroying the enemy's sea-based strategic nuclear forces—that is, against ballistic-missile submarines (SSBNs) and their supporting elements. Such operations might be focused on destruction of the enemy SSBNs, either at their basing areas, during their transits to or from them, or in patrol zones.

F. Major Operations in Support of Ground Forces on the Coast

One's naval forces can cooperate with ground forces operating along the coast in both the offense and defense. In the broadest terms, naval forces support troops on the coast by providing cover, support, and supply. Cover means preventing enemy air, missile, or gunnery strikes, or amphibious landings on the flank or in the rear of friendly ground troops. These tasks are accomplished by destroying forces that threaten friendly ground troops from the sea, as well as enemy amphibious forces at their embarkation areas, in transit, and in their landing areas. Support by naval forces encompasses a range of tasks, from destroying important targets on the coast and in the depth of the enemy defenses to attacking maritime traffic in coastal waters flanking troops on the coast. Supply includes transport of troops and materiel; seizure of crossings over water obstacles (straits/narrows, river estuaries, etc.) for friendly troops; and evacuation of troops from beaches, naval bases, or ports.

IV. Aegis Ballistic Missile Defense (BMD)

Ref: www.mda.mil/system/aegis_bmd.html

Aegis Ballistic Missile Defense (BMD) is the sea-based component of the Missile Defense Agency's Ballistic Missile Defense System (BMDS). Aegis BMD builds upon the Aegis Weapon System, Standard Missile, Navy and joint forces' Command, Control and Communication systems. The Commander, Operational Test and Evaluation Force formally found Aegis BMD to be operationally effective and suitable. The Navy embraces BMD as a core mission. In recognition of its scalability, Aegis BMD/SM-3 system is a keystone in the Phased, Adaptive Approach for missile defense in Europe.

BMD Capabilities

- Defeats short- to intermediate-range, unitary and separating, midcourse-phase, ballistic missile threats with the Standard Missile-3 (SM-3), as well as short-range ballistic missiles in the terminal phase with the SM-2.
- Flight tests are conducted by Fleet standard warships, operated by fleet Sailors and Officers. Each test increases the operational realism and complexity of targets and scenarios and is witnessed by Navy and Defense Department testing evaluators.
- Aegis BMD ships on Ballistic Missile Defense patrol, detect and track ballistic missiles of all ranges including Intercontinental Ballistic Missiles and report track data to the missile defense system. This capability shares tracking data to cue other missile defense sensors and provides fire control data to Ground-based Midcourse Defense interceptors located at Fort Greely, Alaska and Vandenberg Air Force Base, California and other elements of the BMDS including land-based firing units (Terminal High Altitude Area Defense, Patriot) and other Navy BMD ships.

There are 21 Aegis BMD combatants (5 cruisers [CGs] and 16 destroyers [DDGs]) in the U.S. Navy. Of the 21 ships, 16 are assigned to the Pacific Fleet and 5 to the Atlantic Fleet. The Secretary of Defense announced earlier this year that 6 more DDGs would be BMD equipped. These additional six DDGs will be from Fleet Forces in the Atlantic. The MDA and the Navy, working together, will increase the number of BMD capable ships to 32 by end of 2013.

V. Cyber Warfare

Ref: www.stratcom.mil; www.fcc.navy.mil; www.netwarcom.navy.mil

On June 23, 2009, the Secretary of Defense directed the Commander of U.S. Strategic Command (USSTRATCOM) to establish USCYBERCOM. Initial Operational Capability (IOC) was attained on May 21. USCYBERCOM will fuse the Department's full spectrum of cyberspace operations and will plan, coordinate, integrate, synchronize, and conduct activities to: lead day-to-day defense and protection of DoD information networks; coordinate DoD operations providing support to military missions; direct the operations and defense of specified Department of Defense information networks and; prepare to, and when directed, conduct full spectrum military cyberspace operations. The command is charged with pulling together existing cyberspace resources, creating synergy that does not currently exist and synchronizing war-fighting effects to defend the information security environment.

U.S. Cyber Command (USCYBERCOM)

USCYBERCOM plans, coordinates, integrates, synchronizes, and conducts activities to: direct the operations and defense of specified Department of Defense information networks and; prepare to, and when directed, conduct full-spectrum military cyberspace operations in order to enable actions in all domains, ensure US/Allied freedom of action in cyberspace and deny the same to our adversaries.

VI. Maritime Operational Threat Response (MOTR) Plan

Ref: NWC Maritime Component Commander's Guidebook.

MOTR is a presidentially-directed plan to achieve a coordinated U.S. Government response to threats against the United States and its interests in the maritime domain. The MOTR Plan includes operational coordination requirements to ensure quick and decisive action to counter maritime threats/hazards.

National Security Presidential Directive (NSPD) 41 / Homeland Security Presidential Directive (HSPD) 13 establish U.S. policy, guidelines and implementation actions to enhance U.S. national security and homeland security by protecting U.S. maritime interests. The Maritime Operational Threat Response (MOTR) plan, approved in 2006, was one of the plans directed by NSPD 41/HSPD 13. There is no geographic scope to the MOTR process; it applies to maritime homeland security/defense in the vicinity of the homeland as well as maritime threats against the U.S. and its interests overseas.

The MOTR process directs the integration of national-level maritime command and operations centers to ensure coordinated whole-of-government responses. The Office of the Secretary of Defense (OSD) represents the Department of Defense (DOD) during MOTR coordination activities and policy discussions. The Joint Staff J3, on behalf of OSD, and with support from affected combatant commanders and corresponding navy component commander staffs, serves as the DOD action agent for MOTR coordination activities.

The MOTR plan has specified triggers/criteria for when agencies shall initiate co-ordination activities. Based upon authority, jurisdiction, capability, competency and partnerships, the plan has pre-designated lead and supporting agencies depending on the activity and area. Through MOTR coordination activities, these roles are further refined to the desired USG outcome and the maritime threat that is being addressed. Lead MOTR agencies are those that have the most direct role and responsibility with respect to a specific MOTR; the designated lead MOTR agency will coordinate with all other MOTR agencies throughout the event/response. Supporting MOTR agencies will provide expertise and assistance to the lead MOTR agency in support of the desired national (USG) outcome.

The Global MOTR Coordination Center (GMCC) is an interagency organization established in 2010 to serve as the executive secretariat and facilitator. The GMCC serves as the "honest broker" and supports the facilitation of the MOTR coordination process among the USG agencies/organizations.

MOTR teleconferences, VTCs and/or emails are used as tool to facilitate interagency coordination. Issues raised during the MOTR coordination process can lead to the initiation of JFMCC conference calls for coordination of DOD maritime actions. (JFMCC conference calls may also include U.S. interagency and certain multinational commands or organizations.) Convening a MOTR conference is not a prerequisite for responding to a threat. Neither the GMCC nor the MOTR process supplants or replaces existing agency authority.

MOTR addresses the full range of threats including unlawful or hostile acts by state and non-state actors, terrorism and piracy. When requested or required, MOTR can also be a coordination enhancement tool in support of existing WMD counter-proliferation protocols. Thus, MOTR is a whole of government coordination process in response to a range of threats against the U.S. and its interests in the maritime domain. From the maritime operational level of war perspective, it is a top down or bottom up process that is a tool for the maritime component commander and combatant commander to address issues in the maritime domain where DOD either lacks or overlaps with other USG agencies/organizations in the authority, capacity and/or competency to fully resolve the maritime threat.

USCYBERCOM will centralize command of cyberspace operations, strengthen DoD cyberspace capabilities, and integrate and bolster DoD's cyber expertise. Consequently, USCYBERCOM will improve DoD's capabilities to ensure resilient, reliable information and communication networks, counter cyberspace threats, and assure access to cyberspace. USCYBERCOM's efforts will also support the Armed Services' ability to confidently conduct high-tempo, effective operations as well as protect command and control systems and the cyberspace infrastructure supporting weapons system platforms from disruptions, intrusions and attacks.

USCYBERCOM is a sub-unified command subordinate to USSTRATCOM. Service elements include:

- USA – Army Forces Cyber Command (ARFORCYBER)
- USAF – 24th USAF
- USN – Fleet Cyber Command (FLTCYBERCOM)
- USMC – Marine Forces Cyber Command (MARFORCYBER)

U.S. Fleet Cyber Command (U.S. Tenth Fleet)

The mission of Fleet Cyber Command is to direct Navy cyberspace operations globally to deter and defeat aggression and to ensure freedom of action to achieve military objectives in and through cyberspace; to organize and direct Navy cryptologic operations worldwide and support information operations and space planning and operations, as directed; to direct, operate, maintain, secure and defend the Navy's portion of the Global Information Grid; to deliver integrated cyber, information operations cryptologic and space capabilities; and to deliver global Navy cyber network common cyber operational requirements.

Tenth Fleet was reactivated January 29, 2010 as U.S. Fleet Cyber Command/U.S. Tenth Fleet at Fort Meade, Maryland. The command has both joint and service responsibilities. The split name alludes to these dual responsibilities.

As Fleet Cyber Command, it is the Naval component to U.S. Cyber Command, the sub-unified cyber commander. As U.S. Tenth Fleet, the command provides operational support to Navy commanders worldwide, supporting information, computer, electronic warfare and space operations. In addition to joint and service reporting, the command also serves as the Navy's cryptologic commander, reporting to the Central Security Service. Tenth Fleet has operational control over Navy information, computer, cryptologic, and space forces. Components include Naval Network Warfare Command, Navy Cyber Defense Operations Command (NCDOC), Naval Information Operation Commands (NIOC), Combined Task Forces (CTF), and NIOC Suitland (Research, Development, Test & Evaluation).

Naval Network Warfare Command (NETWARCOM)

NETWARCOM's mission includes operations to secure the Navy's portion of the Global Information Grid and deliver reliable and secure net-centric and space warfighting capabilities. This is performed on behalf of Commander, 10th Fleet for Navy and Joint operations worldwide, leveraging decades of cyber experience and expertise. As the Navy's Cyber warfare force, it is the Navy's central authority for the delivery of cyber forces and capabilities to warfighters and is the commander of the Navy's service cryptologic element (SCE).

In 2002, some 23 organizations from several commands, including the former Naval Space Command, Naval Computer and Telecommunications Command, Fleet Information Warfare Center, and Navy Component Task Force - Computer Network Defense were brought together to form Naval Network Warfare Command, emphasizing the organization's focus on the operation and defense of the Navy's networks.

In 2009, Naval Network Warfare Command was reorganized and it's mission revised to operate and defend the Navy's portion of the Global Information Grid and deliver reliable and secure Net-centric and Space warfighting capabilities in support of strategic, operational and tactical missions across the Navy.

III. Objectives of Naval Operations

Ref: NWC 4020A, On Naval Warfare (Dr. Milan Vego, Sept '08).

War at sea is an integral part of war as a whole. It has almost never taken place alone but has been conducted in conjunction with war on land and, in the modern era, with war in the air. The objectives of naval warfare have been and continue to be an integral part of war's objectives. These, in turn, are accomplished by the employment of all the services of a country's armed forces. In contrast to war on land, the objectives in war at sea are almost invariably physical in character. The principal strategic or operational objective for a stronger side is to obtain sea control in the whole theater or a major part of it, while the weaker side tries to obtain sea denial. Normally, a relatively strong but initially weaker side at sea aims to ultimately obtain sea control for itself. When operating in an enclosed sea theater, a blue-water navy would aim to obtain choke-point control, while the weaker side would conduct counter–choke-point control. Another operational objective for both the stronger and weaker sides at sea is to establish, maintain, and, if possible, expand control of their respective basing and deployment areas for their naval forces and aircraft, thereby creating prerequisites for planning, preparing, and executing major operations. Navies are also for accomplishing diverse objectives in low-intensity conflict and operations short of war. These objectives might range from sea control or sea denial in support of an insurgency or counterinsurgency, to the enforcement of a cease-fire and arms embargoes in peace operations.

Strategic Offensive vs. Defensive Posture

The overall strategic posture in a war at sea is inextricably linked to the strategic situation in general and the war on land in particular. One's navy can generally be on a strategic offensive or defensive. Depending on the course of the war, a strategic offensive could be followed by a strategic defensive or vice versa. Whether a war at sea is conducted primarily offensively or defensively depends upon, among other things, the country's geostrategic position on land and at sea, the initial balance of strength at sea and in the air, and the overall war objectives to be accomplished. The progress of the war on land, however, ultimately determines whether one's navy is employed predominantly offensively or defensively. Such a decision cannot be made without full consideration of the factors of space, time, forces and their combinations, and how they relate to the accomplishment of the overall war's objectives.

I. Sea Control

In the past, the principal objective of any fleet was to obtain and maintain what is commonly called "command of the sea" in a certain part of the open ocean or the sea. In the aftermath of World War I, the term "command of the sea" was replaced in the West by the term sea control, while the Soviets (and the Russians today) continued the use of the equivalent of the older term "sea mastery." The use of the term "sea control" was a result of the gradual realization that the new technological advances, specifically mines, torpedoes, submarines, and aircraft, made it difficult, even for the stronger navy, to obtain full command of the sea for any extended time over a large part of the theater. In fact, the term "sea control" more accurately conveys the reality that in a conflict between two strong opponents at sea, it is not possible, except in the most limited sense, to completely control the seas for one's use or to completely deny their use to an opponent. It essentially means the ability of one's fleet to operate with a high degree of freedom in a sea or ocean area, but only for a limited time.

Operational Factors in Naval Warfare

Ref: NWC 4020A, On Naval Warfare, Dr. Milan Vego (Sept '08), pp. 13 to 19 and fig. 1.

In contrast to the war on land, sea control is not obtained by occupying or capturing a certain ocean or sea area. It is not obtained through the permanent presence of one's fleet. There are no front lines at sea and no fortified positions by which one can control the territory. The sea or ocean area is invariably abandoned by the victorious side regardless of whether the opponent was completely defeated or not. Sea control means that more or less by the destruction of the enemy fleet one can accomplish tasks without serious opposition from the enemy.

In theory sea control and "disputed" (or contested) sea control can be strategic, operational, and tactical in their scale. Strategic sea control pertains to the entire maritime theater, while control of a major part of a maritime theater represents "operational" sea control. Tactical control refers to control of a maritime combat sector or zone but sometimes can encompass a maritime area of operations. However, in practical terms, the focus should be invariably on strategic or operational sea control or disputed control, not tactical sea control.

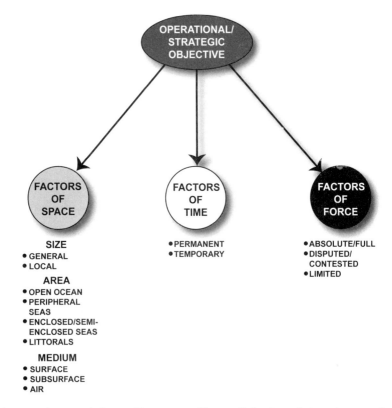

In general, sea control cannot be expressed in quantitative terms but can be recognized only in its effects. Sea control is always relative in spatial terms. It pertains only to the specific part of the theater in which a certain degree of control must be obtained. Sea control is also relative in terms of the factor of time. Once sea control is obtained, it must be continuously maintained. It is also relative in terms of the factor of force.

Factors of Space

In the sailing era, command of the sea was limited to control of only the sea surface. It was obtained by defeating the enemy fleet in a single or several major battles. The naval battle was the main method for obtaining command of the sea. However, even after the decisive battle, the victorious side did not enjoy absolute command of the sea. By the late nineteenth century, it was increasingly more difficult for a stronger fleet to obtain command of the sea because of the advent of torpedoes and mines. In the decade prior to World War I, submarines and aircraft made that task even more complicated.

In the modern understanding of the term, **general sea control** can be understood to mean that the stronger side can carry out its main operational tasks in terms of scope, time, and place without significant challenge from its opponent.

Local sea control exists when one side possesses superiority in the part of the sea or ocean area that is operationally significant for executing one or several specific tasks. Sometimes local control of such an area must be obtained to conduct amphibious landings or to bombard the enemy's coastal installations/facilities. Drastic changes to a situation are a common occurrence. In general, local sea control is usually temporary.

Factors of Time

In terms of the factor of time, sea control can be either permanent or temporary. **Permanent sea control** exists when the stronger fleet completely dominates a given theater, either because the other side does not have any means to deny that control or because its fleet has been destroyed. In practice, however, it is more common that the weaker side still has some means at its disposal to challenge the stronger side's control.

Temporary sea control often results from the inability of either side to obtain a decision. When one side loses the initiative, for whatever reason, the abandonment may be permanent or temporary. The weaker side at sea then usually falls back on the defensive and keeps a major part of its fleet in bases, avoiding any decisive action at sea. If a weaker opponent succeeds in obtaining superiority in the air, this in itself could be sufficient to allow him the use of the sea for a specific purpose and for a limited time.

Factors of Force

In terms of the factor of force, sea control can range from absolute to contested. It can also mean the free use of particular types of ships but not others. **Absolute sea control** means, in practice, that one's fleet operates without major opposition while the enemy fleet cannot operate at all. It aims in general to obtain sea control of the entire theater, or the major part of the theater, so that one can employ one's fleet whenever and wherever required without threat from the enemy. In other words, absolute sea control equates to maritime supremacy or maritime mastery. The weaker side then cannot employ its submarines, aircraft, or sometimes even mines. In practice, control of large ocean or sea areas cannot be absolute in terms of either space or time in the presence of an undefeated and strong opponent. The only exception is when one side possesses a fleet and the other does not and has no other means to dispute control.

Limited sea control (also called conditional or working control) is usually the consequence of the drastic shift in the operational or strategic situation when the initiative passes from one side to the other. Then one side in the conflict has a high degree of freedom to act, while the other operates at high risk. In a case where absolute control cannot be obtained, the stronger side would usually try to secure temporary control of limited sea or ocean areas for conducting operations necessary to the successful progress of the war on land.

Disputed (or contested) sea control is usually the principal objective of a weaker but relatively strong navy in the initial phase of a war at sea. Disputed sea control occurs when the opposing sides possess roughly equal capabilities and opportunities to obtain sea control in a theater as a whole (or in one of its parts) and there is no significant change in the ratio of forces, nor a change of the initiative to either side.

Obtaining, maintaining, and exercising sea control are related but not identical terms. They differ in terms of time and the efforts of one's naval forces. Sea control is obtained primarily by the employment of one's maritime forces in the form of major naval operations. In the littorals, these operations will be joint or combined; that is, not only naval forces but combat arms/branches of other services will take part. The result of one's sea control should ensure that one's forces can carry out their main tasks without significant interference by the opponent. After sea control is obtained, it must be maintained. In operational terms, this phase equates to consolidation of strategic or operational success. The navy's main operational tasks should determine the degree of one's sea control to be obtained and maintained. Exercising sea control is carried out through a series of operational tasks aimed to exploit strategic or operational success. The successful execution of operational tasks should expand and reinforce the degree of one's sea control obtained in a certain sea or ocean area in terms of time and space.

Choke-Point Control

The struggle for control of the straits and narrows, or "choke points," is a unique feature of war for control of a typical narrow sea. For a blue-water navy, general sea control is hardly possible without establishing not only control on the open ocean but also direct or indirect control of several critical passages of vital importance to the world's maritime trade, or by obtaining control of a given enclosed or semi enclosed sea theater. The objective for a weaker side is, then, just the opposite: choke-point-control denial. In either case, but particularly for a weaker side, this objective would normally require the highest degree of cooperation among naval forces and combat arms of other services.

Basing/Deployment Area Control

One of the principal and most important tasks of any navy is to obtain and maintain control of its own basing and deployment areas. Without securing control of one's basing and deployment area first, it is difficult, if not impossible, to prepare and execute major naval operations. This objective is especially critical for naval forces operating in an enclosed or semi enclosed sea. It is aimed to obtain a sufficient degree of security for coastal traffic both in coastal waters and ashore and create preconditions for the successful deployment of one's naval forces.

Basing/deployment area control is an integral part of a broader task, operational protection in a given maritime theater. Control of one's own basing and deployment area is an operational objective in war. The ultimate purpose is to ensure the safety of one's naval and other forces at their bases and deployment areas (close to one's controlled shores) from enemy attacks on the sea, the air, and the ground.

Obtaining Air Superiority

A blue-water navy can obtain control of a typical narrow sea by a combination of methods. The first and most critical operational task is to obtain air superiority in the area. In fact, sea control usually cannot be obtained if air superiority is lacking. The struggle for air superiority in narrow seas is most closely linked to war on land. Both naval and air forces should take part in the struggle for sea control. Because of the short distances, the effectiveness of air strikes against enemy ships and forces and installations/facilities on the coast is considerably higher in a typical narrow sea than on the open ocean. Land-based aircraft can fly more sorties within a given time frame than can carrier-based aircraft on the open ocean. In a narrow sea with many offshore islands and islets, land-based aircraft can strike from bases flanking the transit routes of the enemy ships. Aircraft represent a constant threat to the survivability of all ships, but especially to one's surface forces.

Open Ocean vs. Narrow Seas

Ref: NWC 4020A, On Naval Warfare (Dr. Milan Vego, Sept '08), pp. 38 to 39.

Sea control on the open ocean cannot be isolated from sea control in semi enclosed and enclosed sea theaters, especially where land and sea boundaries are contiguous. The situation on land, especially in the coastal area, greatly affects the struggle for sea control and air superiority. Hence, in such theaters, the struggle for sea control would require the closest cooperation among all services. Normally, attaining sea control in an enclosed sea or a marginal sea of an ocean should result in a drastic change in the operational situation. This criterion is a determining factor for deciding in which parts of the theater and in what sequence sea control should be obtained. The objective should be to obtain complete sea control in the entire theater. This does not exclude, especially at the beginning of hostilities, establishing areas with a different degree of control by one's forces. The physical size of these areas should be such as to have an operational impact on the entire situation in the respective theater.

Obtaining and maintaining sea control in a typical narrow sea depends on both the progress of the war in the littoral area and the possession of air superiority. The withdrawal of enemy forces from a given part of an enclosed sea theater would not necessarily mean that one's forces would obtain absolute sea control. One's sea control is usually relative, because the enemy could dispute it by using aircraft, missiles, submarines, and even coastal antiship missiles and guns.

The place, the time, and the required degree of one's sea control in a typical narrow sea depend on the objective and operational scheme of a major operation conducted by ground forces on the coast. Normally, sea control obtained in part or all of the theater should facilitate attaining the operational objectives on land. One's naval forces accomplish this through continuous actions in support of the ground forces' coastal flank. The purpose is to neutralize the enemy's threat from the sea or to prevent the enemy's forces from operating within one's own area of sea control. If successful, such actions make it possible to release considerable forces on the ground to be employed in the selected sector of main effort. A lack of one's sea control on the army flank would otherwise require significant defensive forces on the coast. Also, by exercising sea control the opponent would be able to inflict high losses on one's own forces. This, in turn, could jeopardize the overall objective of a major ground operation. Obtaining and maintaining sea control is the first, the most important, and the most effective way of providing support to friendly ground forces on the coast.

The actions of ground forces and the general situation on the land front usually have a significant effect on attaining and maintaining sea control. Seizing the enemy's coastal area and nearby naval bases and installations makes it impossible for the opponent to maintain his sea control. One's sea control cannot be obtained as long as the enemy still retains control of his bases and ports. This does not exclude one's forces from attacking enemy forces. Yet the opponent would then have considerable advantages, specifically better surveillance and reconnaissance, shorter distances from his own coast, and good opportunities to repair ships and evacuate personnel. Then the presence of one's naval forces in an area under enemy control would be sporadic, not permanent. The enemy could defend such an area not only with naval forces but also with ground forces such as coastal missile/gun batteries and field artillery.

A. Obtaining Sea Control

Ref: NWC 4020A, On Naval Warfare (Dr. Milan Vego, Sept '08), pp. pp. 41 to 72.

Sea control on the open ocean and in large narrow seas can be obtained quickly and decisively by conducting a series of major naval and joint/combined operations. Obtaining sea control normally requires the sequential and/or simultaneous execution of a number of operational tasks. Some operational tasks, such as destroying an enemy fleet at sea, can be carried out predominantly by one's naval forces. Other operational tasks, for example, seizing the enemy's naval basing areas, largely require the employment of one's ground forces and air forces. The tasks of interdicting or cutting off enemy maritime trade and defending and protecting one's own cannot usually be successfully accomplished without the employment of the combat arms of several services.

Besides major operations, tactical actions are often conducted in the struggle to obtain sea control on the open ocean or a typical narrow sea. Such a situation might occur when a blue-water navy, due to long distances and lack of naval bases and airfields, cannot employ its forces at all or is limited to only certain types of forces. In another case, one's forces might enjoy such a large numerical superiority that the weaker side limits its actions to sporadic ambushes and attacks. Then a naval battle can accomplish an operational task and not be an integral part of a major operation.

Sea control is obtained through a number of methods. Specifically, these include destroying or annihilating the enemy's fleet at sea or its bases, neutralizing the enemy's fleet, seizing choke points and other operationally significant positions within a given enclosed sea theater, and physically seizing the enemy's naval basing areas.

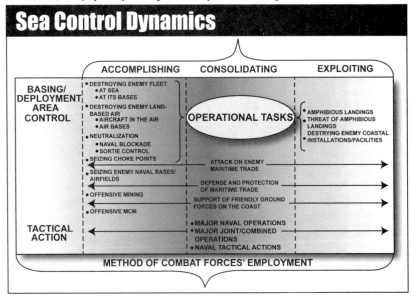

Ref: NWC 4020A, On Naval Warfare, fig. 2, p. 41.

Destroying the Enemy Fleet at Sea

Usually, the quickest and most effective way to obtain sea control in a major part of the open ocean or a narrow sea is by destroying or annihilating the enemy fleet at sea at the very outset of the hostilities. In the past, the most common method used to destroy the enemy fleet was by seeking a decisive battle. The objective for both sides then was to destroy, incapacitate, or capture the opponent's main body and thereby accomplish the principal strategic objective of the war at sea.

Destroying the Enemy Fleet at Its Base
Another option for a stronger side is to mount a major operation to destroy a major part of the opponent's coastal navy either at sea or at its bases or both. In the past, four basic methods of destroying an enemy fleet in its bases were used: launching attacks from beyond the range of the enemy base's defenses, penetrating the enemy naval base by one's surface forces, conducting raids by special force teams, and blockading the sea approaches of enemy naval bases.

Attrition
Most often the enemy's fleet is significantly reduced—attrited—over time through a series of minor tactical actions conducted in the course of the entire conflict. In practice, a stronger side should combine attrition with major operations to destroy a major part of the enemy fleet and thereby obtain sea control in a certain part of the maritime theater. Neither side should remain long in a position where attrition is likely to be successful against it. Generally, in a conflict between two strong opponents at sea, attrition-based warfare requires much more time and larger forces than does the conduct of major naval operations. Moreover, the initiative is often not with the stronger fleet but with its much weaker opponent. The exception to this is when the stronger fleet possesses overwhelming strength over its much weaker opponent.

Neutralization
A stronger fleet can neutralize or contain the weaker force in a typical narrow sea by blockading it in its bases or restricting its operating area. In the past, a stronger side usually declared a naval blockade at the outset of hostilities. A naval blockade is primarily aimed to contain enemy naval or military forces. The general objective of a blockade is to prevent the enemy from interfering in any substantial way with the stronger navy's capacity to use the sea as it wishes.

Offensive Mining
Mines are one of the most effective weapons for blockading enemy naval forces. They are used tactically to inflict damage or losses, delay or hamper enemy naval activities and commercial shipping in a certain combat zone or sector, and reduce the space for tactical maneuver of the enemy's forces. The operational employment of mines is intended to have an operational impact on the course and outcome of a major operation or campaign.

Choke-Point Blockade
Often the best place to engage the weaker fleet is in a geographical bottleneck through which it must pass, or by conducting choke-point control. The main purpose of blockading a choke point is to disrupt enemy maritime communications and to prevent deployment and maneuver of his naval forces. Blockading a sea's only exit is one of the most effective methods of creating the key prerequisite for neutralizing and subsequently destroying the enemy fleet.

Seizing Choke Points
General control of a narrow sea can be accomplished in wartime by physically controlling or seizing one or both shores of a sea's exits and operating along short divergent lines of operations against any hostile force approaching the strait zone.

Seizing Enemy Naval Basing Areas
One's ground forces could be employed to capture choke points, a large part of the mainland coast, and key offshore islands with naval/air bases and ports, thereby completing a major part of the task of obtaining sea control in a certain enclosed sea theater. As one's troops advance along the coast, the enemy's maritime position is also steadily reduced. Full control of the sea is obtained after the enemy's entire coast or archipelago has been seized by one's troops. With this, the enemy fleet would be either destroyed or forced into internment in neutral ports.

B. Maintaining Sea Control

After obtaining sea control, a stronger fleet must consolidate its operational or strategic success. Therefore, obtaining and then maintaining sea control is not the same thing. The task of maintaining sea control encompasses similar but progressively reduced operational tasks, carried out sequentially and/or simultaneously until the conflict on land ends. Many methods used for obtaining control are used in maintaining that control. However, the intensity of one's effort in maintaining control is usually significantly lower than in obtaining control. There is no clearly delineated line in terms of time between maintaining and exercising sea control. If the first phase ends in a drastic reduction of the enemy threat, then maintaining and exercising sea control by the stronger fleet would be conducted almost simultaneously.

In maintaining control of the sea, the stronger navy should be capable of concentrating enough of its strength to ensure that the remaining enemy forces cannot present a serious threat to one's own and friendly forces. Full and effective control of an ocean or sea area may, however, require considerable dispersion of the available strength of a stronger navy. It is by the deployment of one's fleet elements in the areas where they can effect a superior concentration that the stronger fleet would be able to maintain the desired degree of sea control. This would usually require that the most essential part of one's fleet be deployed in such areas and at such distances from its base of operations to ensure a superior concentration of strength prior to the arrival of the hostile forces.

C. Exercising Sea Control

A stronger fleet normally begins exercising sea control in a certain part of the open ocean or narrow sea as soon as a certain degree of sea control is obtained in a specific part of the theater. In operational terms, this phase pertains to exploiting operational or strategic success. There is no clear separation between maintaining and exercising sea control in terms of time or space. In many cases, the tasks of maintaining and exercising sea control are carried out simultaneously. The situation in a given theater and the balance of the opposing forces would be primary factors in determining in which sequence and in which areas any of these tasks would be carried out.

Projecting Power Ashore

For a blue-water navy, projecting power ashore is one of the principal strategic tasks in exercising one's sea control. For a coastal navy, projecting power on the opposite shore might well be one of the main operational tasks, especially in a strategic offensive. The littoral is the area of the sea or ocean that must be controlled by a blue-water navy to ensure support of operations ashore.

To operate successfully in a typical narrow sea, a blue-water navy should significantly reduce the threat posed by the enemy's land-based aircraft, antiship missiles, submarines, and mines. Therefore, a blue-water navy should possess sufficient capabilities in antiair, antisubsurface, antisurface, and antimine warfare areas. A particularly acute problem is the effective defense against the enemy's long-range coastal antiship missiles and theater ballistic missiles in operating in a typical narrow sea.

In projecting power ashore, a stronger navy could be assigned a number of operational tasks, including large-scale amphibious landings, posing the threat of amphibious landings, destroying the enemy's coastal installations/facilities, and destroying the enemy's political and military-economic sources of powers in the strategic depth of the enemy's territory. Exercising sea control is usually accomplished by executing major naval and joint/combined operations and tactical actions.

- In a typical narrow sea, **amphibious landings** usually are conducted across much shorter distances and are smaller than those mounted against a coast bordering an open ocean. Today, the hazards of operating large amphibious forces in confined waters and under constant threat from the air, submarines, and mines are too great in some narrow seas, such as the Baltic, the Black Sea, and the Arabian (Persian) Gulf. A great benefit of possessing a credible amphibious threat is to tie up enemy forces in defending the coast and offshore islands. In this way, the success of a large-scale attack by one's troops on the coast can greatly be enhanced.

- **Attacks on coastal installations and facilities** are usually conducted to either tie up or divert enemy strength from one's main sector of effort. These actions can have a significant, although temporary, psychological effect by raising morale among one's own and friendly forces and population and depressing the opponent. The main methods of combat employment are attacks and raids. A major naval operation aimed to destroy enemy coastal installations/facilities can also be conducted as a preliminary to an amphibious landing or as an integral part of a naval blockade. It can be conducted either with naval or air forces or jointly. Major navies today have a much greater ability than in the past to attack a wide range of targets deep in the enemy's operational and even strategic depth by using shipboard guns, attack aircraft, and cruise missiles.

- The enormous advances in technology since the end of World War II allow a blue-water navy to **attack and destroy diverse military, economic, and other targets** hundreds and thousands of miles deep into enemy territory. This capability, exemplified by the ever-greater effective range and endurance of carrier-based aircraft and land attack cruise missiles, allows a blue-water navy to exert greater influence on the course of war on land.

II. Sea Denial

Sometimes the terms sea control and sea denial are used interchangeably, as if they mean the same thing. All too often it is contended that the stronger navy, by virtue of obtaining sea control, also somehow conducted sea denial. However, sea denial is primarily (though not exclusively) the option for the weaker side in war at sea. A weaker side at sea would normally try to dispute control in certain sea or ocean areas for the duration of the conflict. After the stronger navy is sufficiently attrited, the objective for a relatively weaker fleet might be to obtain and maintain sea control in a certain part of the theater. Because no navy has unlimited resources, a blue water navy might be forced to go on the defensive, that is, conduct sea denial, until the conflict in the main theater has ended, allowing sufficient forces to be brought into the secondary theater so it can go on the offensive. In a typical narrow sea, a stronger fleet can sometimes be forced to contest sea control by a much weaker force. This situation could also occur if the weaker side controls one or both shores of an international strait or controls the principal exits from the naval basing areas of a stronger fleet. In a typical narrow sea, a weaker navy would have greater opportunities to inflict significant losses on the stronger adversary. Then the threat of submarines, land-based aircraft, coastal missile batteries, and mines would make the extended stay of large surface combatants too risky. This threat is sometimes taken too lightly or dismissed entirely by blue-water navies. A multitude of islands and islets and a highly indented coastline allow many small navies to challenge the dominance of a much stronger fleet operating within a narrow sea. For a stronger navy, complete control of a narrow sea cannot be obtained as long as its weaker adversary exists and is active.

The principal methods used by a weaker side in contesting control in narrow seas range from avoiding the stronger fleet, instituting a counterblockade, and employing strategic diversions, to defending the coast, supporting major defensive operations of ground forces on the coast, and exercising offensive mining and offensive mine countermeasures (MCM).

See following pages (pp. 2-28 to 2-29) for discussion of the sea denial dynamics.

Sea Denial Dynamics

Ref: NWC 4020A, On Naval Warfare, (Dr. Milan Vego, Sept '08), pp. 73-92.

The principal methods used by a weaker side in contesting control in narrow seas range from avoiding the stronger fleet, instituting a counterblockade, and employing strategic diversions, to defending the coast, supporting major defensive operations of ground forces on the coast, and exercising offensive mining and offensive mine countermeasures (MCM). In generic terms, most of these methods would be assigned as operational tasks and would be carried out predominantly by one's naval forces. Other operational tasks would require employing multiservice combat arms. In the littorals, land-based air would often have a decisive role in the successful execution of all operational tasks. The principal methods of combat force employment in disputing enemy sea control would be tactical actions. Some major defensive operations in support of one's own ground forces on the coast and in defense of maritime trade could also be conducted in a typical narrow sea.

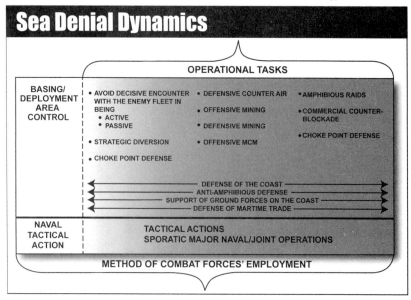

Ref: NWC 4020A, On Naval Warfare, fig. 3, p. 74.

Avoiding Engagement with the Stronger Fleet

A weaker fleet whose ultimate objective is to obtain control in a certain sea or ocean area should normally try to avoid any decisive encounter with the stronger fleet until the balance of forces turns in its favor. Historically known as the "fleet-in-being," this method of disputing sea control has often been applied by a weaker navy.

The stronger fleet must watch over enemy bases, and its main force must stay in a high state of combat readiness in order to attack the weaker fleet when it comes out. As long as the weaker fleet is undefeated, its very existence constitutes a permanent menace to the stronger fleet's control of the sea. In theory, an "active" fleet-in-being can eventually defeat a stronger fleet and thereby secure sea control for itself.

Naval Counterblockade

A weaker side bordering a narrow sea usually resorts to establishing a counterblockade to neutralize or sometimes to nullify the effects of a blockade imposed by a stronger power. Some theoreticians consider a counterblockade a form of "sporadic warfare" at sea.

In general, the actions of a blockaded country are aimed at lifting either a naval or a commercial blockade. Success against the enemy naval blockade means the resolution of both tasks. By conducting a naval counterblockade, a weaker fleet is usually unable to obtain the necessary degree of sea control in a given area. This is a more difficult problem for a fleet operating within the confines of a narrow sea because of its usually unfavorable base of operations regarding free access to another sea or open ocean. Actions to break through a naval blockade are primarily directed toward allowing one's naval forces access to the open ocean so that they can accomplish their assigned tasks. Another objective is to reestablish the country's links to overseas commerce. One's naval forces are normally employed in cooperation with the other services. A blockaded fleet can choose to contest an enemy naval blockade by breaking through the blockading line with part of its surface forces or submarines. A weaker fleet can also hope, under favorable circumstances, to force the straits or narrows defended by a blockading fleet.

Commerce Raiding
In contrast to attacks against merchant shipping, commerce raiding (guerre de course) is purely directed against enemy merchant shipping. This is often the most logical course of action for a weaker fleet. A fleet conducting commerce raiding essentially prolongs the conflict. The main objective of commerce raiding is not to cause the economic ruin of the stronger opponent at sea, but to inflict as much damage as possible on his maritime trade. Commerce raiding succeeds even when conducted on a very small scale, by pushing up marine insurance rates and freight costs. It also might lead the adversary to abandon some of his economic activity.

Defense of the Coast
Defense of one's coast is one of the operational tasks conducted by the weaker side at sea. This task is closely related to the navy's other tasks in war at sea: establishing and maintaining basing/deployment area control. Defense of one's coast is generally much more critical and more difficult to accomplish for a side contesting control because of the stronger side's ability to project its power ashore at the place and time of its own choosing. Generally, this task is not vital for a forwardly deployed blue-water navy when operating in the littorals far from its country's shores.

Defensive Mining
The defense of the coast in general, including one's naval bases and ports, and sea traffic in coastal waters depends to a large degree on the defensive use of mines. Mine barriers intended to prevent enemy movements to and from a sea's only exit or to significantly reduce an enemy's ability to use a major part of an enclosed or semienclosed sea represent the operational employment of mines by a weaker navy. Minefields and mine banks, in contrast, are used in smaller parts of one's coastal waters or offshore islands for a specific tactical purpose. Defensive mining is intended to deny the enemy access to the operating areas of one's surface forces and submarines, naval bases and ports, possible landing beaches, and important straits/narrows and channels, and to protect seaward flanks of one's sea routes in coastal waters. Defensive mine barriers are usually larger in physical scope and number of mines laid than offensive mine barriers.

Offensive Mine Countermeasures (MCM)
Offensive MCM is designed to prevent or neutralize enemy minelaying capabilities in a given sea or ocean area. Normally, efforts should be made to destroy or disable enemy mines before they are laid. Specifically, offensive MCM is directed at destroying enemy minelaying platforms at sea or in their bases, and mine production and storage facilities on the coast. One's forces could conduct offensive mining of enemy ports and thereby prevent enemy surface minelaying platforms from existing and laying mines. Offensive MCM also includes strikes by one's land-based or carrier-based aircraft against selected mine-related targets on the coast or in the country's interior. Mining can also be considered as a part of offensive MCM if directly attacking minelaying assets or mine stockpiles is not possible.

III. Attack on Maritime Trade

In the era prior to the advent of aircraft, one of the principal tasks of any navy was to attack enemy shipping at sea and at the same time provide sufficient defense and protection to one's own and friendly shipping. Ports and their facilities were rarely targets of enemy attacks. This changed drastically in World War II, when land and carrier-based aircraft were used in attacking not only shipping but also other elements of maritime trade: ships at ports and port facilities, shipyards and ship repair facilities, storage areas, and even railroad/road traffic in the coastal area. Hence, it is more accurate to refer to offensive and defensive actions as, respectively, attack on and defense and protection of maritime trade.

Attack on enemy and defense and protection of friendly maritime trade, generically called "trade warfare," are an integral part of navies' mission to exercise and contest command of the sea. They are conducted by both stronger and weaker fleets throughout the entire conflict. The difference is that the focus of an inferior fleet in contesting command will be on attacking the enemy's maritime trade, while the stronger fleet will protect its country's and friendly maritime trade. These tasks are an integral part of a much broader task of weakening the enemy's or protecting one's own and friendly military-economic potential. Protection or attack on maritime trade cannot be accomplished quickly or cheaply. Great care should be taken not to give too much emphasis to these tasks to the detriment of others.

In the broader context, one's attack on enemy maritime trade is an integral part of the multi-service strategic objective to weaken the enemy's military-economic potential. In a war with a strong opponent at sea, an attack on enemy maritime trade is usually considered one of the strategic tasks for one's naval forces. This task is aimed to destroy or significantly reduce the enemy's maritime trade in a given theater and thereby weaken his ability to prosecute the war.

Objectives

The objectives of attack on enemy maritime trade are to reduce the traffic volume, usually expressed in percentages, in a given sea or ocean area for a certain period of time. In generic terms, this is accomplished by "interfering" with or "interdicting" enemy maritime trade. The degree to which this affects enemy maritime trade ranges from hampering it to curtailing, interrupting, and cutting it off. The effectiveness of actions by one's own and friendly forces in attacking enemy maritime trade can also be expressed in number of ships or total tonnage sunk and damaged over a certain time and in a given sea or ocean area. "Hampering" means that one's own forces inflict such losses on enemy shipping and shipping-related facilities ashore that the enemy troops become unable to carry out the planned combat actions within the assigned timetables. In practice, this is accomplished if the enemy maritime transport volume is reduced by 25 to 30 percent. The enemy maritime trade is "curtailed" if reduced from 30 to between 50 and 60 percent in terms of volume. "Interruption" means a serious disruption of enemy maritime trade in a given sea or ocean area. This objective is accomplished by reducing the volume of enemy maritime traffic by 60 to 80 percent. "Cutting off" maritime trade means, in practice, at least 80 percent losses in bottoms and a corresponding reduction in the capabilities of loading and off-loading shipping terminals. Then the enemy forces and civilian economy function only by drawing on existing stockpiles of material. This objective can be normally accomplished only by instituting a full naval and air blockade.

Tenets

The most effective way to attack enemy merchant shipping is to use diverse naval combat arms and weapons against all elements of the enemy's maritime communications. This requires smooth and effective operational and tactical cooperation among all the participating combat arms. Past experience clearly shows that success in protecting maritime trade can be achieved only using diverse combat arms and weapons. No single combat arm can accomplish that task, but each arm should complement the other.

Offensive Mining

Mines are one of the most effective weapons against merchant shipping. They can be used to maintain a continuous threat in the sea areas where bombs and torpedoes might pose a threat when ships or aircraft are present. They are also the only naval weapons capable to some extent of altering geographical circumstances, by making certain areas impassable to enemy ships. Mining large sea areas greatly complicates the enemy's ability to counter the mine threat, and sometimes it can lead to a serious strain or even breakdown of the entire naval defense effort in a given area. Generally, mines can be used to blockade straits or narrows, on coastal shipping routes, and at the approaches to enemy ports before the arrival or departure of convoys. Once laid, mine barriers must be continuously renewed to maintain the same degree of threat to enemy shipping. Mines not only create an extended threat to enemy shipping but also restrict the maneuverability of ships and their escorts on shipping routes. The purpose of minelaying can also be to force an enemy's maritime traffic to abandon the use of more protected routes and come into the open, where it is more vulnerable to attack by submarines or aircraft.

The best results are achieved if offensive mining is coordinated with other types of attack. Hence, mining should be combined with attacks against merchant shipping at sea and against ports and ship-related facilities ashore. The most effective method is the use of several types of mines in order to exert the greatest possible strain on enemy minesweeping forces.

Commercial Blockade

Commercial blockade is the most effective method for weakening the enemy's economic potential. It is almost invariably an integral part of a naval blockade. Its immediate objective is to stop the flow of both the outbound and inbound flow of enemy seaborne trade, whether carried in his own or neutral bottoms. Commercial blockade can aim to stop only some or all goods coming from or into the blockaded country.

A maritime country rich in resources is less likely to rely heavily on maritime trade for its economic well-being and therefore is usually less vulnerable to commercial blockade. In general, an island nation poor in natural resources, such as Japan or Britain, is much more vulnerable to any prolonged interruption of overseas trade. In the past, a large continental country rich in resources was largely impervious to commercial blockade. However, this situation has drastically changed in the modern era, because even the largest countries depend on the import of some critical strategic materials. Also, countries that rely almost exclusively on the export of a single commodity are vulnerable; for example, all the oil-producing countries in the Arabian Gulf are extremely vulnerable to a stoppage of the flow of oil, either by tankers or via pipelines to port terminals.

IV. Defense and Protection of Maritime Trade

One of the principal operational tasks of both the stronger and the weaker fleet is the defense and protection of maritime trade. However, this task is much more critical, and more complicated, for the weaker fleet because of its lack of the necessary strength to secure a sufficient degree of sea control where one's own traffic moves. Regardless of all the difficulties and expected losses, the defense and protection of one's own and friendly maritime trade must be carried out throughout the entire conflict. This task is much easier to accomplish if one's ground troops are largely successful in defending the littoral area, because naval bases and airfields remain available. On the other hand, the withdrawal or retreat of one's own troops will result in the loss of airfields from which air cover could be provided to one's own and friendly ships at sea.

One's maritime trade is made secure by organizing the defense and protection of not only commercial shipping at sea but also all other elements of trade: port terminals, cargo storage depots, shipbuilding and ship repair facilities, railway/road junctions, and railway/road traffic in the littoral area. In the littorals, this operational task cannot be successful without the closest cooperation between the navy and other services. Methods and procedures for defense and protection of maritime trade should be fully practiced in peacetime through war games and exercises. This task is not "defensive," and therefore is somewhat less important than other areas of naval warfare.

All maritime countries depend to a large extent on the continuous flow of trade over coastal routes for the everyday functioning of their economy. This is especially the case for a country with undeveloped land communications in the littoral area. Generally, the importance of maritime trade depends on a country's geographic location and its level of economic self-sufficiency. Obviously, an island nation depends much more on continuous seaborne trade than does a continental state bordering a narrow sea. Likewise, countries whose main export commodity is oil depend considerably on the uninterrupted functioning of their oil-exporting industries and shipping. Defense and protection of one's maritime trade sea are usually conducted almost continuously once hostilities begin. The maintenance of uninterrupted seaborne trade can often be of strategic importance for a major power in time of war. A failure to protect maritime traffic in one sea area often has adverse strategic consequences in the adjacent sea or ocean area. A failure to protect maritime traffic in one sea area often has adverse strategic consequences in the adjacent sea or ocean area.

Defense of merchant shipping depends chiefly on the number of and mutual distances between ports of loading and offloading, the number of ships to be protected, the number and positions of naval bases and airfields, and the number and location of enemy naval bases and airfields. A prerequisite to ensure the safety of one's maritime shipping is to possess a sufficient degree of sea control in a certain part of the sea/ocean. Consequently, too much emphasis on the need to secure sea control might easily lead to a lack of attention to the difficult business of exercising that control. Moreover, one should never set objectives that are too high. Temporary local command of the sea is usually adequate to ensure the safety of one's own shipping, and sometimes it is sufficient to obtain only the command of a focal area of maritime trade.

Defense of Maritime Trade
Destruction or neutralization of the enemy forces that pose a threat to one's maritime trade can best be achieved through a combination of offensive and defensive actions. Ideally, the enemy surface ships and aircraft should be destroyed at their bases or during transit to their respective operating areas. These objectives can best be accomplished by mounting a series of offensive major naval operations at the very outset of hostilities. Subsequently, offensive tactical actions and, occasionally, major naval operations will be conducted when the operational situation in the theater is favorable.

Protection of Maritime Trade
The main methods of protecting one's shipping are convoying, short-range independent sailing, and evasive routing. The principal advantages of convoys are that they provide concentration of forces, as well as economy of force, and often bring the enemy to a decisive action. In general, convoying improves the morale of merchant seamen. On the open ocean, sailing in a convoy offers individual ships the greatest mathematical chance of avoiding detection and attack. Individual ships can escape after the first ships are attacked.

IV. Levels of Maritime Command

Ref: NWP 3-32, Maritime Operations at the Operational Level of War, chaps. 3 & 4.

Operational vs. Administrative Control

The President and Secretary of Defense (SECDEF) exercise authority and control of the Armed Forces through two distinct branches of command, operational and administrative. The operational branch includes the combatant commanders (CCDR's) for missions and forces assigned or attached to their commands. The administrative branch is used for purposes other than operational direction of forces assigned or attached to the CCDR's. Operational-level Navy commanders are responsible and accountable to both branches. See also pp. 1-2 to 1-3.

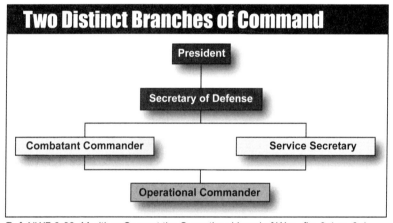

Ref: NWP 3-32, Maritime Opns at the Operational Level of War, fig. 3-1, p. 3-1.

Navy commanders at the operational level prepare for and execute major operations and campaigns. A campaign is a series of related major operations aimed at achieving strategic and operational objectives within a given time and space. A major operation is a series of tactical actions (battles, engagements, strikes) conducted by combat forces of a single or several services, coordinated in time and place, to achieve strategic or operational objectives in an operational area (OA). Navy operational-level commanders will be assigned or attached units or personnel to conduct the campaign/major operation. The Joint Publication, Doctrine for the Armed Forces of the United States, discusses in detail the authorized command relationships and authority for the military commander and provides guidance for the exercise of that military authority. Commanders and their staffs must understand the different levels of authority and the impact each has on the commander's ability to control assigned and attached forces.

I. Command Relationships

Ref: NWP 3-32, Maritime Operations at the Operational Level of War (Oct '08), p. 3-2 to 3-7.

The Doctrine for the Armed Forces of the United States, defines four types of command relationships. The specific command relationship (COCOM, OPCON, TACON, and support) will define the level of authority a commander has over assigned or attached forces. Joint doctrine also defines three other types of authority (outside of command authority): administrative control (ADCON), coordinating authority, and direct liaison authorized (DIRLAUTH).

U.S. Joint Doctrine Command Relationships

Combatant Command (Command Authority)
(Unique to Combatant Commander)
- Planning, Programming, Budgeting and Execution Process Input
- Assignment of Subordinate Commanders
- Relations with Department of Defense Agencies
- Directive Authority for Logistics

When OPERATIONAL CONTROL is delegated
- Authoritative Direction for All Military Operations and Joint Training
- Organize and Employ Commands and Forces
- Assign Command Functions to Subordinates
- Establish Plans and Requirements for Intelligence Surveillance, and Reconnaissance Activities
- Suspend Subordinate Commanders from Duty

When TACTICAL CONTROL is delegated: Local direction and control of movements or maneuvers to accomplish mission

When SUPPORT relationship is delegated: Aid, assist, protect or sustain another organization

1. Combatant Command (COCOM - Command Authority)
COCOM is the command authority over assigned forces vested only in the commanders of combatant commands by Title 10, United States Code (USC), Section 164, or as directed by the President in the Unified Command Plan (UCP), and cannot be delegated or transferred.

2. Operational Control (OPCON)
OPCON is inherent in COCOM and is the command authority over assigned or attached forces. OPCON is the authority of a commander to perform those functions over subordinate forces involving organizing and employing commands and forces, assigning tasks, designating objectives, and giving authoritative direction necessary to accomplish the mission.

3. Tactical Control (TACON)
TACON is inherent in OPCON and is the command authority over assigned or attached forces or commands, or military capability or forces made available for tasking. It is limited to the detailed direction and control of movements or maneuvers within the OA necessary to accomplish assigned missions or tasks. TACON may be delegated to and exercised by commanders at any echelon at or below the level of CCDR. Tactical control provides sufficient authority for controlling and directing the application of force or tactical use of combat support assets within the assigned mission or task.

4. Support
Support is a command authority with four categories: general, mutual, direct, and close. A support relationship is established by a superior commander between subordinate commanders when one organization should aid, protect, complement, or sustain another force.

Allied/multinational maritime tactical instructions and procedures define five types of command relationships: full command, OPCOM, OPCON, TACOM, and TACON.

Allied/Multinational Maritime Command

Full Command (National Command Authority)
- National military commander who has national authority to attach forces to an allied/multinational commander using either OPCOM or OPCON
- Logistics responsibility
- Administrative command

Operational Command (OPCOM)
- Assign missions or task
- Reassign forces
- Deploy units
- Retain/delegate OPCON
- Retain/delegate TACOM/TACON

Operational Control (OPCON)
- Direct forces to accomplish specific mission or task
- Deploy units
- Retain/delegate TACOM/TACON

Tactical Command (TACOM)
- Assign and conduct tasks pertaining to mission
- General safety of assigned units
- Retain/delegate TACON

Tactical Control (TACON)
- Detailed and usually local direction and control of movements or maneuvers neccessary to accomplish missions or assigned tasks

1. Full Command (National Command Authority)
National military commander who has national authority to attach forces to an allied/multinational commander using either OPCOM or OPCON. Includes logistics responsibility and administrative control.

2. Operational Command (OPCOM)
OPCOM is the command authority granted to an Allied/multinational maritime commander by a national commander with full command to assign missions or tasks to subordinated commanders, to deploy units, to reassign forces, and to retain or delegate OPCON, TACOM, or TACON as may be deemed necessary. It does not in itself include administrative command or logistical responsibility. OPCOM is a unique authority for allied/multinational forces.

3. Operational Control (OPCON)
As defined in allied/multinational maritime tactical instructions and procedures, OPCON is subordinate to OPCOM. OPCON is a command authority granted to an allied/multinational maritime commander by a national commander with full command or an allied/multinational maritime commander with OPCOM to direct forces assigned so that the commander can accomplish specific missions or tasks that are usually limited by function, time, or location; to deploy units concerned; and to retain or assign tactical command and/or control of those units. It does not include the authority to assign separate employment of the units concerned. Neither does it, of itself, include administrative command or logistic responsibility.

4. Tactical Command (TACOM)
As defined in allied/multinational maritime tactical instructions and procedures, TACOM is subordinate to OPCOM and OPCON. It is a command authority granted to an Allied/multinational maritime commander by an Allied/multinational maritime commander with either OPCOM or OPCON. TACOM is authority delegated to an Allied/multinational commander to assign subordinate forces for the accomplishment of the mission assigned by higher authority.

5. Tactical Control (TACON)
As defined in allied/multinational maritime tactical instructions and procedures, TACON is subordinate to TACOM. TACON is a command authority granted to an allied/multinational maritime commander by an allied/multinational maritime commander with OPCOM, OPCON, or TACOM command authority. TACON is the detailed and usually local direction and control of movements or maneuvers necessary to accomplish missions or assigned tasks.

Command Authority

Command is the authority that a commander in the Armed Forces lawfully exercises over subordinates by virtue of rank or assignment. Command includes the authority and responsibility for effectively using available resources, and for planning the employment, organizing, directing, coordinating, and controlling military forces for the accomplishment of assigned missions. It also includes responsibility for health, welfare, morale, and discipline of assigned personnel.

Inherent in command is the authority that a military commander lawfully exercises over subordinates, including authority to assign missions and accountability for their successful completion. Although commanders may delegate authority to accomplish missions, they may not absolve themselves of the responsibility for the accomplishment of these missions. Authority is never absolute. The extent of authority is specified by the establishing authority, directives, and law.

Unity of command (one of the principles of war) means all forces operate under a single commander with the requisite authority to direct all forces employed in pursuit of a common purpose. However, unity of effort requires coordination and cooperation among all forces toward a commonly recognized objective, although they are not necessarily part of the same command structure.

Command relationships define the interrelated responsibilities between and among commanders, as well as the operational authority exercised by commanders in the chain of command.

Impact Of Level Of Authority On Command at the Operational Level

Navy commanders at the operational level may be delegated any level of authority (except COCOM) for assigned or attached forces. Normally Navy commanders at the operational level will have OPCON over assigned and attached forces. Unlike the other Services, the Navy has a long tradition of exercising TACOM as defined in allied/multinational maritime tactical instructions and procedures. This provides the Navy operational-level commander flexibility when specifying a tactical commander's command authority over tactical forces.

Navy commanders at the operational level organize or assign tactical forces to maintain the span of control they desire. Each Navy tactical force established typically has a designated officer in tactical command (OTC). Normally the senior task force commander is the OTC. The operational commander will always consider the mission, nature, and duration of the operation, force capabilities, and command and control (C2) capabilities when selecting the OTC. The operational-level commander designates the OTC's command authority over assigned forces. The operational-level commander can designate OPCON, TACOM (inherent in OPCON), or TACON (inherent in TACOM). Typically, the OTC is assigned TACOM over forces made available by the operational-level commander.

With TACOM the OTC can organize the force but does not have the broad authorities of OPCON. OTCs can organize assigned and attached forces to accomplish missions or tasks. Each subordinate commander is assigned either TACOM or TACON command authority over forces made available by the OTC. Typically OTCs will assign TACON command authority to subordinate commanders (e.g., composite warfare commander (CWC), sector CWC, or warfare commander). The operational-level commander also can designate a support command authority between two or more OTCs. An OTC who has OPCON can designate a support command authority between two or more subordinate force commanders.

The Navy's concurrent use of command authorities from U.S. joint doctrine and allied/multinational maritime tactical instructions and procedures can cause confusion. Operational-level commanders and staffs need to ensure all commanders in the force understand their delegated command authorities over forces made available and the applicable governing reference(s).

II. Joint Force Components

Joint forces are established at three levels: unified commands, subordinate unified commands, and joint task forces (JTF's).

A. Unified Command

A unified command is a command with broad continuing missions. The President, through the Secretary of Defense (SECDEF) and with the advice and assistance of the Chairman of the Joint Chiefs of Staff (CJCS), establishes combatant commands for the performance of military missions and prescribes the force structure of such commands. Combatant commanders (CCDR's) are either geographic or functional.

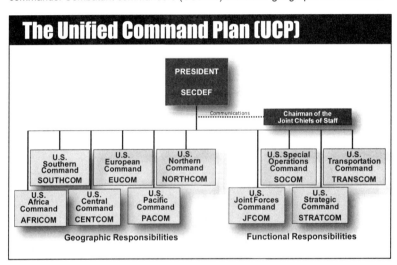

B. Subordinate Unified Command

When authorized by the SecDef through the CJCS, CCDR's may establish subordinate unified commands (also called sub-unified commands) to conduct operations on a continuing basis in accordance with the criteria set forth for combatant commands. CCDR's exercise COCOM over assigned forces. Forces are assigned in accordance with guidance contained within the SecDef "Forces for Unified Commands Memorandum" contained within the SecDef's Global Force Management Implementation Guidance (GFMIG). Forces are allocated for crisis action planning (CAP) or execution through the Joint Operation Planning and Execution System (JOPES) crisis action procedures. CCDRs may assign or attach assigned forces to subordinate commanders. The CCDR may designate operational control (OPCON) and/or tactical control (TACON) command authority for subordinate commanders to exercise over assigned or attached forces.

Refer to The Joint Forces Operations & Doctrine SMARTbook (Guide to Joint, Multinational & Interagency Operations) for further discussion of joint force components. Topics include joint doctrine fundamentals; joint operations; joint operation planning; joint logistics; joint task forces (JTFs); information operations; multinational operations; and interagency, intergovernmental, and nongovernmental organization coordination.

C. Joint Task Force (JTF)

A joint task force (JTF) is a joint force that is constituted and so designated by a JTF establishing authority (i.e., the Secretary of Defense, a combatant commander [CCDR], a subordinate unified commander, or an existing commander, joint task force [CJTF]) to conduct military operations or support to a specific situation. It usually is part of a larger national or international effort to prepare for or react to that situation. In most situations, the JTF establishing authority will be a CCDR.

III. Joint Force Commander

The CCDR, commander subunified command, and JTF commander are joint force commanders (JFCs) who normally exercise OPCON over assigned (and normally over attached) forces. For attached forces, the JFC's authority will be designated in the establishing directive. OPCON allows the JFC to organize forces to best accomplish the assigned mission. JTFs can be established by the SecDef, a CCDR, or an existing JTF commander. The JFC will establish subordinate commands, assign responsibilities, establish or delegate appropriate command relationships, and establish coordinating instructions for the component commanders.

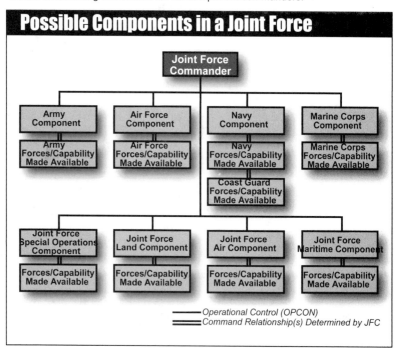

Ref: JP 1, Doctrine for the Armed Forces of the United States, fig. V-1, p. V-3.

Forces, not command authority, are transferred between commands. Forces are assigned to a joint force when the transfer will be permanent, or for an unknown but long period of time. Forces are attached to a joint force when the transfer will be temporary. Included in the directive that transfers forces is the command authority of the gaining command.

The CCDR may conduct operations through the Service component commanders; subordinate JFCs may conduct operations through Service force commanders, e.g., Navy TF commanders. This relationship is appropriate when stability, continuity, economy, ease of long-range planning, and the scope of operations dictate organizational integrity of Service forces for conducting operations.

2-38 (Maritime Operations) IV. Levels of Maritime Command

Availability of Forces for Joint Operations

Ref: NWP 3-32, Maritime Operations at the Operational Level of War, pp. 4-1 to 4-2.

Forces, not command authority, are transferred between commands. Forces are assigned to a joint force when the transfer will be permanent, or for an unknown but long period of time. Forces are attached to a joint force when the transfer will be temporary. Included in the directive that transfers forces is the command authority of the gaining command.

Availability of Forces

1. **Assigned**
2. **Attached**
3. **Apportioned**
4. **Allocated**

The CCDR, commander subunified command, and JTF commander are joint force commanders (JFCs) who normally exercise OPCON over assigned (and normally over attached) forces. For attached forces, the JFC's authority will be designated in the establishing directive. OPCON allows the JFC to organize forces to best accomplish the assigned mission. JTFs can be established by the SecDef, a CCDR, or an existing JTF commander. The JFC will establish subordinate commands, assign responsibilities, establish or delegate appropriate command relationships, and establish coordinating instructions for the component commanders.

The terms assigned, apportioned, and allocated are defined in the "**Forces for Unified Commands Memorandum**" contained within the SecDef's GFMIG as follows:

- **Assigned**: Those forces and resources that have been placed under the COCOM of a unified commander by direction of the SecDef in the "Forces for Unified Commands Memorandum."
- **Attached**: The term "attached" is defined in JP 1-02, DoD Dictionary of Military and Associated Terms, as the placement of units and personnel in an organization where such placement is relatively temporary.
- **Apportioned**: Those forces and resources assumed to be available for contingency planning as of a specified date. They may include those assigned, those expected through mobilization, and those programmed. Apportionment tables are included in part IV of the GFMIG.
- **Allocated**: Those forces and resources provided by the President and SecDef for CAP or execution. Forces are allocated through the JOPES crisis action planning procedures.

Global Force Management (GFM)

Global force management (GFM) is the process by which military forces and capabilities are assigned, apportioned, and allocated to the various combatant commanders. Since these forces are typically employed by subordinate joint task force, service, or functional component commanders it is imperative for the fleet commander or JFMCC to understand how the GFM process works. In particular, the commander needs to understand GFM allocation since this is how required forces and capabilities will be requested and sourced to fulfill emergent mission requirements.

JFCs may conduct operations through functional component commanders. This relationship is appropriate when forces from two or more military departments must operate within the same mission area or geographic domain, or there is a need to accomplish a distinct aspect of the mission. Levels of command in the joint force that naval commanders routinely occupy are shaded. These roles are JFC, Navy component commander (NCC), Marine component commander (MCC), and joint force maritime component commander (JFMCC).

A. Commander, Joint Task Force (CJTF)

A commander, joint task force (CJTF) is a joint force commander (JFC). Navy commanders may be designated a JFC by the SecDef, a CCDR, a subordinate unified commander, or an existing CJTF. CJTFs normally are operational-level commanders. The authority that establishes the joint command will state the forces that are made available and also include the overall mission, purpose, and objectives for the directed military operations.

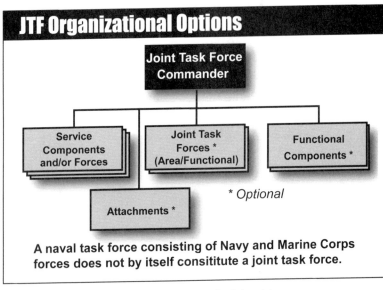

Ref: JP 3-33, Joint Task Force Headquarters, fig. I-1, p. I-1.

The CJTF will organize those forces over which OPCON is exercised to ensure sufficient flexibility to meet the planned phases of the contemplated operation and any development that may necessitate a change in plan. The CJTF should allow service tactical and operational groupings to function generally as they were designed. The intent is to meet the CJTF's needs while maintaining the tactical and operational integrity of the service organizations. The CJTF assesses tasking, develops plans, and then directs and monitors plan execution. This cycle of assess, plan, direct, and monitor continues until the CJTF mission is complete.

While assigned CJTF responsibilities, the Navy commander's staff will be augmented to include representatives from each service with significant forces assigned to the joint command. The Navy commander will retain service responsibilities (e.g., support the SECNAV's execution of United States Code (USC) Title 10 responsibilities), held before being designated a JFC. The commander must clearly define staff roles and responsibilities to ensure fleet management and joint command responsibilities are executed correctly and efficiently.

2-40 (Maritime Operations) IV. Levels of Maritime Command

Joint Task Force Establishing Authority Responsibilities

Ref: JP 3-33, *Joint Task Force Headquarters* (Feb '07), fig. I-2, pp. I-1 to I-2.

CJTFs have full authority to assign missions, redirect efforts, and direct coordination among subordinate commanders. CJTFs should allow Service tactical and operational groupings to function generally as they were designed. The intent is to meet the needs of CJTFs, while maintaining the tactical and operational integrity of Service organizations. The manner in which CJTFs organize their forces directly affects joint force operational responsiveness and versatility.

- Appoint the commander, joint task force (CJTF), assign the mission and forces, and exercise command and control of the joint task force (JTF)
  - In coordination with the CJTF, determine the military forces and other national means required to accomplish the mission
  - Allocate or request forces required
- Provide the overall mission, purpose, and objectives for the directed military operations
- Define the joint operations area (JOA) in terms of geography or time. (Note: The JOA should be assigned through the appropriate combatant commander and activated at the date and time specified)
  - Ensure freedom of action, communications, personnel recovery, and security for forces moving into or positioned outside the JOA
- Ensure the development and approval of rules of engagement or rules for the use of force tailored to the situation
- Monitor the operational situation and keep superiors informed through periodic reports
- Provide guidance (e.g., planning guidelines with a recognizable end state, situation, concepts, tasks, execute orders, administration, logistics, media releases, and organizational requirements)
- Promulgate changes in plans and modify mission and forces as necessary
- Ensure administrative and sustainment support
- Recommend to higher authority which organizations should be responsible for funding various aspects of the JTF
- Establish or assist in establishing liaison with US embassies and foreign governments involved in the operation
- Determine supporting force requirements
  - Prepare a directive that indicates the purpose, in terms of desired effect, and the scope of action required. The directive establishes the support relationships with amplifying instructions (e.g., strength to be allocated to the supporting mission; time, place, and duration of the supporting effort; priority of the supporting mission; and authority for the cessation of support).
- Approve CJTF plans
- Delegate directive authority for common support capabilities (if required)

Normally, joint forces are organized with a combination of Service and functional component commands with operational responsibilities. The CJTF may elect to retain some or all component commander responsibilities; however, care must be exercised when a single commander executes multiple roles in the joint force.

B. Service Component Commander

All joint forces include Service components. Service component commanders have administrative and logistic support responsibilities for their Service forces that are part of the overall joint force. Service forces may be assigned or attached to subordinate joint forces without the formal creation of a respective Service component command of that joint force. The JFC also may conduct operations through the Service component commanders or, at lower echelons, Service force commanders. This relationship is appropriate when stability, continuity, economy, ease of long-range planning, and the scope of operations dictate organizational integrity of Service forces for conducting operations.

A Navy component command assigned to a CCDR consists of the NCC and the Navy forces (NAVFOR) (such as individuals, units, detachments, and organizations, including the support forces) that have been assigned to that CCDR. A Marine component command assigned to a CCDR consists of the MCC and the Marine forces (such as individuals, units, detachments, and organizations, including the support forces) that have been assigned to that CCDR. When a command is designated as the NCC or MCC to multiple CCDRs, the NCC/MCC and only that portion of the NCC/MCC's assets assigned to a particular CCDR are under the command authority of that particular CCDR.

Unless otherwise directed by the CCDR, the NCC/MCC will communicate through the combatant command on those matters over which the CCDR exercises COCOM. On Navy-specific matters, such as personnel, administration, and unit training, the NCC will normally communicate directly with the CNO, informing the CCDR as the CCDR directs. On Marine Corps–specific matters, such as personnel, administration, and unit training, the MCC will normally communicate directly with the CMC, informing the CCDR as the CCDR directs.

Subject to the directive authority of the CCDR, the NCC/MCC will retain and exercise the operating details of the naval logistic support system in accordance with instructions of the Navy Department. Joint force transportation policies will comply with the guidelines established in the Defense Transportation System.

Service Component with Operational Control

The JFC may conduct operations through the Service component commanders or, at lower echelons, Service force commanders. This relationship is appropriate when stability, continuity, economy, ease of long-range planning, and the scope of operations dictate organizational integrity of Service forces for conducting operations. When this is applicable, the JFC will normally establish an OPCON command authority between some or all Navy and Coast Guard forces assigned to the JFC and the NCC or Navy force commanders. Those Navy and Coast Guard forces not under the OPCON of the NCC will remain under the OPCON of the CCDR or be assigned to either a sub-unified commander, JTF commander, or functional component commander.

Typically, the Service component will conduct a low-end range of military operations; i.e., military engagement, security cooperation, and deterrence. These activities shape the operational environment and keep the day-to-day tensions between nations or groups below the threshold of armed conflict while maintaining U.S. global influence. These ongoing and specialized activities establish, shape, maintain, and refine relations with other nations and domestic civil authorities (e.g., state governors or local law enforcement). The fluid and long-term nature of these operations, as well as the impact unplanned events can have on force availability, complicates the operational assessment, planning, direction, and monitoring tasks of the NCC/Navy force commanders. In addition, the NCC/Navy force commander must ensure maritime actions are synchronized with those of other services that are also supporting the JFC's low-end range of military operations missions. When authorized by the JFC, the NCC/Navy force commander should also coordinate with other government agencies (DOS) where appropriate to ensure unity of action.

Service Component Commands

Ref: NWP 3-32, Maritime Operations at the Operational Level of War, pp. 4-6 to 4-7.

The JFC may provide NCCs and MCCs command authority over Navy and Marine forces assigned or attached to the JFC. In addition to fulfilling any designated command authorities, Service component commanders have responsibilities that derive from their Services' support function. NCCs/MCCs are always responsible for the following Navy/Marine Corps–specific functions.

- Makes recommendations to the JFC on the proper employment of the Navy/Marine Corps forces

- Accomplishes such operational missions as may be assigned

- Selects and nominates specific units of the Navy/Marine Corps component for attachment to other subordinate commands. Unless otherwise directed, these units revert to the NCC/MCC's control when such subordinate commands are dissolved.

- Conducts joint training, including, as directed, the training of components of other Services in joint operations for which the NCC/MCC has or may be assigned primary responsibility, or for which the Navy/Marine Corps component's facilities and capabilities are suitable

- Informs their JFC (and their CCDR, if affected) of planning for changes in logistic support that would significantly affect operational capability or sustainability sufficiently early in the planning process for the JFC to evaluate the proposals prior to final decision or implementation. If the CCDR does not approve the proposal and discrepancies cannot be resolved between the CCDR and the NCC/MCC, the CCDR will forward the issue through the CJCS to the SecDef for resolution. Under crisis action or wartime conditions, and where critical situations make diversion of the normal logistic process necessary, NCC/MCCs will implement directives issued by the CCDR.

- Develops program and budget requests that comply with CCDR guidance on war-fighting requirements and priorities. The NCC/MCC will provide to the CCDR a copy of the program submission prior to forwarding it to the Chief of Naval Operations (CNO)/Commandant of the Marine Corps (CMC). The NCC/MCC will keep the CCDR informed of the status of CCDR requirements while Navy/Marine programs are under development.

- Informs the CCDR (and any intermediate JFCs) of program and budget decisions that may affect joint operation planning. The NCC/MCC will inform the CCDR of such decisions and of program and budget changes in a timely manner during the process in order to permit the CCDR to express the command's views before a final decision. The NCC/MCC will include in this information Navy/Marine Corps rationale for nonsupport of the CCDR's requirements.

- As requested, provides supporting joint operation and exercise plans with necessary force data to support missions that may be assigned by the CCDR

NCC/MCC or other Navy/Marine Corps commanders assigned to a CCDR are responsible to the CNO/CMC for the following:

- Internal administration and discipline

- Training in joint doctrine and Navy/Marine Corps doctrine, tactics, techniques, and procedures

- Logistic functions normal to the command, except as otherwise directed by higher authority

- Navy/Marine Corps intelligence matters and oversight of intelligence activities to ensure compliance with the laws, policies, and directives

D. Functional Component Commander

The multiple complex tasks confronting the JFC may challenge the JFC's span of control and ability to oversee and influence each task. Functional components allow control of joint forces and capabilities on a functional level and enhance component interaction. JFCs may decide to establish a functional component command to integrate planning, reduce span of control, and/or significantly improve combat efficiency, information flow, unity of effort, weapon systems management, component interaction, or control over the scheme of maneuver.

The JFC can establish functional component commands to conduct operations. Functional commands are the joint force air component commander (JFACC), joint force land component commander (JFLCC), joint force special operations component commander (JFSOCC), and JFMCC. Functional component commands are appropriate when forces from two or more military departments must operate within the same mission area or geographic domain, or when there is a need to accomplish a distinct aspect of the assigned mission.

Functional component commanders are component commanders of a joint force and do not constitute a "joint force command" with the authorities and responsibilities a JFC, even when employing forces from two or more military departments.

The JFC establishing a functional component command has the authority to designate its commander. Normally, the Service component commander with the preponderance of forces to be tasked and the ability to command and control those forces will be designated as the functional component commander. However, the JFC will always consider the mission, nature, and duration of the operation, force capabilities, and the C2 capabilities in selecting a commander.

Joint Force Maritime Component Commander (JFMCC)

The JFMCC is the JFC's maritime functional component commander. The JFMCC's forces/capabilities may consist of subordinate commanders and forces from any Service and may include multinational forces. The JFMCC's subordinate commanders use the JFMCC's operational-level plans to develop their specific maritime tactical plans.

See following pages (pp. 2-45 to 2-60) for discussion of the JFMCC.

V. Joint Force Maritime Component Cmdr (JFMCC)

Ref: JP 3-32 (Chg 1), Command and Control for Joint Maritime Operations (May '08) and NWC Maritime Component Commander Handbook (Feb '10), chap. 6.

The JFMCC is the JFC's maritime warfighter. The JFMCC reports directly to the JFC and advises the JFC on the proper employment and joint integration of attached and assigned forces to accomplish the portion of joint operations that occurs predominately in the maritime domain, which is defined as "the oceans, seas, bays, estuaries, islands, coastal areas, and the airspace above these, including the littorals."

The manner in which the JFC organizes forces directly affects the responsiveness and versatility of joint force operations. The JFC organizes assigned forces based on the mission assigned and the commander's vision of how to accomplish that mission. The JFC normally has OPCON of subordinate commanders. Most often, JFCs organize forces with a combination of service and functional components. Functional component commands may be appropriate when forces from two or more military departments operate in the same domain, location, or medium or when there is a need to accomplish a distinct aspect of the assigned mission. These conditions apply when the scope of operations requires that similar capabilities and functions of forces from more than one service be directed toward closely related objectives.

JFMCC Functions

A joint force maritime component commander can provide the joint force commander with focused maritime expertise.

1. Command and Control
2. Coordination and Deconfliction
3. Communication System Support
4. Intelligence, Surveillance and Reconnaissance (ISR)
5. Movement and Maneuver
6. Fires
7. Force Protection (FP)
8. Logistics Support
9. Planning

Ref: JP 3-32, fig. II-2, p. II-5. See p. 2-47 for an overview of these functions.

I. Establishing a JFMCC

Combatant commanders, commanders of subordinate unified commands, and JFCs have the authority to establish functional components to conduct military operations. Functional component command may be established across the range of military operations (ROMO) to perform operational missions that may be of short or extended duration. The JFC designates the forces and/or military capabilities that are available for tasking by the JFMCC and the corresponding command authority. The size of the JFMCC staff and forces is tailored and scaled to the scope of the joint operation. The functional component commander normally exercises OPCON over that commander's parent service forces and tactical control (TACON) over other services' forces made available for tasking. The JFC may also establish support relationships between functional component commanders and other component commanders to facilitate operations.

The JFC normally designates a commander from the service whose forces are principally represented as the functional component commander (e.g., a Navy commander serving as JFMCC; an Army or Marine Corps commander as joint force land component commander (JFLCC); an Air Force commander or a Navy commander as JFACC; and an appropriate special operations forces commander as joint force special operations component commander (JFSOCC)). The JFC augments the designated JFMCC staff as necessary to complete the following:
- Reflect the composition of assigned maritime forces
- Properly execute the JFMCC staff mission
- Fulfill joint, multinational, and interagency requirements

A. JFMCC Area Of Operations

When a JFMCC is designated, normally the JFC will designate an area of operations, which may include air, land, and sea. The AO may not encompass the entire littoral area; however, it should be large enough for the JFMCC to accomplish the mission and protect the maritime force. Based on the nature of the operations, the geography, the adversary, and the operations and activities of other elements of the joint force, this AO can be dynamic and evolving as the operation or campaign matures. The AO should be of sufficient size to allow for movement and maneuver and employment of weapons systems as well as other force projection and inherent war fighting capabilities. The AO must also provide the operational depth for required logistics and force protection (FP). Within the AO, the JFMCC establishes the geometry that allows for independent yet supporting operations of subordinate elements while enabling the synchronization of employment of forces across all components.

When the JFC designates a JFMCC AO, the JFMCC is the supported commander within the AO. As a supported commander, the JFMCC integrates and synchronizes maneuver, fires, and interdiction. To facilitate this integration and synchronization, the JFMCC has the authority to designate target priority, effects, and timing of fires within the JFMCC AO.

B. Integration with Joint Campaign Planning

Joint planning at the JFMCC headquarters is predominantly focused on the operational level of war. Maritime force planning links the tactical employment of maritime forces to operational and strategic objectives. Focus at the command level is on operational art — the use of military forces to achieve strategic objectives through the design, organization, and execution of strategies, campaigns, major operations, and battles.

JFMCC planning must be in consonance with the guidance of senior commanders and must support the JFC CONOPS and should support other component commanders as well. The JFMCC's operational concept is typically built upon the following missions:
- Sea control
- Maritime power projection and projection of defense from sea to land
- Deterrence
- Strategic sealift
- Forward maritime presence
- Seabasing operations

Other specified and implied tasks can involve planning and directing naval operations (e.g., undersea operations, mine operations, strike operations, interdiction, amphibious and expeditionary operations, maritime interception operations, foreign humanitarian assistance and noncombatant evacuation operations, and civil support operations), as well as providing communications system support and FP.

See chap. 5, Naval Planning, for discussion of the Navy Planning Process.

II. JFMCC Functions

Ref: JP 3-32 (Chg 1), Cmd & Control for Joint Maritime Operations (May '08), pp. II-4 to II-8.

1. Command and Control. The JFMCC commands assigned and attached forces, prepares OPLANs, and executes operations in support of the assigned tasks and strategic goals as directed by the JFC. Upon JFC approval of the JFMCC's plan, the JFMCC exercises specified authority and direction over forces/capabilities in the accomplishment of the assigned mission. The JFMCC publishes daily targeting orders for the execution of maritime operational activity and special procedures. The JFMCC must also maintain liaison with other functional/Service components and agencies and provide representation on boards, groups, and cells.

2. Coordination and Deconfliction. The JFMCC must ensure that forces/capabilities are coordinated within the maritime force and with other component commanders. Where appropriate, the JFMCC may make coordination and deconfliction recommendations to the JFC, to include, but not limited to airspace, land space and waterspace management; fire support; and interagency, intergovernmental and NGOs.

3. Communications System Support. The JFMCC is responsible for planning and activating all validated joint maritime communications links that are consistent with the overall campaign plan and allow accomplishment of the JFC directives.

4. Intelligence, Surveillance & Reconnaissance (ISR). The JFMCC provides the operational requirements and continuous feedback to the JFC to ensure optimum maritime and littoral ISR support. Maritime forces typically bring a rich complement of sensors and sensor fusion capability. Close coordination with other component commanders and the JFC's communications system directorate (J-6) early in joint planning are essential to creating architectures and sensor employment plans that provide the best mix of ISR services throughout the joint force.

5. Movement and Maneuver. The JFMCC, responsible at the operational level for movement and maneuver of assigned forces, directs subordinate commanders in the execution of force level operational tasks, advises the JFC of their movement, and coordinates with other functional/Service components and other agencies in supporting JFMCC missions. *See pp. 2-52 to 2-55.*

6. Fires. Joint fires can be in support of gaining sea control/maritime superiority/maritime supremacy, in support of projecting power ashore/interdiction/fire support, projecting defense inland from the sea, and other missions. The JFMCC is responsible for the planning and employment of operational fires within the assigned AO, both in terms of developing and integrating multidimensional attacks on the adversary's centers of gravity (COGs) and in terms of shaping the JFMCC AO. The JFMCC should provide guidance for the employment of JFMCC forces' fires. *See pp. 2-56 to 2-57.*

7. Force Protection (FP). The JFMCC is responsible to the JFC for all aspects of maritime FP. The JFMCC creates force protection plans and sets priorities for JFMCC forces. FP is part of each mission assigned as a function routinely conducted by maritime forces.

8. Logistic Support. Each Service is responsible for the logistic support of its own forces, except when logistic support is otherwise provided for by agreement with national agencies, multinational partners or by assignments to common, joint or cross-servicing. The supported combatant command can determine whether or not common servicing would be beneficial within the theater or designated area. The JFMCC makes recommendations concerning the distribution of material and services commensurate with priorities developed for JFMCC operations. A CCDR may delegate responsibility for a common support capability to the JFMCC. The JFMCC will usually assume logistic coordination responsibilities for all Services and forces operating from a sea base.

9. Planning. The JFMCC assists the JFC in long-range or future planning, preparation of campaign and joint OPLANs, and associated estimates of the situation.

III. JFMCC Roles and Responsibilities

Ref: NWC Maritime Component Commander Handbook (Feb '10), pp. 4-9 to 4-10.

The JFMCC is the JFC's maritime functional component commander. The JFMCC's forces/capabilities may consist of subordinate commanders and forces from any Service and may include multinational forces. The JFMCC's subordinate commanders use the JFMCC's operational-level plans to develop their specific maritime tactical plans.

The JFMCC's role is to delineate the objectives, plan in support of the JFC, and conduct maritime operations. In setting the objectives for the force, the commander provides focus. The JFMCC contributes to unity of effort by integrating maritime action across all components to support the JFC's objectives and intent. The JFMCC shall plan and execute JFC assigned missions. Planning, the act of envisioning and determining effective ways of achieving a desired end state, serves to direct and coordinate actions. Planning also develops a shared SA among the staff and commanders and generates expectations about how actions will evolve and how those actions will affect the desired outcome. The JFMCC issues planning guidance to all subordinate and supporting elements and analyzes proposed COAs. The intent is to concentrate combat power at the right time and place to accomplish operational or strategic goals. The JFMCC influences outcomes through decisions made in assigning missions to subordinates, prioritizing their efforts, allocating resources, assessing risks, and directing necessary changes.

JP 3-32, Command and Control for Joint Maritime Operations, provides detailed descriptions of the JFMCC's authority, functions, and staff organization. JP 3-32 lists the JFMCC responsibilities as follows:

- Develop a joint maritime operations plan to best support joint forces
- Provide centralized direction for the allocation and tasking of forces/capabilities made available
- Request forces of other component commanders when necessary for the accomplishment of the maritime mission
- Make maritime apportionment recommendations to the JFC
- Provide maritime forces to other component commanders in accordance with JFC maritime apportionment decisions
- Control the operational-level synchronization and execution of joint maritime operations, as specified by the JFC to include adjusting targets and tasks for available joint capabilities/forces. The JFC and affected component commanders will be notified, as appropriate, if the JFMCC changes the planned joint maritime operations during execution.
- Act as supported commander within the assigned area of operations (AO)
- Assign and coordinate target priorities within the assigned AO by synchronizing and integrating maneuver, mobility and movement, fires, and interdiction. If the JFMCC cannot service targets within the maritime AO with organic maritime forces, the JFMCC may nominate those targets to the joint targeting process that may potentially require action by another component commander's assigned forces
- Evaluate results of maritime operations and forward combat assessments to the JFC in support of the overall effort
- Support JFC information operations (IO) with assigned assets, when directed
- Function as supported and/or supporting commander as directed by the JFC
- Perform other functions as directed by the JFC

JFMCC responsibilities include, but are not limited to, planning, coordination, allocation, tasking, and synchronization of joint maritime operations based on the JFC's CONOPS and maritime apportionment decisions. Specifically, the JFMCC's responsibilities include but are not limited to the following:

Responsibilities to the Joint Force Commander

- Advise the JFC on the proper employment of all assigned and attached maritime forces
- Recommend to the JFC the apportionment of the joint, multinational, coalition, and interagency maritime–related forces against assigned missions
- Command and control of assigned and attached forces
- Position forces through movement and maneuver to support JFC objectives and missions
- Advise the JFC on issues pertaining to logistic support
- Provide and assist the JFC with future planning, including preparation of campaign plans, operational plans, and estimates of the situation
- Evaluate the results of maritime operations and forward results to the JFC to support the combat assessment effort
- Participate in the campaign assessment process and other duties as designated by the JFC
- Assist the JFC in development and verification of the time-phased force and deployment data
- Nominate targets and designate priorities, effects, and timing within the AO as a member of the joint targeting coordination board
- Recommend the use of and employ space-based, air breathing, and terrestrial ISR assets as well as other applicable space capabilities within the JFMCC AO

Responsibilities to other Component Commanders

- Provide maritime forces to other component commanders in accordance with JFC apportionment decisions. Employ other component forces in accordance with JFC apportionment decisions for accomplishment of JFMCC's assigned missions
- Coordinate the planning and execution of maritime operations with the other components, JTF commanders, and other supporting agencies
- Provide effective liaison to other functional components
- Acting as the supported commander within maritime AO, exercise general direction of the supporting effort. (General direction includes the designation and prioritization of targets or objectives, timing and duration of the supporting action, and other instructions necessary for coordination and efficiency.)

Responsibilities within the JFMCC

- Develop a supporting maritime concept plan, OPLAN, and/or OPORD that supports the operational objectives of the JFC and optimizes the operations of task-organized maritime forces
- Direct the execution of the maritime OPORD as specified by the JFC, which includes making timely adjustments to the tasking of assigned/attached forces. The JFMCC coordinates changes with affected component commanders as appropriate.
- Synchronize and integrate command and control; movement and maneuver; fires; intelligence, surveillance, and reconnaissance; logistics; and force protection in support of maritime operations
- Synchronize JFMCC forces and functions

IV. Notional Joint Maritime Operations Organization and Processes

Ref: JP 3-32 (Chg 1), Cmd & Control for Joint Maritime Operations (May '08), app. G.

Provided below is a notional model for the organization and associated processes through which the JFMCC may command and control the maritime component of a joint force. The JFMCC staff organization uses a common planning process to support high level operational planning of the JFC, the lower level maritime specific operational planning at the JFMCC level, and subordinate commander tactical planning. The JFC staff uses a synchronization process to ensure coordination between component commanders. Similarly, the JFMCC staff uses a synchronization process to ensure coordination between subordinates. All levels of command have processes for analysis and assessment when execution occurs.

Notional Joint Force Maritime Component Commander Planning and Operations Process

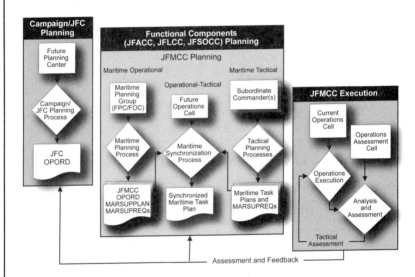

FOC	future operations cell
FPC	future plans cell
JFACC	joint force air component commander
JFC	joint force commander
JFLCC	joint force land component commander
JFMCC	joint force maritime component commander
JFSOCC	joint force special operations component commander
MARSUPPLAN	maritime support plan
MARSUPREQs	maritime support requests
OPORD	operation order

Collaboration is critical to the integration of planning, synchronization, and execution processes and allows multiple echelons of C2 to work together. Analysis and feedback are additional key elements that allow elements of the joint organizations and processes to respond to emerging events on the battlefield. The products associated with the processes allow modification of the desired outcome in response to emergent battlefield events, or alteration of plans to adapt to opportunities.

Maritime Operations Center (MOC)

The JFMCC staff is divided into organizational elements called centers and cells that are collectively called the maritime operations center. Centers are manned 24 hours a day during the planning and execution of maritime operations. Cells meet as required for specific tasks. There are six centers dedicated to major maritime functions: knowledge management, intelligence and analysis, operations, logistic coordination, future planning, and support.

See chap. 3 for discussion of the maritime headquarters (MHQ) and the maritime operations center (MOC). See pp. 3-6 to 3-9 for specific discussion of MOC centers and B2C2WG (boards, bureaus, centers, cells, and working groups) structure.

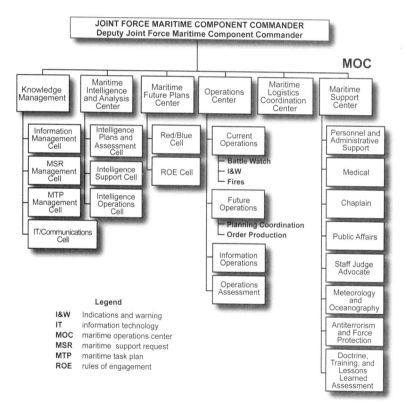

Notional Joint Force Maritime Component Commander Functional Organization

The JFMCC is involved in campaign planning, contingency planning, and CAP. The JFMCC planning process interfaces with the joint contingency planning process during the development of supporting plans. Supporting plans are developed once the CCDR's concept has been approved and a plan has been developed. Maritime supporting plans address the tasks identified for JFMCC forces and define the planned actions of assigned and augmenting forces. The JFMCC planning process interfaces with the CAP process beginning in situation development and throughout the process as JFMCC planners develop new plans, or expand or modify existing plans.

C. Movement and Maneuver of Maritime Forces

Movement and maneuver of maritime forces, when combined with the element of surprise, provide a significant advantage over adversaries and allow the commander to rapidly concentrate forces where necessary. They are also key attributes upon which the JFMCC CONOPS is built, applying directly to the key maritime areas such as control of the sea and maritime AO, power projection from sea to land, deterrence, strategic sealift, forward maritime presence, and other specified and implied tasks as assigned.

The JFMCC plans, controls, and coordinates operational level movement and maneuver to gain a positional advantage over the adversary. The objective for operational maneuver is usually focused on an adversary's COG.

A prime attribute of maritime forward presence is the ability to, on short notice and with no host-nation notification or approval, conduct a show of force in an area of increased tension or a show of support to friends and allies. While these are typically peacetime operations conducted by a Service component commander, a JFMCC can conduct them by using maneuver and movement.

Maneuver and movement of maritime forces are integral aspects of joint operations. Examples of maneuver and movement include undersea warfare (USW), strategic sealift, movements of other military forces, convoy operations, and amphibious operations. During maritime operations, the use of information, initiative, and decisive force is applied through maneuver to dominate specific regions of the environment, in all dimensions, at the chosen time and place. Maintaining awareness in the transition from the open ocean to littoral areas is key to the continuing conduct of maneuver on and from the sea, allowing the JFMCC to focus on the adversary's COG.

Whether at sea or from the sea, maneuver can guide application of power projection in support of the land battle. The concepts for maneuver are articulated in JFC and JFMCC tasking documents, and include timing, sequencing, method, and location of entry into the assigned AO.

The notional JFMCC staff organization discussed in JP 3-32, Appendix E, "Joint Force Maritime Component Commander's Staff," provides for the integration of staff from the appropriate Services into the JFMCC staff. It is essential that staffs from each Service participate in the planning process for all operational level movement and maneuver to ensure that all Service-specific capabilities are considered.

Maritime Interception Operations (MIO)

Maritime interception operations, which restrict a vessel's ability to move and maneuver, isolates a vessel of interest from outside support, enhances free use of the sea lines of communications (SLOCs) for friendly operations, assures law and sanction enforcement, and enforces and provides security for maritime operations.

Amphibious Operations

A key to maximizing capabilities of the JFMCC force is in understanding the requirements of each assigned command. One example of requirements is a joint amphibious operation where the commander of the landing force (e.g., Marine expeditionary unit, Marine expeditionary brigade, or Marine expeditionary force) requires a sufficient AO or amphibious objective area to effectively employ his aviation combat element, requiring coordination with the JFACC, the airspace control authority (ACA), and area air defense commander (AADC), with careful consideration of fire support coordinating measures and boundaries. The inherent mobility of JFMCC units must also be preserved in deconfliction plans worked out between components.

Refer to JP 3-02, Joint Doctrine for Amphibious Operations, for more information.

Antisubmarine Warfare (ASW)

US joint forces maintain the capability to destroy enemy naval forces, suppress enemy sea commerce, protect vital sealanes, and establish maritime superiority in support of a joint or multinational operation. Control of the undersea portion of the operational area is vital to the success of joint operations. A principal threat in the maritime domain comes from enemy submarines. To counter this threat, the JFC will coordinate, and when required, integrate assets from the joint force to conduct ASW during all phases of the joint operation or campaign. Although often viewed as a Navy-only mission, the JFMCC may utilize a variety of joint forces and capabilities (air, land, maritime, space, and special operations) to facilitate or conduct ASW.

See following pages (2-54 to 2-55) for a discussion of antisubmarine warfare.

Operational- and Tactical-level Movement and Maneuver

Operational and tactical level maneuver and movement of maritime forces, both significant counter-targeting measures, are integral in defense planning considerations such as JFMCC support to, or execution of, assigned ACA or AADC responsibilities. As described in JP 3-0, Joint Operations, maneuver must be synchronized with fires and interdiction. Such synchronization provides one of the most dynamic tools available to the JFC.

Waterspace Management (WSM)

Coordination of waterspace is a requirement unique to the joint maritime component. The JFMCC will coordinate the operations of forces/capabilities in order to prevent mutual interference between maritime elements as well as with commercial shipping, multinational forces, and other units. The JFMCC will provide necessary information to the JFC and other component commanders, specifically JFACC, to prevent inadvertent engagement of friendly forces and physical interference among maritime forces.

See also p. 4-9.

Recommendations Concerning the Employment of Forces

The JFC normally tasks the JFMCC to make recommendations concerning the employment of forces for maritime operations within the JFMCC's AO. This includes:

- Organizing for combat
- Developing a JFMCC force scheme of maneuver and a fire support plan to support the JFC's campaign plan
- Identifying interdiction targets or objectives within the JFMCC's AO
- Establishing priorities of effort
- Coordinating, integrating, and synchronizing operational reconnaissance
- Coordinating/planning operational fires that impact maneuver
- Integrating multinational forces in maritime operations

D. Intelligence, Surveillance, and Reconnaissance (ISR)

Like the JFC, JFMCC is both a broker of intelligence support for those in subordinate commands as well as the executive agent for intelligence support tasks that may be required by the JFMCC maritime CONOPS or directed by higher authority. The JFMCC has primary responsibility for coordinating maritime intelligence efforts. Based on JFC guidance, the JFMCC develops intelligence and reconnaissance plans for component operations and provides feedback to the JFC on maritime-related issues affecting joint operations. The JFMCC defines intelligence responsibilities and prioritized maritime-related intelligence requirements of tactical forces. Finally, the JFMCC provides representation for the maritime component at the JFC's joint targeting coordination boards.

Normally, the JFMCC needs access to national, theater, and tactical intelligence systems/data. The JFMCC also requires core analysis capability, ability to provide indications and warnings, ISR collection management skills, targeting capability, and systems and administrative support.

V. Undersea/Antisubmarine Warfare (USW/ASW)

Ref: JP 3-32 (Chg 1), Cmd & Control for Joint Maritime Operations (May '08), app. J.

Control of the maritime domain is typically critical to joint force operations. US joint forces maintain the capability to destroy enemy naval forces, suppress enemy sea commerce, protect vital sealanes, and establish maritime superiority in support of a joint or multi-national operation. Control of the undersea portion of the operational area is vital to the success of joint operations. A principal threat in the maritime domain comes from enemy submarines. To counter this threat, the JFC will coordinate, and when required, integrate assets from the joint force to conduct ASW during all phases of the joint operation or campaign.

- **Undersea Warfare (USW).** USW operations are conducted to establish dominance in the underwater environment, which permits friendly forces to accomplish the full range of potential missions and denies an opposing force the effective use of underwater systems and weapons. It includes offensive and defensive submarine, antisubmarine, and mine warfare operations. *For more information regarding mine warfare operations, see JP 3-15, Barriers, Obstacles, and Mine Warfare for Joint Operations.*

- **Antisubmarine Warfare (ASW).** ASW is operations conducted with the intention of denying the enemy the effective use of submarines. ASW is a subset of USW.

Although often viewed as a Navy-only mission, the JFMCC may utilize a variety of joint forces and capabilities (air, land, maritime, space, and special operations) to facilitate or conduct ASW. At the operational level of war, ASW will have joint implications. In particular, given the nature of the operating environment, the size of the area to be covered, and the requirement to find, fix, track, target, and engage enemy submarines, the use of persistent national and joint ISR is one of the essential resources to ASW mission accomplishment. For example, the monitoring, tracking, and engagement of enemy submarines in port or transiting on the ocean surface may be effectively accomplished by non-Navy aircraft, satellites, special forces, or other joint assets.

While the JFC is responsible for ASW planning inside the JOA, coordination of ASW plans and activities with commands outside the JOA will be essential and may require close coordination with other government agencies, multinational partners, and host nations.

ASW Planning Considerations

ASW missions will be centrally planned, typically under the direction of the JFMCC or a Navy component commander and executed in a decentralized manner in support of the JFC's concept of operations. ASW is extremely complex, requiring the coordination and integration of multiple platforms and systems in order to mitigate the risks posed by enemy submarines. ASW planning should include consideration of the submarine threat, operational environment, force planning, ISR, communications systems, and command and control.

A. Understanding the Operational Environment

Because it is difficult to detect and track submarines operating underwater, a thorough understanding of the operational environment is a key tenet of success. Only after through analyses of the physical environment and adversary systems will the JFC be able to properly develop the concept of operations.

- **Physical Environment.** The physical characteristics of the maritime domain have a significant impact on ASW execution. The highly variable acoustic properties of the underwater environment will impact the ability to detect identify, track, and engage enemy submarines. Factors that may affect these properties include surface shipping (including that of the joint force and commercial shipping), inherent environmental noise and oceanographic properties, and seasonal weather patterns.

- **Adversary Forces.** A thorough understanding of the adversary's ability to conduct submarine warfare is essential. Intelligence efforts must focus on the both the physical attributes of specific enemy platforms, their supporting physical and C2 infrastructure, and past and anticipated employment patterns.

B. Force Planning

ASW may require joint and combined forces and capabilities. Maritime forces must be identified early to account for long transit times. Initial force planning considerations should include utilization of prepositioned capabilities, early deployment of surface and subsurface forces, and reassignment of forward deployed forces to the ASW operation. Early presence of joint forces may be essential in seizing the initiative.

C. Concept of Operations

The objective of ASW operations is to assist in the establishment or maintenance of maritime superiority by denying enemy submarines influence in the operational area. This is accomplished through detection, identification, tracking, and engagement of enemy submarines. Unlocated enemy submarines often have the most influence in the JOA, possibly affecting fleet maneuver and commercial shipping operations. A single unlocated submarine could result in a significant operational, political, or economic impact. The JFC should designate enemy submarines as time-sensitive targets and develop and implement a comprehensive plan to reduce this influence.

- **Focused Operations.** The operational key to limiting the influence of unlocated submarines is to focus the ASW effort to hold enemy submarines at risk and secure friendly maneuver areas.
- **ISR**. Detection of submarines can have a significant impact on maritime operations. Even if engagement of enemy submarines is prevented by ROE or other considerations, the ability to track enemy submarine movement will shrink the area of influence from all possible areas of enemy submarine operations to the known location of the submarine. The integration of intelligence and operations through a comprehensive ISR concept of operations is essential to the conduct of ASW. The theater ASW commander (TASWC) should maintain direct liaison with the Joint Intelligence Center (JIC).
- **Communications Planning.** A survivable, networked joint communication system is essential to facilitate ISR, coordinate multi-platform execution, manage the waterspace, and prevent fratricide. Information connectivity, exchange, and integration at all levels can help maximize maritime domain awareness and mission accomplishment.

Antisubmarine Warfare Organization

Each geographic combatant commander (GCC) operates theater ASW commanders (TASWCs), through the NCC. Each NCC appoints submarine operating authorities (SUBOPAUTHs). The TASWC and SUBOPAUTH closely coordinate submarine operations. In some cases, the TASWC and SUBOPAUTH responsibilities may be shared by a single commander.

A. Theater ASW Commander (TASWC)

The TASWC is the Navy commander assigned to develop plans and direct assigned assets to conduct ASW within an assigned operational area. The TASWC may exercise either operational or tactical command and control of assigned assets. The TASWC is normally designated as a task force or task group commander subordinate to a navy component commander or joint force maritime component commander.

B. Submarine Operating Authority (SUBOPAUTH)

The SUBOPAUTH is the Navy commander responsible for ensuring safety, PMI, providing WSM, and controlling the submarine broadcast for assigned submarines within a designated operational area.

(Maritime Operations) V. JFMCC 2-55

VI. Maritime Fires Support

Ref: JP 3-32 (Chg 1), Cmd & Control for Joint Maritime Operations (May '08), app. D.

Operational fires accomplish specific missions and create conditions for success in the AO. Such operations can focus on accomplishing JFMCC objectives or serve as support to other components. In addition to providing fires, the JFMCC must also prepare to direct a multidimensional fires operation using assigned forces/capabilities.

Maritime fires consist of strike and fire support. The extended range, speed, and versatility of maritime forces and the associated fires capabilities they provide, combine to make strike fires the salient operational capability of JFMCC forces. With assets that include fires-capable units, the JFMCC is a significant fires provider for various operations of the JFC campaign.

Types of Fires

Fires from maritime platforms can produce a full range of effects and are a critical component of maritime power and defense projection from sea-to-land missions. Examples of maritime missions employing fires at targets ashore or over land include interdiction, close air support, suppression of enemy air defenses, counterair (offensive and defensive), and naval surface fire support (direct and general).

Shaping the AO

The use of fires is one of the principal means of shaping the JFMCC AO. IO, employed to affect adversary information and information systems, are also integral to this process. The JFMCC's interests are those theater-wide adversary forces, functions, facilities, and operations that impact JFMCC plans and operations. As a significant strike and fire support provider, the JFMCC works closely with the JFC and component commanders to coordinate fires for maximum operational effect.

Interdiction

Joint interdiction operations are a key focus for JFMCC fires. Fires from maritime assets may be major active elements of interdiction. The key attributes in the JFC's joint interdiction operations are the flexibility, maneuverability and speed of JFMCC fires assets and freedom from host nation permissions.

Joint Force Maritime Component Commander Fires Assets

Concentrated fires, even from dispersed JFMCC forces, are possible because of the maneuverability of JFMCC forces and the extended range of their fires. JFMCC resources for fires encompass forces/capabilities assigned by the JFC and may include:

- Sea- or shore-based aircraft including fixed- or rotary-wing assigned to theater naval forces, Marine air-ground task force, or other aircraft made available for tasking
- Armed and attack helicopters
- Surface- and subsurface-launched cruise missiles and torpedoes
- Surface gunnery, including naval surface fire support (NSFS)
- Surface-, subsurface- and air-launched mines
- Air, land, maritime, and special operations forces
- Unmanned vehicles

Fires Synchronization and Coordination

The JFMCC must synchronize and coordinate fires in support of the JFC's objectives. Because of the highly specialized nature of some assigned operations, maritime fires require a high degree of coordination between component and subordinate commanders. Fires need to be synchronized with maneuver and interdiction.

The JFMCC synchronizes operational fires and C2 by the active participation of the JFMCC strike, NSFS cell, supporting arms coordination center, and landing force fire support planners, where available, and as required. *Refer to JP 3-09, Joint Fire Support, for additional information.*

JFMCC Targeting Functions

Once the JFC provides targeting guidance, subordinate component commanders can recommend to the JFC how best to use their assigned forces/capabilities to achieve the JFC's objectives. The JFC may retain central targeting authority or delegate it to the subordinate component commander best able to accomplish it. Key functions include synchronization, integration, deconfliction, fratricide avoidance targeting, and force allocation. Components typically will nominate targets to the JFC's designated targeting authority for centralized servicing and will provide the direct support sortie requirements plan for central targeting support. *See also following pages (pp. 2-58 to 2-59).*

JFMCC targeting functions and responsibilities include:

- Advise on application of operational fires
- Identify fires requirements from other components
- Provide apportionment recommendations
- Recommend JFMCC assets for JFC allocation
- Advise on fires asset distribution (priority) of JFMCC forces
- Develop JFMCC priorities, timing, and effects for interdiction within the JFMCC AO
- Develop JFMCC targeting guidance and priorities.
- Develop a prioritized target nomination list for submission to the JFC.
- Integrate/deconflict fires activity with the JFC and other component commanders or forces
- Plan, coordinate, and supervise the execution of JFMCC deep supporting fire operations
- Coordinate with designated airspace control authorities for all planned airspace requirements
- Staff and man the JFMCC time-sensitive strike branch in the assigned operations cell

Air Tasking Order/Airspace Control Order (ATO/ACO)

The air tasking order (ATO) delineates tasked sorties allocated by the JFACC after the JFC's air apportionment decision in support of the JFC's campaign. While intended to be a methodically planned document, operations frequently force dynamic changes which may be received only a few hours before launch. The ATO codifies tasked sorties, capabilities and/or forces, assigned targets and specific missions to components, subordinate units, and tactical C2 agencies. The airspace control order provides airspace coordination measures to deconflict the employment of air assets throughout the joint force operational area. It is published either as part of the ATO or as a separate document and provides the details of approved requests for airspace control measures.

Refer to JP 3-30, Command and Control for Joint Air Operations, JP 3-52, Doctrine for Joint Airspace Control in the Combat Zone, and Navy Warfare Publication (NWP) 3-56.1, Naval Air Operations Center Organization and Processes, for additional details.

Combat Assessment (CA)

CA addresses the effectiveness of overall joint targeting in light of the JFC's objectives, guidance and intent. CA gives both the JFMCC and the JFC a broad perspective on the total effects of joint targeting against the adversary at both the operational and strategic levels. Pre- and post-mission reconnaissance are required as part of CA. For example, the product of the CA effort directly impacts the survivability of air assets, by avoiding unnecessary restrikes on targets, hitting key targets again which were not destroyed, and freeing assets to take on additional tasking. *See pp. 4-22 to 4-23.*

VII. Joint Maritime Operations Targeting

Ref: JP 3-32 (Chg 1), Cmd & Control for Joint Maritime Operations (May '08), pp. III-3 to III-8.

Joint Publication 3-60, Joint Targeting, provides standard terminology and procedures that govern JFC targeting responsibility and execution. These basic tenets are amplified by JFC guidance that adapts them to specific missions and force capabilities for each component.

Commanders must clearly understand the joint targeting process in order to arrange for fires that support their own objectives while satisfying the tasking of the JFC. Effective coordination, deconfliction, and synchronization maximize the strategic, operational, and tactical effects of joint targeting. The JFMCC and other components must employ joint targeting procedures that ensure:

- Compliance with JFC objectives, guidance and intent, rules on the use of force, rules of engagement (ROE), and collateral damage concerns
- Coordination, synchronization, and deconfliction of targets
- Fratricide avoidance
- Rapid response to time-sensitive targets (TSTs)
- Minimal duplication of effort
- Expeditious combat assessment (CA)
- Common perspective of all targeting on the adversaries' COG

Targeting Process

Once the JFC provides targeting guidance, subordinate component commanders can recommend to the JFC how best to use their assigned forces/capabilities to achieve the JFC's objectives. The JFC may retain central targeting authority or delegate it to the subordinate component commander best able to accomplish it. Key functions include synchronization, integration, deconfliction, fratricide avoidance targeting, and force allocation. Components typically will nominate targets to the JFC's designated targeting authority for centralized servicing and will provide the direct support sortie requirements plan for central targeting support. Typically, LNOs acting in support of their component commander actively participate in the targeting. For example, if the JFACC is designated the targeting authority, the JFMCC provides target input to the joint guidance, apportionment, and targeting team directly via collaborative tools or JFMCC LNOs. The JFMCC is charged with establishing procedures and mechanisms to manage its part of the joint targeting function. The JFMCC must be prepared to coordinate and employ alternative procedures for any aspect of the joint targeting process should established procedures require change to keep pace with execution. The theater joint intelligence center (JIC) normally provides targeting support to the JFC and components. This is supplemented by continental United States reach-back services. In addition to these targeting measures, there is a growing need for real time operational targeting capability, which reflects the very dynamic environment that grows from taking advantage of our force agility and highly trained personnel.

For additional information, refer to JP 3-60, Joint Targeting.

Targeting for JFMCC Operations/JFMCC Distributed Targeting

The JFMCC directs the maritime segment of the joint targeting process within an organizational framework tailored to the scope of the maritime targeting requirements. A combination of maritime force processing capabilities resident in maritime assets (e.g., aircraft carriers, large-deck amphibious ships, command ships, surface combatants, submarines, and aircraft) offer the JFMCC a potentially significant distributed targeting arrangement. SOPs for employing these capabilities must be consistent with the targeting direction from the JFC and with specific arrangements for the shared use of sensors.

The JFMCC's targeting structure must be agile enough to react to rapidly changing events at sea, in the seaward littoral, and the landward littoral, as well as to project power and defense inland in support of other component commanders while providing for efficient and continuous execution of all phases of the joint targeting process. Deliberate targeting supports the standard targeting cycle established by the JFC. Additional dynamic targeting procedures are required to support time-sensitive targeting requirements. The JFMCC's targeting process should mirror the standard joint process as closely as possible. The JFMCC provides long-term, top level planning guidance that highlights the commander's intent for fires. Subordinate units submit targeting requests for servicing either by the JFMCC's own targeting capability or for consideration by the JFC's targeting authority, if necessary.

OPS normally is responsible for organizing and executing the JFMCC's targeting responsibilities. The joint targeting process cuts across traditional functional and organizational boundaries. Operations, plans, and intelligence specialists are the primary active participants, but other functional areas, such as logistics, weather, law, and communications may also support the process. Close coordination, cooperation, and communications are essential. Depending on technical capabilities resident inside and outside the JFMCC staff, portions of the targeting effort may need to be delegated to subordinate units best equipped and manned for it.

The servicing of RFIs is a key element of the targeting process. Standard format RFIs are essential for compiling information necessary for all aspects of the joint targeting process. In order to service joint force RFIs in a timely manner, the JFMCC's process must be synchronized with that used by the JFC targeting authority.

Component Targeting Board

The JFMCC typically organizes a maritime component targeting board to function as an integrating center for maritime targeting oversight and review. Primary direction is provided by OPS. This maritime targeting board must be a joint activity with representatives from the JFMCC staff, all component LNOs, and major subordinate units. It provides a forum for review of joint targeting guidance and joint apportionment that advises the JFMCC on alternatives for achieving the JFC's theater plans and campaign objectives. The board acts to consider the capabilities of the JFMCC force and other component forces. Specific responsibilities include:

- Assigning maritime target priorities
- Developing a general asset/platform allocation plan to service the prioritized target list for refinement by the strike cell
- Assisting subordinate units with translating JFMCC objectives, guidance, and desired effects into the targeting effort
- Providing targeting guidance based on major plans and priorities
- Providing pre-execution review of JFMCC's major plans
- Specifying the desired effects of joint targeting in the maritime AO
- Recommending supplemental ROE to enhance the effectiveness of fires missions
- Recommending changes to the JFC's restricted target and no-strike lists
- Developing a prioritized target list to submit to the JFC for inclusion in the joint integrated prioritized target list
- Considering potential nonscheduled fires that can be held in reserve for direct (close battle) support
- Identifying requests for supporting fires from other component commanders.
- Coordinating with other component targeting to avoid duplication of effort and reduce the risk of fratricide

(Maritime Operations) V. JFMCC 2-59

VIII. Considerations for Employing a JFMCC

Ref: JP 3-32 (Chg 1), Cmd & Control for Joint Maritime Operations (May '08) pp. I-4 to I-5.

JFCs can conduct operations through subordinate JTFs, Service components, functional components, a combination of Service and functional components, or, in operations of limited scope and duration, the JFC may retain control and use the joint staff to direct and execute maritime operations.

JFCs may decide to establish a functional component command to integrate planning; reduce their span of control, and/or significantly improve combat efficiency, information sharing, unity of effort, weapons system management, component interaction or control over the scheme of maneuver. When the JFC designates a JFMCC, the JFMCC's authority and responsibility is also defined by the JFC.

When designated, a JFMCC is the single maritime voice regarding maritime forces and requirements and makes recommendations to the JFC regarding prioritization and allocation of joint maritime force assets and synchronization of maritime operations with overall operations. The following are some considerations for employing a joint force maritime component commander:

Planning
The need for detailed, coordinated, concurrent, and parallel planning is a consideration when deciding whether the JFC should establish a JFMCC. While JFMCC-integrated planning is focused primarily on employment, the JFMCC also may be tasked to integrate planning of multiservice maritime forces for deployment, transition, and redeployment/reconstitution at a level subordinate to that of the JFC.

Duration
The length of an operation may be sufficient to warrant the establishment of a single maritime commander. The decision to establish a JFMCC should consider the time required for personnel and staff sourcing and training, the establishment of C2, and communications system support architecture.

Maritime Perspective
A JFMCC provides the JFC with focused maritime expertise to enhance the detailed planning, coordination, and execution of joint operations.

JFC Span of Control
A JFMCC resolves joint maritime issues when task or organizational complexities limit the JFC's effective span of control.

Multinational Operations
A JFMCC (or a CFMCC) integrates multinational maritime forces into the overall operation on a level commensurate with the capabilities they provide.

Timing
The decision to establish and designate a JFMCC ideally occurs during the concept development phase of the campaign plan, permitting the JFMCC to fully participate and to maximize unity of effort.

VI. Navy Tactical HQs and Task Organization

Ref: NWP 3-32, Maritime Operations at the Operational Level of War (Oct '08), p. 6-3.

Operations at the tactical, operational, and strategic levels are conducted concurrently. Tactical commanders fight engagements and battles understanding their relevance to operational objectives and goals. Operational commanders utilize operational art and design to set conditions for battles within a major operation or campaign to achieve military strategic and operational objectives. Combatant commanders (CCDRs) integrate theater strategy and operational art while remaining acutely aware of the impact of tactical events. Because of the inherent interrelationships between the various levels of war, commanders cannot be concerned only with events at their respective echelon, but must also understand how their actions contribute to the military end state.

I. Notional Relationship of Command Level to Level of War

Joint doctrine does not establish explicit linkage between command levels and levels of war. These documents also outline a notional relationship between command level and level of war. This notional relationship was developed by reading joint/Navy doctrine and reviewing historical records. As discussed earlier, the boundaries between the tactical, operational, and strategic levels of war overlap and frequently are pictorially displayed as three circles with sides overlapping. Similarly the notional relationships between command level and level of war, depending on the mission, reality may, and often does, result in a command level operating at a level either above or below the one at which it normally operates.

Notional Relationship of Command Level

Commander	Strategic	Operational	Tactical
Combatant Commander	•		
Subunified Commander	•		
JTF Commander (CJTF)		•	
Navy Component Commander (NCC)		•	
Major Fleet Commander		•	
Commander, Navy Forces (COMNAVFOR)		•	
Joint Force Maritime Component Commander (i.e., JFMCC)		•	
Numbered Fleet Commander		•	
Principal Headquarters Commander		•	
Commander, Task Force (i.e., CTF)			•

Ref: NWP 3-32, Maritime Opns at the Operational Level of War, table 6-1, p. 6-3.

Today's information systems allow for the rapid exchange of information between commanders at the various levels of war. Strategic- and operational-level commanders are now provided insight into the operational environment that was not historically possible. This insight allows these commanders to provide situational awareness (SA) and apply resources not previously available at the tactical level to the tactical problem. Similarly, the enhanced information systems provide the tactical commander means to influence strategic and operational assessments of the operational environment. It is important that while various echelons may be aware of events at other levels, the commanders and their staffs must focus on the appropriate level of war.

See pp. 1-2 to 1-3 for discussion of Navy organization and command structure to include numbered fleets, principal headquarters and task forces.

II. Navy Task Organization
(Task Forces, Groups, Units and Elements)

Ref: NWP 3-32, Maritime Operations at the Operational Level of War (Oct '08), p. 4-14 to 4-18.

The operational campaign/major operation plan (OPLAN) synchronizes, integrates, and coordinates forces at the tactical level. The Navy operational-level commander commands forces at the tactical level organized from individual platforms. Task organizing enables the operational-level commander to exercise a more reasonable span of control. Individual platforms are assigned and/or attached to a **task force (TF)**. Each TF is assigned a commander, and only the commander reports to the operational commander. TFs also allow an operational commander to subdivide subordinate forces and delegate authority and responsibility to plan and execute based on mission, platform capability, geography, or other issues and challenges. The commander, task force (CTF) further subdivides the TF into **task groups, units and elements** to ensure span of control at the tactical level is maintained. These subdivisions may be organized based on capabilities, missions, geography, or a hybrid of all three; they produce tactical and operational results to accomplish operational and strategic objectives to satisfy strategic goals toward fulfillment of national policy.

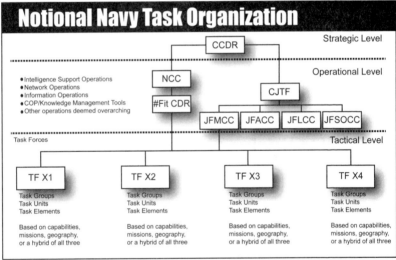

Ref: NWP 3-32, Maritime Opns at the Operational Level of War, fig. 4-9, p. 4-16.

The Navy operational-level commander serves as the Navy coordinator for intelligence, network operations, establishment/maintenance of knowledge repositories, and the establishment and maintenance of a common operational picture (COP). These activities are required by the operational-level commander and must be shared with all of the TFs in order to help them execute their missions. Having the maritime commander at the operational level responsible for these activities ensures that they are common across the maritime force. When operations require two or more TFs or their subordinates to occupy the same maritime AO and/or have the same/complementary focus, either the CTFs or the common superior will establish the relationship to ensure unity of effort.

Establishment of TFs ensures unity of command and unity of effort across the multi-mission platforms assigned for an operational mission or tasking. TF categories (Logistics, Navy Special Warfare, etc.) are established by doctrine. Fleet or other operational commanders (i.e., JFC/JFMCC) are not required to assign or attach forces to all TF categories.

Fleet and other commanders who may assign or attach forces to a TF should consider the following before establishment of a TF:

- **Need: Is a TF needed?** The purpose of task organizing is to produce tactical and operational results that enable accomplishment of operational and strategic objectives. What missions does the current operational effort require? Task forces execute discrete groups of missions that when coupled will accomplish a tactical or operational result. Can a discrete task or set of tasks that is not applicable to the entire maritime force be identified? Do the scale, scope, complexity, and duration of expected operations necessitate the creation of multiple TFs? Are there missions that all TFs will be required to execute? If yes, can the mission be executed by the establishing command or should a separate TF be created?

- **Mission:** What operational and tactical effects will the missions expected of each TF accomplish? Are all missions expected to be performed by the maritime force assigned to TFs? Are the units assigned to TFs capable of accomplishing missions assigned to the TF? Who has the preponderance of forces necessary, the ability to characterize the operational environment, and the ability to command and control the forces assigned to achieve success?

- **Span of Control/Command.** There are limits as to how many subordinates a commander can effectively control. At some point, managing the incoming information will exceed the abilities of the commander and staff. Expanding beyond this threshold taxes the cognitive abilities and may cause a loss of proper focus or, in extreme instances, lead to organizational paralysis. One can reduce the number of subordinate commands by narrowing the span of control. This approach deepens the fighting organization by adding intermediate command echelons. It may increase the number of individual staffs and the amount of time for information to flow to the top — that is, it may reduce speed of command. Flattening increases speed of vertical dissemination and may increase tempo, but requires widening the span of control. Trade-offs exist between organizational width (number of TFs) and depth (fewer TFs, with each having multiple subordinate commanders).

- **TF Relationships:** What will be the relationship between TFs operating concurrently in the maritime AO? With multimission ships it is expected that ships in one TF have capabilities applicable to mission(s) assigned to another TF. *NWP 3-56 (Rev. A), Composite Warfare Commander's Manual and allied/multinational instructions and procedures identify three* **"support situations" (SUPSITs):**
  - **Situation A.** The support force is to join and integrate with the other force. The senior officer present or the officer to whom he has delegated tactical command is to become the OTC of the combined force.
  - **Situation B.** The support force does not integrate. Unless otherwise ordered, the senior OTC of the two forces is to coordinate the tactical operations of the two forces
  - **Situation C.** The support force cdr has discretion how to best provide support

- **Flagship:** TFs typically require a flagship or shore headquarters facility with sufficient communications and computer networks to effectively provide situational awareness (SA) to the operational commander, other TF commanders, and subordinate forces.

Officer in Tactical Command/Composite Warfare Commander

Navy tactical forces composed of one or more units will be designated a task organization component and have an OTC. For task components comprising ships, a composite warfare commander (CWC) may be established. The OTC is the senior officer present eligible to assume command, or the officer to whom the senior officer has delegated tactical command.

See pp. 1-2 to 1-3 for discussion of Navy organization and command structure to include numbered fleets, principal headquarters and task forces. See following page (p. 2-64) for discussion of Navy tactical headquarters to include Navy strike groups (CSG/ESG/SSG) and expeditionary strike forces. See pp. 2-65 to 2-70 for discussion of the Officer in Tactical Command/Composite Warfare Commander.

III. Navy Tactical Headquarters

Ref: NWP 3-32, Maritime Operations at the Operational Level of War (Oct '08), p. 4-13 to 4-14.

Navy fleet commanders task organize their tactical platforms to create force packages able to execute various maritime missions. Within each tactical organization there is an officer in tactical command (OTC). Navy task force commanders are the Navy's primary tactical warfighting commanders; they are also typically commanders of Navy strike groups. Navy strike group commands are normally associated with an aircraft carrier, large amphibious ship, or major surface combatant. With appropriate augmentation, a Navy strike group command can be the JTF or Navy force command headquarters for small contingencies. A Navy strike group force package may include any mix of ships, aircraft, submarines, detachments, other units, and personnel. OPNAVINST 3501.316A describes three notional types of strike groups from which Navy tactical headquarters operate:

Carrier Strike Group (CSG)
The CSG provides the full range of operational capabilities for sustained maritime power projection and combat survivability. It is a flexible, heavy strike group that can operate in any threat environment, in the littorals or open ocean. CSG capabilities support initial crisis response missions and the ability to operate in non-permissive environments characterized by multiple threats, including, but not limited to, antiship missiles, ballistic missiles, fighter/attack aircraft, electromagnetic jammers, cruise missile–equipped surface combatants, submarines (nuclear and diesel), and terrorist threats.

Expeditionary Strike Group (ESG)
Currently, the ESG has organic air defense (provided through the air defense–capable surface combatant ships), expeditionary warfare capability (provided through amphibious shipping and Marine expeditionary unit (MEU) (special operations capable), when so equipped), and strike capability (provided through cruise missile land attack–capable ships and submarines) required for operating independently in low-to-medium threat environments. An ESG is a flexible strike group that can operate in the littorals or open ocean. ESG capabilities support initial crisis response missions that may be undertaken in limited nonpermissive environments characterized by multiple threats, including, but not limited to, antiship missiles, ballistic missiles, fighter/attack aircraft, electromagnetic jammers, cruise missile–equipped surface combatants, submarines (nuclear and diesel), and terrorist threats.

Surface Strike Group (SSG)
The SSG provides combat effectiveness by providing fire support to allies and joint forces ashore. It is a surface group that can operate independently or in conjunction with other maritime forces. SSGs support crisis response missions or sustained missions and may be employed in limited nonpermissive environments characterized by multiple threats. These Tomahawk land attack missile (TLAM)/standard missile–equipped surface support groups provide deterrence and immediate contingency response while maintaining the ability to conduct maritime interception and other tasks. SSGs are primarily designed to be an independent, sea-based, mobile group that can provide sea control and strike power to support joint and allied forces afloat and ashore. SSG capabilities include passive surveillance and tracking, passive defense and early warning, strike operations, and sea control, as well as the multiwarfare platform capabilities inherent within the SSG.

* Expeditionary Strike Force
When an operational commander combines the capabilities of more than one strike group (CSG/ESG/SSG), the tactical organization is designated an expeditionary strike force. An expeditionary strike force has increased striking power, enhanced flexibility, and improved responsiveness to permit operations in any threat environment.

See also pp. 1-4 to 1-5 for composition and capability of major deployable elements.

IV. Navy Composite Warfare Commander (CWC)

Ref: NWP 3-56 (REV A), Composite Warfare Commander's Manual (Aug '01) and NWC 3153K, Joint Military Operations Reference Guide (Jul '09), App. C.

The Navy Composite Warfare Commander (CWC) Concept was developed in concert with the "defense in depth" concept during the Cold War. It was designed to permit the Carrier Battle Group (CVBG) Commander to fight "fleet on fleet" engagements across multiple warfare areas simultaneously against the Soviet Union. The key operational assumption was that the CVBG would fight simultaneously in all three dimensions (air, surface, subsurface) within roughly 200 to 300 miles of the aircraft carrier (CV). The prevailing thought was that the CVBG Commander would be overwhelmed with so much simultaneous information that it would impact speed of command and tactical proficiency. Therefore, the Navy developed the CWC Concept and implemented its two central tenets: command by negation, and decentralized control. The CVBG has evolved into the Carrier Strike Group (CSG) and integrated into the joint operational domain.

The Navy's CWC Concept shares similarities with the joint force command and control construct, albeit with a more tactical focus. Under the CWC architecture, the Officer in Tactical Command (OTC) delegates command authority in tactical warfare areas to subordinate commanders within his organization. Subordinate commanders then execute assigned tasks within those warfare areas with assigned assets based on the OTC's guidance. The CWC concept is the Navy's C2 construct for tactical execution. This construct, though designed for tactical level maritime operations, is not unlike the construct utilized by Joint Force Commanders (JFC), who employ decentralized execution and a form of command by negation in directing the operational execution of subordinate component commanders. The relationship between the JTF Commander and his subordinate component commanders is similar to the relationship between the CWC/OTC and his subordinate warfare commanders.

The CWC retains responsibility and accountability for the conduct of Carrier Strike Group (CSG) operations and delegates authority for employing forces to his principle warfare commanders. In order to decide the assignment and location of subordinate warfare commanders and coordinators, the CWC must take into account the tactical situation, force size, and capabilities required to cope with an expected threat.

Principle warfare commanders employ forces assigned to them by the CWC and respond to threats according to the CWC's guidance. The CWC's guidance is promulgated via a naval message called the OPGEN. The OPGEN describes the operational environment, assigns warfare commanders and warfare areas, allocates assets, outlines mission priorities, defines command relationships, and provides commander's guidance.

The CWC Concept and Amphibious Operations

When a CSG is tasked to support an amphibious operation the command and control environment is challenging. Conducting amphibious operations in a joint environment with coalition partners adds another layer of complexity. The command and control challenges are both positional and sequential. The first challenge is the command relationship between the CSG Commander and the Amphibious Ready Group – Marine Expeditionary Unit (ARG-MEU) Commander. The second challenge is within the ARG-MEU between the Commander Amphibious Task Force (CATF) and the Commander of the Landing Force (CLF).

A. Officer in Tactical Command (OTC)/ Composite Warfare Commander (CWC)

Ref: NWP 3-56 (REV A), Composite Warfare Commander's Manual (Aug '01), chap. 2 and NWC 3153K (Jul '09), p. C-2.

The Composite Warfare Commander is the central command authority and overall commander. The CWC is usually a Rear Admiral (O7/O8) in command of a CSG (Carrier Strike Group Commander) who is embarked in the aircraft carrier (CVN) along with a support staff. Under the CWC architecture, the Officer in Tactical Command (OTC) delegates command authority in particular warfare areas to subordinate commanders within the CWC organization. Usually, the OTC is also the CWC, but in certain situations the CWC and OTC may be separate officers. The OTC has the option to delegate overall coordination of defensive warfare areas to a CWC and can retain direct command in any one (or more) warfare area(s) if desired. Because this rarely occurs, the terms CWC and OTC are usually interchangeable. However, great care must be exercised when applying the CWC or OTC distinction to coalition operations because of the possibility of misunderstanding the CWC/OTC relationship.

Officer in Tactical Command (OTC)

In maritime usage, the OTC will be the senior officer present eligible to assume command, or the officer to whom the senior officer has delegated TACOM. The OTC will always be responsible for accomplishing the mission of the forces assigned. He may delegate authority for the execution of various activities in some or all warfare areas to designated subordinate warfare commanders and/or coordinators.

> **TACOM** — The authority delegated to a commander to assign tasks to forces under their command for the accomplishment of the mission assigned by higher authority. This term is used primarily in maritime operations. It is narrower in scope than operational command but includes the authority to delegate or retain tactical control (TACON). (MMOPS-1)

The five principal areas of warfare the OTC is responsible for are air defense (AD), antisubmarine warfare (ASW), information warfare (IW), strike warfare (STW), and surface warfare (SUW). Dependent on the situation, the ASW and SUW areas may be combined into a singular warfare area called sea combat (SC). Requirements for air and surveillance coordination concern more than one area of warfare and shall be taken into consideration when developing command structure.

The OTC is always responsible for formulating and promulgating policy. Other OTC functions including warfare functions may be delegated to subordinates within the constraints of the Rules of Engagement (ROE) in force and stated policy. For the five principal warfare areas, the OTC has the following options:

1. The OTC retains command in the principal areas of warfare by retaining all warfare functions.
2. The OTC delegates to one subordinate commander one or more warfare functions.
3. The OTC delegates to more than one subordinate commander several warfare functions.
4. The OTC delegates to subordinates within geographic areas (or sectors) warfare functions relevant to that area, but may retain any part of the overall function for himself. This form of delegation can be used by a principal warfare commander (PWC) as well, if so assigned.
5. A special form of delegation as described in item (2), is that the OTC of a large force may allocate all his warfare functions for the defense of a force to a CWC while retaining overall responsibility for the mission. The CWC may, in turn, delegate some or all warfare functions as described above.

Delegation of Authority

The OTC may retain TACON authority or he may delegate some of that authority to subordinate commanders and coordinators. Such delegation does not mean that the OTC relinquishes authority over subordinates. It does mean that the subordinate is given some or all of that same authority over forces assigned to him by the OTC. If the OTC elects to designate a CWC and/or warfare commander/coordinators, he may delegate TACON required to carry out assigned tasks to them by using ATP 1, Vol. II, Table D (Duty Table) and NWP 3-56, Appendix A.

TACON—Command authority over assigned or attached forces or commands, or military capability or forces made available for tasking, that is limited to the detailed and usually, local direction and control of movements or maneuvers necessary to accomplish missions or tasks assigned. TACON is inherent in operational control (OPCON). TACON may be delegated to, and exercised at any level at or below the level of combatant command. (JP 1-02)

The OTC has overall responsibility for accomplishing the mission of the force. The OTC's policy and procedure for succession of command authority as well as designation of the standby OTC, should be specified in advance of the operation in the OTC's orders. The OTC shall specify the chain of command between himself and, when designated, the CWC, PWCs, functional commanders, coordinators and the force under their TACON. This may be done by task number designation or by stipulating which task groups (TGs), units (TUs), or elements (TEs) are designated for each commander.

Composite Warfare Commander (CWC)

The CWC is the commander to whom the OTC has delegated specific authority for force direction and control of warfare functions. The OTC is normally the CWC. However, the CWC concept allows an OTC to delegate TACOM to the CWC. The CWC wages combat operations to counter threats to the force and to maintain tactical sea control with assets assigned; while the OTC retains close control of power projection and strategic sea control operations.

If the OTC is also the CWC, TACON remains with the OTC. When the OTC has designated another commander as the CWC, the OTC must specify the assigned forces over which the CWC has TACON. The OTC, through the CWC, exercises overall responsibility for C2 of the force. In this example, the OTC of a naval force has chosen to define an organization in which the aircraft CVBG commander has established a CWC organization using BG assets. This commander serves as the CWC for forces assigned to the CVBG. Concurrently, and independently, the ATF commander has established a CWC organization using ATF assets and is serving as the CWC for forces assigned to the ATF. The ATF and the CVBG may be operating independently in this case, or they may be operating under a supported/supporting arrangement established by the OTC. The relationship established by the OTC between the two organizations will be the result of that commander's estimate of the mission, threat, and the forces assigned.

In deciding where to locate warfare commanders, functional commanders, and coordinators, the CWC should take into account the TACSIT, size and capabilities of the force, the C2 capabilities of available assets to cope with the expected threat, possible threat courses of action (COA), and meeting the mission objectives. This analysis may lead the CWC to retain control of one or more of the warfare areas. When appropriate, designated commanders may be assigned alternate/supporting functions in addition to their primary responsibility.

Subordinate to the CWC are principal warfare commanders to include: Sea Combat Commander (SCC), Surface Warfare Commander (SUWC), Undersea Warfare Commander (USWC), Air Warfare Commander (AWC), Information Warfare Commander (IWC), Strike Warfare Commander (STWC), and Mine Warfare Commander (MIWC).

See following pages (pp. 2-68 to 2-69) for discussion of the principle warfare commanders and p. 2-70 for discussion of functional coordinators.

B. Principle Warfare Commanders (PWCs)

Ref: NWP 3-56 (REV A), Composite Warfare Commander's Manual (Aug '01), pp. 31 to 32 and NWC 3153K (Jul '09), pp. C-2 to C-4.

Subordinate to the OTC and CWC are five PWCs: Air Defense Commander (ADC), Antisubmarine Commander (ASWC), Information Warfare Commander (IWC), Strike Warfare Commander (STWC) and Surface Warfare Commander (SUWC). Dependent on the situation, the ASWC and SUWC can be linked and put under a SCC. Principle warfare commanders are usually individual ship Commanding Officers (CO's) or embarked Commanders (Destroyer Squadron (DESRON) or Carrier Air Wing (CAG). They usually exercise tactical control (TACON) of all assets assigned to their warfare area. Each principle warfare commander reports directly to the CWC.

- **Functional Warfare Commanders.** The OTC or CWC may form temporary or permanent functional groups within the overall organization. Functional groups are subordinate to the OTC and CWC and are usually established to perform duties which are generally more limited in scope and duration than those acted upon by PWCs. Examples of functional commands include: MIOC, MIWC, Operational Deception Group Commander, Screen Commander (SC), and Underway Replenishment Group (URG) Commander.

- **Sector Warfare Commanders.** When the force is dispersed the CWC or, when authorized, the warfare commander, may appoint sector warfare commanders. Sector warfare commanders need not be assigned in all warfare areas simultaneously; for example, the CWC may assign sector ADCs while designating only one force ASWC and SUWC. Sector warfare cmdrs report to their primary force warfare commander.

1. Antisubmarine Warfare Commander (ASWC)

For TG operations in a multi-threat environment ASWC responsibilities should be assigned to a commander served by a staff augmented with representatives of the ASW air communities (e.g., fixed-wing patrol and reconnaissance squadron (VP), helicopter ASW squadron (HS), helicopter antisubmarine squadron light (HSL)). When ASW is assigned as the primary mission of submarines operating with a naval force and if the ASWC is also designated as SOCA, a qualified submarine officer may be assigned to the ASWC to act as submarine element coordinator (SEC) in order to achieve the required coordination.

2. Surface Warfare Commander (SUWC)

The SUWC is an incorporated responsibility of the SCC, but may be executed separately as the situation dictates. The mission of the SUWC is the conduct of both offensive and defensive maritime operations to defeat enemy surface threats (primarily ships). The SUWC; if divested from the SCC will normally have an alternate warfare commander assigned. Candidates include the CO of the CVN, CAG, or a coalition partner. Both the SCC and the SUWC (when divested) place a high demand signal on limited aviation assets in order to support surface surveillance and anti-surface operations. The Force over the Horizon Coordinator (FOTC) support function is usually associated with this warfare area.

3. Sea Combat Commander (SCC)

The SCC is an optional position within the CWC structure that, when activated, is used to combine the SUWC, ASWC, and other warfare commander and coordinator positions as desired by the OTC/CWC. The SCC is normally the DESRON Commander (a Navy O6) and is embarked on the CVN. The SCC is responsible for conducting operations in multiple warfare areas. Originally conceived as a combination of Surface Warfare (SUW) and Undersea Warfare (USW); SCC's operational reach has expanded to include (MIWC) and other supporting functions as well. At the discretion of the SCC (with the CWC's concurrence), the SCC may divest warfare area responsibilities to other commanders in the CSG or to coalition partners. Some examples include assigning the CO of the CVN as SUWC, or a coalition partner as USWC.

4. Air Defense Commander (AWC/ADC)
The ADC is normally the CO (a Navy O6) of one of two AEGIS guided-missile cruisers (CG). The mission of the ADC is to defend the CSG against air threats (primarily airplanes and missiles). The alternate ADC is normally the other CG Commanding Officer or the CO of a DDG if only one CG is assigned to the CSG. The ADC, CAG and the CVN must coordinate their efforts in order to effectively conduct air operations. Within the joint air defense architecture the ADC will typically be a RADC/SADC (Regional Air Defense Commander/Sector Air Defense Commander) supporting the JFACC AADC functions.

5. Strike Warfare Commander (STWC)
The STWC is normally the Carrier Air Wing (CAG) Commander. The mission of the STWC is power projection. The STWC promulgates strike philosophy, policy, and employs manned aircraft and tactical missiles in accordance with either the Air Tasking Order (ATO) in a joint/combined environment, or the air plan in an exclusively naval environment. The alternate STWC may be a senior officer on the embarked CWC or CVN staff. In rare cases, the CWC may chose to retain this warfare area and augment his staff with officers from the CVN and CAG staffs.

6. Information Warfare Commander (IWC)
The IWC includes the function of Space and Electronic Warfare Commander (SEWC). The IWC is normally apart of the CWC's support staff embarked on the CVN and responsible for all CSG Information Operations. This individual serves as the principle advisor to the CWC for the exploitation of the electromagnetic spectrum by friendly forces and the denial of its use to enemy forces. The IWC promulgates Emissions Control (EMCON) restrictions, monitors intelligence, sensor operations, develops operational deception, and counter-targeting plans for the CSG among other functions. The CO of the non-ADC CG may also be assigned as an alternate.

7. Maritime Interception Operations Commander (MIOC)
The MIOC is normally a DESRON or amphibious squadron (PHIBRON) commander, as the scope of the MIOC's duties requires a dedicated commander and staff who have been trained to execute this specialized function. The MIOC must be embarked in a platform with the capability to develop and maintain a good surface surveillance plot and with adequate communications capability for MIO. An alternate MIOC is not normally designated.

8. Mine Warfare Commander (MIWC)
The MIWC performs those tasks within the CWC organization relating to mine warfare – to include both mining (offensive and defensive) and mine countermeasures (MCM). Usually embarked on a CVN, the MIWC works closely with STWC to plan and conduct mine-laying operations due to the requirement for airborne strike assets as mine layers. The MIWC also works closely with the other principle warfare commanders to ensure that offensive and defensive mine warfare is conducted across the full spectrum of warfare specific areas. Today, the MIWC is a relatively new addition to the CWC organizational staff and has been integrated into the organizational structure due to the asymmetric nature of the mine threat in littoral areas of operations.

9. Undersea Warfare Commander (USWC)
The USWC is also an incorporated responsibility of the SCC responsible for anti-submarine warfare (ASW) operations. It may be executed separately from the SCC as the situation dictates. The mission of the USWC is the conduct of both offensive and defensive maritime operations to defeat enemy subsurface threats (primarily submarines). When divested, the USWC will also have an alternate warfare commander assigned. Candidates include the CO's of either the Guided Missile Cruiser (CG) or Guided Missile Destroyer (DDG); or a coalition partner. The Screen Coordinator (SC) and the Light Air Multi-purpose System (LAMPS) Element Coordinator (LEC- responsible for coordination of ASW helo assets) support functions are also associated with this warfare area.

C. Functional Coordinators

Ref: NWP 3-56 (REV A), Composite Warfare Commander's Manual (Aug '01), pp. 4 to 5 and NWC 3153K (Jul '09), pp. C-4 to C-5.

Functional coordinators play important roles in the CWC concept. Their specific responsibilities cross warfare areas boundaries and require information to be shared between warfare commanders in order for all concerned to be effective. These functional coordinators can be staff members embarked on the carrier, principle warfare commanders who are "dual-hatted", or individual ship CO's. Supporting coordinators differ from the warfare commanders in one very important respect. Warfare commanders have TACON of resources assigned and may employ forces. The supporting coordinators execute policy, but do not employ forces.

1. Air Resource Element Coordinator (AREC)
The AREC is responsible for managing and coordinating the distribution of aircraft and keeping the CWC and other warfare commanders/coordinators apprised of air operations. Normally the AREC is the CVN CO and does not function as a warfare commander. Instead, the AREC's responsibilities include serving as an air operations advisor to the CWC and producing the daily air plan that allocates CSG aircraft.

2. Airspace Control Authority (ACA)
The OTC and/or CWC embarked in the CV, or Commanding Officer of the CV, should retain the ACA function. Alternatively, the ACA function may be delegated to the ADC.

3. Cryptologic Resources Coordinator (CRC)
The OTC/CWC should retain the CRC function. The alternate CRC shall have cryptologic personnel embarked.

4. Force Over-the-Horizon Track Coordinator (FOTC)
The FOTC is either the Destroyer Squadron Commander or a staff officer aboard the CVN. FOTC serves as the chief advisor to the CWC and manages the CSGs common operational picture (COP) or recognized maritime picture (RMP). Therefore, FOTC spend the majority of their time managing, processing, and disseminating all-source contact information for CSG.

5. Force Track Coordinator (FTC)
The ADC is normally designated as the FTC. Accordingly, the alternate ADC is designated as alternate FTC.

6. Helicopter Element Coordinator (HEC)
The role of the HEC in the strike group is to advise the CWC of all non-logistics helicopter operations to support the strike group. Usually the HEC is either embarked on board the carrier or is a senior staff member who also has warfare expertise in helicopter platforms. The HEC is normally apart of the CAG staff.

7. Screen Coordinator (SC)
The SC is responsible to the SCC and AWC for coordinating ship positioning in the carrier strike group. He positions the ships in screening or picket stations around the CVN or defended asset/area to provide protection against expected threats. SCC is normally assigned SC duties, but SC may be assigned to AWC (rare) or the CO of the CVN if the SCC is not embarked.

8. Submarine Element Coordinator (SEC)
The SEC is a usually assigned to the ASWC staff when submarines are assigned in direct support of CWC CDRs. The SEC acts as the executive agent in planning submarine operations, conducting waterspace management and battle space deconfliction.

9. LAMPS Element Coordinator (LEC)
The LEC performs a similar function as the AREC for LAMPS helicopters only in advising the CWC and OTC. Usually the LEC is a staff officer who has expertise in the LAMPS helicopter community (usually assigned to a Destroyer Squadron).

VII. Multinational Operations

Ref: JP 3-16, Multinational Operations (Mar '07), chap. I and NWC Maritime Component Commander Handbook (Feb '10), app. C.

Multinational operations are operations conducted by forces of two or more nations, usually undertaken within the structure of a coalition or alliance. Other possible arrangements include supervision by an intergovernmental organization (IGO) such as the United Nations (UN) or the Organization for Security and Cooperation in Europe. Commonly used terms under the multinational rubric include allied, bilateral, coalition, combined, combined/coalition or multilateral.

Alliance
An alliance is a relationship that results from a formal agreement (e.g., treaty) between two or more nations for broad, long-term objectives that further the common interests of the members.

Coalition
A coalition is an ad hoc arrangement between two or more nations for common action. Coalitions are formed by different nations with different objectives, usually for a single occasion or for a longer period while addressing a narrow sector of common interest. Operations conducted with units from two or more coalition members are referred to as coalition operations.

Nations form partnerships in both regional and worldwide patterns as they seek opportunities to promote their mutual national interests, ensure mutual security against real and perceived threats, conduct foreign humanitarian assistance (FHA) operations, and engage in peace operations (PO).

US commanders should expect to conduct military operations as part of a multinational force (MNF). These operations could span the range of military operations and require coordination with a variety of US Government (USG) agencies, military forces of other nations, local authorities, IGOs, and nongovernmental organizations (NGOs).

A **bilateral agreement** exists between two nations. Operations conducted with units from two or more allies are referred to as **combined operations**. A JTF comprising units from two or more allies is called a combined joint task force. A JFMCC with a similar force is called a combined force maritime component commander. A coalition is a short-term alliance for combined action. A **multilateral agreement** involves more than two nations.

I. Multinational Chains of Command

Forces participating in a multinational operation will always have at least two distinct chains of command: a national chain of command and a multinational chain of command.

A. National Command
The President always retains and cannot relinquish national command authority over US forces. National command includes the authority and responsibility for organizing, directing, coordinating, controlling, planning employment, and protecting military forces. The President also has the authority to terminate US participation in multinational operations at any time.

B. Multinational Command

Command authority for a MNFC is normally negotiated between the participating nations and can vary from nation to nation. Command authority could range from operational control (OPCON), to tactical control (TACON), to designated support relationships, to coordinating authority.

See facing page (p. 2-73) for discussion of command structures of forces in multinational operations.

II. Maritime Multinational Forces

Maritime multinational forces either exist or are included in contingency plans for most of the earth's international waters. All multinational tactical forces are employed within constraints defined at the strategic level. By the time these forces are employed, a clear understanding of constraints and restraints needs to exist. The CFMCC integrates these diverse strategic factors with diverse tactical capabilities to optimally employ the force. The CFMCC accomplishes this through providing tailored operational functions to the subordinate tactical forces. Besides these functions, CFMCC needs to coordinate the participation and formation of the subordinate CTF, form an integrated operational-level headquarters, manage information sharing, coordinate ROE, conduct planning and assessment, and coordinate termination.

Where commonality of interest exists, nations will enter political, economic, and military partnerships. These partnerships can occur in both regional and worldwide patterns as nations seek opportunities to promote their mutual national interests or seek mutual security against real or perceived threats. Cultural, psychological, economic, technological, and political factors all influence the formation and conduct of coalitions.

Coalitions, which are created for limited purposes and for a set time, do not afford military planners the same political resolve and commonality of aim as alliances. Thus, planners must closely study the political goals of each participant as a precursor to detailed planning. Political considerations weigh more heavily with coalitions than with alliance operations. The precise role of military forces in these operations varies according to each political and military situation. One reason nations conduct coalition operations is that rarely can one nation go it alone, either politically or militarily. Coalition operations involve a comprehensive approach that includes other government agencies, NGOs, and international and regional organizations. This blending of capabilities and political legitimacy makes possible certain operations that a single nation could not or would not conduct unilaterally. Almost all coalition operations, regardless of how they are formed, build from common fundamentals. Coalition headquarters have similar components. In addition, all coalitions assess the area of operations.

Maritime coalitions and multinational forces follow these same ideas. Long-term maritime coalition task forces, such as ones found in the Indian Ocean and Arabian Gulf, already have established the shared political resolve and commonality of the members of the force. As naval forces rotate into and out of the CTF, these maritime coalitions share many of the same challenges of long-term land forces. Maritime CTFs, such as CTF 150, require an operational headquarters to provide the higher headquarters control and operational functions support.

Understanding Multinational Forces and Coalitions

A **multinational force (MNF)** is "a force composed of military elements of nations who have formed an alliance or coalition for some specific purpose." **Coalition operations** are conducted by forces of two or more nations, which may not be allies, acting together for the accomplishment of a single mission. Coalition operations cross the spectrum of conflict, from major combat operations to peacetime military engagement. Coalition operations are a subset of multinational operations. These operations can also include various nonmilitary organizations and other services.

Command Structures of Forces in Multinational Operations

Ref: JP 3-16, Multinational Operations (Mar '07), pp. II-6 to II-8.

No single command structure meets the needs of every multinational command but one absolute remains constant; political considerations will heavily influence the ultimate shape of the command structure. Organizational structures include the following:

A. Integrated Command Structure

Multinational commands organized under an integrated command structure provide unity of effort in a multinational setting. A good example of this command structure is found in the North Atlantic Treaty Organization where a strategic commander is designated from a member nation, but the strategic command staff and the commanders and staffs of subordinate commands are of multinational makeup.

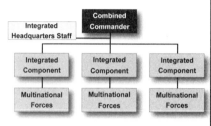

B. Lead Nation Command Structure

A lead nation structure exists when all member nations place their forces under the control of one nation. The lead nation command can be distinguished by a dominant lead nation command and staff arrangement with subordinate elements retaining strict national integrity. A good example of the lead nation structure is Combined Forces Command-Afghanistan wherein a US-led headquarters provides the overall military C2 over the two main subordinate commands: one predominately US forces and the other predominately Afghan forces.

C. Parallel Command Structures

Under a parallel command structure, no single force commander is designated. The coalition leadership must develop a means for coordination among the participants to attain unity of effort. This can be accomplished through the use of coordination centers. Nonetheless, because of the absence of a single commander, the use of a parallel command structure should be avoided if at all possible.

A **coalition action** is a "multinational action outside the bounds of established alliances, usually for single occasions or longer cooperation in a narrow sector of common interest." In the interest of brevity, future references that could be either coalition or alliance will be referred to as coalitions.

Creating a coalition is a political act that sets the conditions for success or failure of a multinational operation. Commanders have an overriding interest in providing advice to assist the political leadership in forming practical military guidance. Further, all national military commanders in a coalition require specific understandings and agreements with the coalition commander and their counterparts if they are to achieve and maintain unity of effort. Establishing these understandings and agreements is a commander's first responsibility. They provide not only the basis for unity of effort but also the foundation for the command guidance needed by staffs when doing campaign planning (political-military-civil).

In coalition operations, consensus building between partners is key to ensure compatibility at the political, military, and cultural levels. While the coalition partners share a commonality of interest, each also has its own strategic objectives, which may not be common. A successful coalition must establish at least unity of effort, if not unity of command. The success of a coalition operation begins with the authority to direct operations of all assigned or attached military forces. The coalition force commander has much to consider in addition to military considerations, including the strategic context within which the operation will be carried out, civil administration, the reestablishment of justice, civil policing, humanitarian assistance, post-conflict development and reconstruction, the possibility of election organization, financial management, and multicultural issues. Commanders must harmonize these considerations to ensure that the operation has the best possible chance of success.

III. Civil-Military Coordination and Dealing with Nonmilitary Agencies (Interagency, Intergovernmental & NGOs)

When dealing with most nonmilitary agencies, the coalition commander focuses on cooperation and coordination rather than command and control. It is important that the military role of the coalition force is coordinated with the roles of **other governmental agencies, NGOs, and intergovernmental and regional organizations**. These agencies have their own missions and goals, and the coalition commander has a limited ability to influence their actions. To ensure that the coalition commander can accomplish the mission and end state while allowing these agencies to do the same requires the commander to seek their cooperation and to coordinate their efforts to prevent interference in one another's missions. Additionally, these agencies may be in a position to help the commander in mission accomplishment. Developing a **civil-military operations center** or coalition coordination center for civil-military cooperation is one way of achieving cooperation and coordination with nonmilitary agencies. It also provides a single point of contact between these agencies and the commander.

Refer to The Joint Forces Operations & Doctrine SMARTbook (Guide to Joint, Multinational & Interagency Operations), chap. 7 for discussion of multinational operations and chap. 8 for discussion of interagency, intergovernmental and nongovernmental (NGO) coordination.

Refer to The Stability, Peace, & Counterinsurgency SMARTbook (Nontraditional Approaches in a Dynamic Security Environment), chap. 4 for discussion of civil-military coordination and the civil-military operations center (CMOC).

Chap 3

I. MHQ / Maritime Operations Center (MOC)

Ref: NWP 3-32, Maritime Operations at the Operational Level of War (Oct '08), chap. 7 and NTTP 3-32.1, Maritime Operations Center (Oct '08), chap. 1.

This chapter will discuss the maritime headquarters with maritime operations center (MHQ with MOC) organization structure and the functions it supports. It will briefly examine the methodology utilized to coordinate the simultaneous actions the maritime operational level command staff may execute to support the commander.

I. The Maritime Headquarters (MHQ)

The staff of an operational maritime command and its associated support infrastructure is collectively known as a MHQ. Today's Navy operational level staffs must continuously balance operational and fleet management (Title 10), routine administrative Navy administrative control (ADCON), responsibilities. Accordingly, the MHQ organization must address fleet management and operational responsibilities. Navy and joint operational organizational constructs are different.

- The Navy organizational construct is based upon a hierarchical command structure (pyramid) with established communication paths from subordinate to senior officers/commands. Each level in the Navy structure is expected to review information, determine its relevance with respect to other information, and provide interpretation and recommendation to higher levels within the organization. This organizational construct is commonly referred to as a Napoleonic organizational construct.

- The joint operational construct is matrix based, with the traditional Napoleonic organizations providing resources to product teams, thereby accelerating the commander's decision processes. Product teams are categorized as boards, bureaus, cells, centers, and working groups (B2C2WGs), depending on the product being provided to the headquarters and the duration that the team will exist.

Adherence to guidance contained herein and MHQ with MOC tactics, techniques, and procedures will ensure standardized staff functions and processes that enable interoperability with the joint community and commonality across all fleet and principal headquarters. To continually address fleet management and operational responsibilities the MHQ with MOC organization must be:

- **Flexible** — capable of adjusting to changing priorities, whether fleet management or operational
- **Tailorable** — capable of smoothly transitioning as the commander is assigned different roles
- **Scalable** — capable of integrating additional capacity or capabilities in response to new missions.

Each MHQ with MOC organizational structure has the same three basic organizational structures: command, fleet management, and maritime operations. Each of these structures is briefly described below, and will be discussed in greater detail in follow-on paragraphs. The commander retains the authority to modify his staff structure as necessary to accomplish the mission. The mission determines what tasks the command is expected to accomplish. These tasks determine how the commander organizes or adapts his staff to support mission accomplishment. The basic staff structure provides the flexibility to make such modifications, while simultaneously maintaining a structure that individuals outside the command can easily understand

and access for interoperability and that allows for transition to support various roles assigned the MHQ commanders — joint force maritime component commander (JFMCC), joint force commander (JFC), Navy force commander (COMNAVFOR), or fleet commander.

- The command structure consists of the MHQ commander, deputy commander (if assigned), chief of staff (COS) and the commander's personal staff. It is supported by the MOC and fleet management directors. It is the focal point of all staff work to collect and process information in order that the commander can make timely, accurate decisions that:
  - Support successful actions at the operational level.
  - Support Secretary of the Navy's (SECNAV) Title 10 responsibilities.
- The fleet management structure provides the commander the information needed to make decisions to execute support of SECNAV's Title 10 responsibilities. These efforts ensure that the Navy is ready (manned, trained, and equipped) today and in the future to successfully execute joint maritime operations. This structure also constitutes the resource pool that is used to man the maritime operations structure with subject-matter experts. Therefore, the MHQ staff can and frequently are required to execute tasks in support of fleet management and maritime operations structures.
- The maritime operations structure provides the commander with the command and control (C2) organization needed to properly assess, plan, direct, and monitor maritime operations from an operational-level perspective.

A. Command Structure of the MHQ

The command structure consists of the commander, deputy commander (if assigned), the COS, and the commander's personal staff.

The commander exercises command across the breadth of the operational maritime force using the C2 processes discussed earlier. Supporting the commander is a carefully designed C2 infrastructure that obtains, analyzes, and submits information to the commander to assist in decision making and manage the execution of the commander's decisions.

The COS is the commander's lead staff officer. The commander normally delegates authority to the COS for the executive management of the entire staff. The COS directs staff tasks, conducts staff coordination, and ensures efficient and prompt staff response. The COS does not normally oversee the commander's personal staff. In major fleet commands, the commander may designate deputy chiefs of staff (DCOS)/directors to direct, coordinate, and supervise the actions of fleet management directorates and/or a MOC director to do the same with the maritime operations components. The COS performs the following functions:

- Keeps the commander informed of current and developing situations
- Receives the commander's decisions and ensures the staff takes appropriate actions to implement those decisions
- Supervises the actions of special assistants, DCOS/directors, and the MOC director
- Serves as the chief information officer
- Monitors the currency, accuracy, and status of commander's critical information requirements (CCIRs)
- Directs and supervises the planning and execution process
- Monitors development of plans, orders, and instructions
- Obtains the commander's approval of and promulgates plans, orders, and instructions
- Monitors, with assistance of the staff, the execution of plans, orders, and instructions

- Organizes, plans, and supervises staff training
- Ensures proper coordination of staff activities internally, vertically (with higher headquarters and subordinate units), and horizontally (with adjacent headquarters)
- Ensures proper staff support to subordinate commanders and staffs
- Prioritizes efforts of fleet management, operations, and shared support.

The commander and COS are supported by the commander's personal staff and fleet management staffs. The commander's personal staff works directly for the commander. The personal staff is normally composed of aides, the command master chief, and personnel secretaries. Specific duties and responsibilities of the personal staff are specified by the commander. The fleet management staff is composed of special assistants and directors who are responsible to the commander and are coordinated and directed by the COS. Directors are responsible for a broad functional area (a directorate) and help the commander coordinate and supervise the execution of plans, operations, and activities in that area. Directors exercise broad coordinating responsibilities over staff sections within their assigned functional area. These responsibilities are intended to facilitate coordination within related areas of staff functioning and to ensure the systematic channeling of information and documents.

Staff officers are accountable for the commander's entire field of responsibilities, except for any areas that the commander may elect to control personally. A staff officer's authority is limited to advising, planning, and coordinating actions within his or her functional area. The commander might also give a staff officer added authority to act on specific matters related to his or her functional area.

Staff officers are responsible for acquiring information and analyzing its implications to provide timely and accurate recommendations to the commander. Staff officers must often request and receive information from staff sections not under their cognizance. For example, the staff judge advocate (SJA) must request, receive, and coordinate information from other staff sections, specifically intelligence and operations, for the purposes of developing and refining rules of engagement (ROE). A clear definition of staff responsibilities is necessary to ensure coordination and eliminate conflict. The command's standard organization and regulation manual (SORM) and standard operating procedures (SOPs) should clearly delineate staff primary responsibilities and requirements for support and coordination.

B. Maritime Operations Center (MOC)

The maritime operations structure of the MHQ consists of the maritime operations center (MOC). Today's information systems, complex operational environment, and need for accelerated decision cycles are not supported by the fleet management hierarchical Napoleonic structure. The fleet management organization with knowledge stovepipes does not facilitate the horizontal and vertical coordination needed to rapidly step through the commander's decision cycle. Therefore, the MOC consists of B2C2WGs in order to facilitate this requirement for rapid decision-making.

The flattened vice hierarchical MOC B2C2WG organizational structure facilitates the ability to accelerate the commander's decision cycle. Standard procedures make staff coordination more routine, increase cross-functional integration, facilitate monitoring, assessment, and planning, provide venues for command decisions, and allow for the management of current operations, future operations, and future plans. The B2C2WGs are staffed with subject-matter experts from the fleet management directorates, thereby providing the cross-directorate knowledge required to address today's complex operational environment. The B2C2WG integrating structure provides the mechanism for bringing together the various expertise of the staff members focused on specific problem sets to provide coherent staff recommendations to the commander.

C. Fleet Management and Operational Structures (Maritime Headquarters)

Ref: NWP 3-32, Maritime Operations at the Operational Level of War (Oct '08), pp. 7-4 to 7-14.

The fleet management structure of the MHQ includes the special assistants to the commander plus the N1 (personnel), N2 (intelligence), N3 (operations), N4 (logistics), N5 (plans and policy), N6 (communications information systems), N7 (training), and N8 (programs) directorates. Based on mission requirements and the nature of the operational environment, the commander commonly establishes additional staff directorates. For instance, an N9 directorate may be responsible for civil-military issues, concept development, and experimentation, or another area that requires the command's attention. The focus of the fleet management module is the readiness of the assigned and attached Navy forces (NAVFOR). This focus covers the operational near-, mid-, and far-time horizons and long-term input to the Navy service as a whole. It requires knowledge of today's operational environment and the projected environment, plus Navy polices and procedures for the manning, equipping, and training of Navy forces.

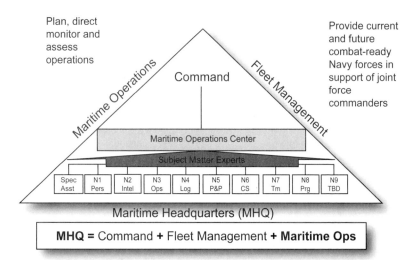

Ref: NWP 3-32, Maritime Opns at the Operational Level of War, fig. 7-3, p. 7-14.

The fleet management module is organized based on the "Napoleonic Structure," with the staff organized into directorates with each directorate having a specific responsibility. In the Navy, directorates are prefaced with the letter "N"; the Marines preface directorate numbers with the letter "G" on staffs whose commander is a general officer. Otherwise, Marine directorate numbers are prefaced with an "S" (regiment level and below). The use of standard directorate assignments enhances external headquarters' communication with the MHQ:

- **N0** Special Assistants to the Commander
- **N1** Personnel
- **N2** Intelligence
- **N3** Operations
- **N4** Logistics
- **N5** Plans and Policy
- **N6** Communications Information Systems
- **N7** Training
- **N8** Programs
- **N9** TBD

The MHQ with MOC ensures Navy operational commanders are able to execute at the operational level. Maintaining the fleet management Napoleonic structure provides the deliberate staffing environment required to properly man, equip, and train Navy forces. Additionally, it provides an organizational structure that external commands understand, allowing staff quick access to expertise to promote interoperability. Organizing the MOC following the B2C2WG flattened organizational construct to plan and execute assigned missions greatly speeds planning and decision making by aligning with the JTF staff model. Additionally, by following this standard model, coordination with other component commanders and transforming to the various joint roles that may be assigned will be made easier. Reach back, or support from or to other similarly organized MOCs, will be facilitated and the utility of schoolhouse training programs maximized through standardization.

MANPOWER & PERSONNEL DIRECTORATE (N-1)	ADMINISTRATIVE AND PERSONNEL CELL LIAISON CELL
INTELLIGENCE DIRECTORATE (N-2)	MARITIME INTELLIGENCE OPERATIONS CENTER INTELLIGENCE PLANS/INTELLIGENCE PREPARATION OF THE OPERATIONAL ENVIRONMENT (IPOE) INTELLIGENCE OPERATIONS COUNTERINTELLIGENCE (CI)/HUMAN INTELLIGENCE (HUMINT) INTELLIGENCE SUPPORT ELEMENT JOINT COLLECTION MANAGEMENT BOARD
OPERATIONS DIRECTORATE (N-3)	FLEET COMMAND CENTER CURRENT OPERATIONS CELL FUTURE OPERATIONS CELL FIRES ELEMENT INFORMATION OPERATIONS CELL MARITIME ASSESSMENT GROUP INFORMATION MGT OFFICER KNOWLEDGE AND INFORMATION MANAGEMENT (KIM) BOARD KIM WORKING GROUP STRATEGIC COMMUNICATION WORKING GROUP METEOROLOGICAL AND OCEANOGRAPHIC (METOC) CELL RULES OF ENGAGEMENT (ROE)/RULES FOR THE USE OF FORCE (RUF) WORKING GROUP PROTECTION WORKING GROUP
LOGISTICS DIRECTORATE (N-4)	LOGISTICS READINESS CENTER DEPLOYMENT AND DISTRIBUTION CELL LOGISTICS PLANS AND READINESS CELL MAINTENANCE CELL FACILITIES AND ENGINEERING CELL SUSTAINMENT AND SERVICES CELL HOST NATION SUPPORT CELL HEALTH SERVICE SUPPORT CELL
PLANS AND POLICY DIRECTORATE (N-5)	MARITIME PLANNING GROUP OPERATIONAL PLANNING TEAMS JOINT OPERATION PLANNING AND EXECUTION SYSTEM (JOPES) CELL RED CELL
COMMUNICATIONS AND INFORMATION SYSTEMS (CIS) DIRECTORATE (N-6)	CIS CENTER COMMUNICATIONS SYSTEM (CS) CURRENT OPERATIONS (COPS) CELL CS PLANS CELL CS MOC/HQ SUPPORT CELL

Ref: NWP 3-32, Maritime Opns at the Operational Level of War, fig. 1-4, p. 1-7.

The MOC's B2C2WG structure, organization, and staffing will vary depending upon the mission assigned, the operational environment, the makeup of existing and potential adversaries or nature of the crisis (e.g., combat operations, tsunami, cyclone, earthquake), and the time available to reach the desired end state.

(Maritime Ops Center) I. MHQ/MOC Organization and Function 3-5

The MHQ with MOC ensures Navy operational commanders are able to execute at the operational level. Maintaining the fleet management Napoleonic structure provides the deliberate staffing environment required to properly man, equip, and train Navy forces. Additionally, it provides an organizational structure that external commands understand, allowing staff quick access to expertise to promote interoperability. Organizing the MOC following the B2C2WG flattened organizational construct to plan and execute assigned missions greatly speeds planning and decision making by aligning with the JTF staff model. Additionally, by following this standard model, coordination with other component commanders and transforming to the various joint roles that may be assigned will be made easier. Reach back, or support from or to other similarly organized MOCs, will be facilitated and the utility of schoolhouse training programs maximized through standardization.

MOC Centers - Intelligence, Operations, and Logistics

A typical MOC is likely to have three centers: maritime intelligence operations, operations, and logistics readiness. Supporting these centers are cells and working groups. NTTP 3-32.1, Maritime Operations Center, provides additional guidance for the formation and functioning of a MOC as a key structure of an MHQ. It provides fundamentals and a generic template to organize the MOC to conduct operations in support of the commander. This organizational template will better facilitate the execution of internal and external (MOC-to-MOC as well as interechelon) processes. It recognizes there are different MOCs, such as tailored MOCs, and that the mission being executed may well dictate organization and process requirements. It is a step toward organizational and process standardization, not a constraint.

See also p. 2-51.

MOC Director

Each MOC will have a director to ensure one individual is focused entirely on the mission and operational tasks, as well as guaranteeing the MOC, as a whole, is functioning as required, ensuring successful fulfillment of the mission and operational tasks. The MOC commander designates the MOC director, who is the officer in charge of accomplishing the main processes of strategy development, combat planning, combat operations, intelligence, surveillance, and reconnaissance (ISR), air and sea mobility, and operational sustainability. The MOC director is charged with running the MOC effectively, based on the commander's guidance, and will report directly to the commander in that role. At each MHQ the individual assigned to this position will depend upon the talent and experience resident in the command and upon the type of operation being conducted. Options for filling the MOC director position might include assignment of a senior officer on a full-time basis or collateral-duty assignment of the DCOS, N3, N5, or other appropriately qualified individual.

B2C2WG (Boards, Bureaus, Centers, Cells, and Working Groups) Structure

The MOC's B2C2WG structure, organization, and staffing will vary depending upon the mission assigned, the operational environment, the makeup of existing and potential adversaries or nature of the crisis (e.g., combat operations, tsunami, cyclone, earthquake), and the time available to reach the desired end state. Each B2C2WG activated for a mission will have principal oversight by a directorate. These B2C2WGs may be physical venues but also support virtual collaboration and participation with other government agencies (OGAs), IGOs, NGOs, and other military headquarters.

MHQ's staffing has not been increased to facilitate maritime operations. Each staff will require a personnel transition plan that provides the necessary scalability and organizational flexibility by augmenting the organization with personnel and equipment, as necessary. Each MHQ will maintain a baseline capability for fleet management and maritime operations that will support the operations tempo required to meet the typical Phase 0 daily demands of the theater, area of operations (AO), or function.

As the tempo and complexity of operations increases across the range of military operations, the MHQ must be able to transition from baseline capabilities to the increased capabilities demanded by the situation. Personnel transition plans should include pre-designated active and reserve joint, multinational and other Service personnel who will augment the headquarters staff to meet the increased demand. This augmentation can occur incrementally as operations increase in complexity, scale, or tempo. Exercising the personnel transition plan will identify potential issues.

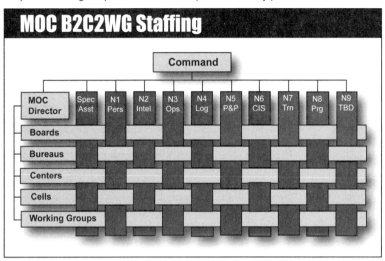

Ref: NWP 3-32, Maritime Opns at the Operational Level of War, fig. 7-4, p. 7-15.

A continuing challenge for commanders when defining the MOC B2C2WG structure is balancing the potentially large number of B2C2WGs necessary for full staff analysis and integration with the limited number of full-time personnel on the staff, time available, and other competing scheduling requirements for the principals and leaders.

One of the means to force discipline on the numbers of B2C2WGs is to require the staff proponent of a B2C2WG to defend its need in terms of what it brings to the decision cycle, e.g., specific inputs, outputs, and recipients of that information. The "seven-minute drill" is a tool to vet B2C2WGs. It is a way the staff proponent summarizes the purpose for the appropriate B2C2WG, its linkage to other B2C2WGs, and its support to decision-making requirements.

See p. 3-17 for discussion of the "seven-minute drill."

The frequency of B2C2WG meetings depends on need. Corresponding B2C2WGs at external staffs, particularly at HHQ, may meet at more frequent and regular intervals; MOC interaction at these sessions, using collaboration tools such as Defense Connect Online (DCO) when appropriate, can improve B2C2WG functioning and cross staff communication. Typically it will be the HHQ that will determine from which B2C2WGs they expect component staff participation, to include required products needed from them. This will in turn drive the JFMCC staff battle rhythm. Some B2C2WGs may not need to meet on a daily basis. For example, during major operations the strategic communication working group (SCWG) may meet every day; otherwise, the SCWG may meet once a month. Meeting unnecessarily wastes staff time and detracts from more productive employment of the staff.

Other publications that may assist in understanding the many facets of a B2C2WG organization and its functions include: JP 3-0, Joint Operations; JP 3-32, Command and Control for Joint Maritime Operations; JP 3-33, Joint Task Force Headquarters; and JP 5-0, Joint Operation Planning.

II. Battle Rhythm

Ref: NWP 3-32, Maritime Operations at the Operational Level of War (Oct '08) p. 7-22 to 7-24

Battle rhythm is defined as the deliberate daily cycle of command, staff, and, unit activities intended to synchronize current and future operations/plans. The battle rhythm serves to optimize the commander's decision making planning and ability to command forces by enabling coordination, and collaboration and accomplishing necessary deconfliction. It is described as the sequencing and execution of actions and events within the operational commander's staff as required by the flow and sharing of information necessary to support the three event horizon commander's decision cycles. A staff battle rhythm supports the commander's battle rhythm. It is synchronized with higher headquarters and peer components, as well as setting a cycle to which subordinate commands can link. The synchronization is calibrated to optimize the most efficient and timely flow of orders, intent, guidance, and information up, down, and between the various levels of headquarters. Battle rhythms are scalable in time cycles based on the pace of operations that range from the normal and routine, which is normally a weekly and monthly cycle, to major combat operations, which is a 24-hour cycle. The battle rhythm is also modulated by components for particular operational circumstances and requirements. The appropriate B2C2WGs are selected for inclusion, which can be done using the seven-minute drill format described.

A challenge for every staff is orchestrating the battle rhythm events for each time horizon ensuring that they also support pertinent information requirements of the other time horizons. Many headquarters minimize the total number of meetings by organizing battle

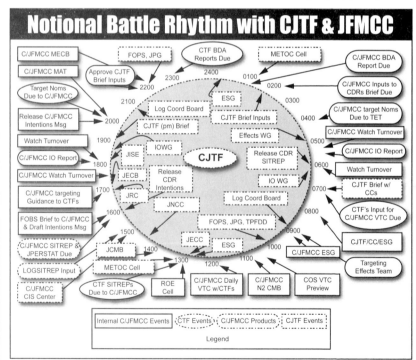

Ref: NWP 3-32, Maritime Opns at the Operational Level of War, fig. 7-8, p. 7-26.

rhythm events by function (e.g., an assessment, planning, or information operations (IO) meeting), and then further setting the agenda of that meeting to satisfy any needed actions for all three time horizons. This reduces the requirement for the leadership to attend three separate meetings, and reduces the time demands on the supporting staff officers.

A battle rhythm is a routine cycle of command and staff activities intended to synchronize the far-, mid-, and near-time horizons to support the commander's planning and execution of an assigned mission. The battle rhythm establishes the time, frequency, and type of meetings, working groups, boards, report/product requirements, and other MOC activities. These activities may be daily, weekly, monthly, or quarterly requirements. Typically, the MOC battle rhythm is managed by the COS. There are several critical functions for a battle rhythm that include, but are not limited to, the following:

- Making staff interaction and coordination within the MHQ routine
- Making commander and staff interaction routine (in so much as it can be)
- Synchronizing centers, groups, bureaus, cells, offices, elements, boards, work groups, and planning team activities
- Facilitating planning by the staff and decision making by the commander

Many factors influence the establishment of a battle rhythm. These include (but are not limited to) the following:

- The higher headquarters battle rhythm and reporting requirements
- Adjacent headquarters battle rhythm and reporting requirements
- The subordinate headquarters battle rhythm requirements
- The duration of the operation
- The intensity of the operation
- The planning requirements within the headquarters, (e.g., future plans, future operations, and current operations)
- Other factors (e.g., operation phase)

The commander is charged with effectively managing operations and establishing the commander's battle rhythm. The commander develops and directs processes to assess, plan, direct, monitor, and coordinate operations in the AO based on the superior commander's intent. Once battle rhythm is ongoing, there are typically five sets of these products being worked at any given time:

- One undergoing assessment (yesterday's plan)
- One in execution (today's plan)
- One in production (tomorrow's plan)
- One in final planning, to include considerations like detailed targeting and deconfliction (the following day's plan)
- One in strategy for development (next phase)

Like all other processes, the commander's battle rhythm must be managed and requires deliberate planning, monitoring during execution, and continuous internal assessment. Battle rhythm management ensures that the commander is responsive to the needs of higher headquarters, the MOC staff, and subordinate commanders. A cell within the MOC typically is the lead to ensure new requirements are incorporated into the battle rhythm, and should establish and manage daily staff battle rhythm, to include daily briefings and meetings. Planning the battle rhythm includes the integration of processes and synchronization of events across the MOC to ensure alignment internally and externally, vertically (higher headquarters and subordinates) and horizontally (joint components, allies, coalition partners, interagency organizations). The battle rhythm of a JFMCC must be synchronized with the battle rhythm of the JFC.

(Maritime Ops Center) I. MHQ/MOC Organization and Function 3-9

III. Operational-Level Functions

Functions are related capabilities and activities that, grouped together, help operational commanders integrate, synchronize, and direct joint operations. Joint doctrine identifies six basic groups of functions that are common to operations at all levels of war: C2, intelligence, fires, movement and maneuver, protection, and sustainment. Some functions, such as C2 and intelligence, apply to all operations. For each level of war the Joint Staff has identified tasks for each functional group and published these tasks in the Universal Joint Task List (UJTL), online at the Joint Doctrine, Education and Training Electronic Information System (JDEIS) Web Portal. The UJTL provides a common lexicon for describing tasks in the joint community and increasingly with interagency partners. Appendix B provides a graphic representation for each functional area of the tasks identified by the UJTL for each level of war. Within the UJTL, each of these tasks has several associated supporting tasks. The UJTL tasks and subtasks identify "what" is to be performed; they do not address "how" or "why" a task is performed (found in joint doctrine or other governing criteria, such as a command SOP), or "who" performs the task (found in the commander's CONOPS and joint doctrine).

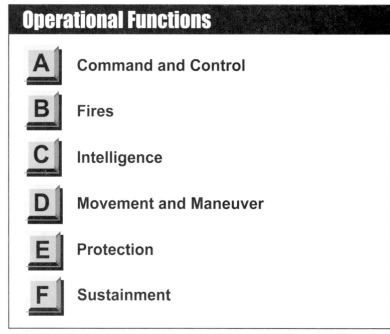

Ref: NWP 3-32, Maritime Opns at the Operational Level of War, chap. 7.

Each operational level headquarters, e.g., MHQ with MOC, utilizes the UJTL to clarify "what" the command will do by identifying a tailored list of UJTL tasks for the command that will result in successful accomplishment of the mission. For each tailored task the reason "why" the task is required should be identified. This provides a means to conduct a post mission analysis and refine future tailored task lists. With an understanding of the "what" and "why" of tasks the command needs to accomplish, the command then can define processes and procedures that define "how" and "who" will be executing the task. Critical in the development of these processes and procedures is identification of their inputs from strategic-level tasks and other operational-level tasks and recipients of their outputs (other operational level tasks or tactical level tasks). This same process is followed by TF commanders using tactical UJTL tasks and subtasks to define processes and procedures that need to be developed.

For each subtask applicable to the command, the staff identifies for each time horizon where in the commander's decision cycle the task will be executed, the inputs required to execute the task, the outputs from the task, and the processes and procedures to generate these outputs.

A. Command and Control (C2)

C2 ties together all the operational functions and tasks and applies to all levels of war and echelons of command across the range of military operations. C2 is the means by which an operational commander synchronizes and integrates force activities in order to achieve unity of command. Unity of effort over complex operations is made possible through decentralized execution of centralized, overarching plans. Unity of command is strengthened through adherence to the following C2 tenets:

- Clearly defined authorities
- Roles and relationships
- Information management (IM)
- Explicit and implicit communication
- Timely decision making
- Coordination mechanisms
- Battle-rhythm discipline
- Responsive, dependable, and interoperable support systems; SA
- Mutual trust

B. Fires

Operational firepower employs kinetic and non-kinetic means to defeat adversary forces or to maintain freedom of movement. By its nature, operational firepower is primarily a joint/multinational task. Firepower refers to the delivery of all types of ordnance to include bombs, rockets, missiles, artillery, and naval gunfire, as well as other lethal means against enemy targets.

The two broad categories of targets are deliberate and dynamic. Deliberate targeting prosecutes planned targets. These are targets that are known to exist in the operational environment with engagement actions scheduled against them to create the results desired to support achievement of JFC objectives. Dynamic targeting prosecutes targets of opportunity that are identified too late, or not selected for action in time to be included in deliberate targeting but, when detected or located, meet criteria specific to achieving objectives. Dynamic targets are often referred to as targets of opportunity. Deliberate and dynamic targeting processes are discussed in the Joint Targeting section.

There are three types of deliberate targeting processes described in JP 3-60, Joint Targeting.

- The joint targeting process
- Six-stage air targeting and tasking process: Air component
- Four-phase targeting process: Land and maritime components.

Land and maritime force commanders normally use a four-phase targeting process known as the decide, detect, deliver, and assess (D3A) for fires planning and execution and to interface with the joint targeting cycle. D3A incorporates the same fundamental functions of the joint target cycle. The D3A methodology facilitates synchronizing maneuver, intelligence, and fire support. D3A is a more responsive deliberate targeting cycle and is not constrained by the 72-hour cycle. Components may strike

(Maritime Ops Center) I. MHQ/MOC Organization and Function 3-11

any target within their AO with organic capabilities that support their objectives and desired results. The AO may include: above, on, and beneath the sea, the land that affects the maritime missions, and the information dimension. If the maritime force has insufficient organic assets to strike a target within the maritime AO, or if a maritime target is outside the maritime AO, the targets can be nominated for joint targeting and prosecution by another component's assigned forces. Likewise, the maritime force can offer strike assets not needed for maritime missions as excess sorties for use in joint missions required by another component. As part of the deliberate targeting process, the maritime operational commander will coordinate target nominations for the joint target list, no strike list, restricted target list, maritime prioritized target list for organic strikes in the maritime AO, and maritime target nomination list for joint missions.

Using responsive organic capabilities can help to ensure the maritime operational commander's decision cycle is inside the adversary's. Dynamic targets are usually fleeting with very small windows for weapon engagement. Commanders and their staffs, in coordination with joint components and other agencies, develop dynamic targeting guidance, which should include: time-sensitive target (TST) type and description; TST prioritization; desired result; approval authority; and acceptable risk. The commander should articulate risk tolerance sufficiently to let on-scene commanders understand his intent when dynamic targeting requires accelerated coordination. The maritime commander and staff at the operational level must ensure the dynamic targeting processes are understood and rehearsed. Components will nominate candidate TSTs, high-payoff targets, and high-value targets during the deliberate targeting process. These targets will be managed by the CCDR-directed dynamic targeting manager.

See pp. 2-56 to 2-57 for discussion of maritime fires support. See pp. 2-58 to 2-59 for discussion of joint maritime operations targeting.

C. Intelligence

Operational intelligence focuses on adversary military capabilities and intentions. Operational intelligence helps the JFC and component commanders keep abreast of events within their areas of interest. It also helps commanders determine when, where, and in what strength the adversary might stage and conduct campaigns and major operations. During counterinsurgency and counterterrorism operations, operational intelligence is increasingly concerned with stability operations and has a greater focus on political, economic, and social factors.

Within the OA, operational intelligence addresses the full range of military operations, facilitates the accomplishment of theater strategic objectives, and supports the planning and conduct of joint campaigns and subordinate operations. Operational intelligence focuses on providing the commander information required to identify adversary centers of gravity (COGs), critical vulnerabilities and decisive points (DPs), and provides relevant, timely, and accurate intelligence and assessments. Operational intelligence also includes monitoring terrorist incidents and natural or man-made disasters and catastrophes.

The senior intelligence officers of the maritime commander at the operational level must know not only their command's intelligence and information requirements, but also must be aware of the priority intelligence requirements (PIRs) of the higher, adjacent, and supporting and subordinate commands, as well as national-level intelligence requirements.

A critical commander's decision cycle assessment phase tool for all event horizons is the intelligence preparation of the operational environment (IPOE) process. IPOE discussed in Chapter 6 is used to define the operational environment, describe the impact of the operational environment on adversary and friendly forces, evaluate the capabilities of adversary forces operating in the operational environment, and determine and describe potential adversary objectives, COG, CVs, DPs, and COAs and civilian activities that might impact military operations.

Analysts use the IPOE process to analyze, correlate, and fuse information pertaining to all relevant aspects of the operational environment (e.g., military, economic, political, and social, information and infrastructure systems). The process is also used to analyze adversary capabilities, identify potential adversary COAs, and assess the most likely and most dangerous adversary COAs. The process can be applied to the full range of joint military operations (to include civil considerations) and to each level of war.

See pp. 5-8 to 5-9 for discussion of the IPOE process.

D. Movement and Maneuver

Operational movement and maneuver focuses on the disposition of joint and/or multinational forces, to impact the conduct of a campaign or major operation by either securing positional advantages before battle is joined or exploiting tactical success to achieve operational or strategic results. This activity includes moving or deploying forces for operational advantage within a joint operations area (JOA) and conducting maneuver to operational depths for offensive or defensive purposes. It also includes enhancing the mobility of friendly forces and controlling the operational environment on land, on and under the sea, in the air, or in space.

Maneuver is defined as the employment of forces in the OA through movement in combination with fires to achieve a position of advantage in respect to the adversary in order to accomplish the mission. At the operational level, maneuver is a means by which operational commanders set the terms of battle by time and location, decline battle, or exploit existing situations. Operational maneuver usually takes large forces from a base of operations to an area where they are in position to achieve operational objectives. The objective for operational maneuver is usually a COG, CV, or DP. Movement and maneuver gains and maintains maritime and air superiority.

The JFMCC directs subordinate commanders in the execution of force-level operational tasks, advises the JFC of its movement, and coordinates with other functional/ Service components and other agencies in supporting JFMCC missions. Operational, and tactical-level maneuver and movement of maritime forces are integral in defense-planning considerations, such as JFMCC support to, or execution of, assigned airspace control authority (ACA) or area air defense commander (AADC) responsibilities. As described in JP 3-0, Joint Operations, maneuver must be synchronized with fires and interdiction.

Movement and maneuver encompasses organizing for and disposing forces to conduct campaigns, major operations, and other contingencies by securing positional advantages before combat operations commence and by exploiting tactical success to achieve operational and strategic objectives.

See also pp. 2-52 to 2-53.

E. Protection

The protection function focuses on conserving the joint force's fighting potential in four primary ways:

- Active defensive measures that protect the joint force, its information, its bases, necessary infrastructure, and lines of communications (LOCs) from an adversary's attack.
- Passive defensive measures that make friendly forces, systems, and facilities difficult to locate, strike, and destroy.
- Applying technology and procedures to reduce the risk of fratricide.
- Emergency management and response to reduce the loss of personnel and capabilities due to chemical, biological, radiological, nuclear, and high-yield explosives (CBRNE) attack, accidents, health threats, and natural disasters. As the mission requires, the protection function also extends beyond force protection to encompass protection of U.S. noncombatants, the forces, systems, and civil population.

There are protection considerations that affect planning in every operation. The greatest risk to the total force — and therefore the greatest need for protection — occurs during campaigns and major operations that involve large scale combat against a capable enemy. Typically, these will require the full range of protection tasks, thereby complicating planning and execution. Although the OA and joint force may be smaller for a crisis response or limited contingency operation, the mission can still be complex and dangerous, with a variety of protection considerations. Permissive operating environments associated with military engagement, security cooperation, and deterrence still require that planners consider protection measures commensurate with potential risks. These risks may include a wide range of threats such as terrorism, criminal enterprises, environmental threats/hazards, and computer hackers. Thus, continuous research and access to accurate, detailed information about the operational environment, along with realistic training, can enhance protection activities.

F. Sustainment

The strategic, operational, and tactical levels of sustainment function as a coordinated whole, rather than as separate entities. Sustainment is the provision of logistics and personnel services necessary to maintain and prolong operations until mission accomplishment. The focus of sustainment in joint operations is to provide the JFC with the means to enable freedom of action and endurance and extend operational reach. Effective sustainment determines the depth to which the joint force is able to conduct decisive operations, allowing the JFC to seize, retain, and exploit the initiative. It is a Service responsibility to provide logistics support to forces assigned to a JFC/JFMCC.

Operational logistics links tactical requirements to strategic capabilities to accomplish operational and tactical objectives. Operational logistics normally supports campaigns and major theater operations by providing theater-wide logistic support. Operational logisticians coordinate the apportionment, allocation, and distribution of resources within theater. They coordinate closely with tactical operators to identify theater shortfalls and communicate these shortfalls to the appropriate theater or strategic source and/or ration supplies to support operational priorities. Operational logisticians coordinate the flow of strategic capabilities into a theater based on the commander's priorities. The concerns of the logistician and the operator are interrelated. The NCC/commander, Navy forces (COMNAVFOR) is responsible for planning, coordinating, and supervising operational logistics. The NCC/COMNAVFOR may designate a logistics group to be a Navy logistics command/coordinator (NLC) to coordinate the execution of operational logistics.

Operational-level logistics includes deployment, sustainment, resource prioritization and allocation, and requirements identification activities to sustain the force in a campaign or major operation. These fundamental decisions concerning force deployment and sustainment are key for the commander to provide successful logistical support.

At tactical and operational levels, commanders and their staffs forecast the drain on resources associated with conducting operations over extended distance and time. They respond by generating enough military resources at the right times and places to enable achievement of military strategic and operational objectives before reaching culmination. If the commanders cannot generate the resources, then the plan is not supportable and needs to be revised.

See chap. 6, Naval Logistics.

Refer to The Joint Forces Operations & Doctrine SMARTbook (Guide to Joint, Multinational & Interagency Operations), chap. 2 for complete discussion of the six joint (operational-level) functions.

II. Forming & Transitioning the MOC Staff

Ref: NWC Maritime Component Commander Guidebook (Feb '10), App. A.

The forming and transitioning of an operational-level fleet staff from normal operations with a focus on area of responsibility–wide issues to a JTF or JFMCC with a specific mission is critical to mission accomplishment. How this occurs is ultimately the prerogative of the commander and depends on the span and scope of the mission. The staff must be structured based on mission requirements, organized as a fully integrated staff at the operational level of war, and synchronized with higher and adjacent commands.

The commander must determine how to allocate the staff's time to meet competing demands from enduring and emergent mission requirements. Staff organization and processes should be designed to support the commander, specifically by synchronizing the staff and streamlining integration of planning, execution, and assessment efforts with higher headquarters, other component commanders, and subordinates. Early focus on forming and transitioning the MOC staff, to include B2C2WGs and battle rhythm refinement, is key to maintaining support to the commander in different roles.

The process of forming a JFMCC headquarters from an existing MOC staff and transitioning from normal to crisis operations is critical to mission accomplishment. To do this correctly, prior planning by the MOC must occur; this will shorten transition time and enable the staff to meet both its enduring and emergent mission requirements. An analysis based on mission requirements will enable the staff to decide how it will be organized and manned. Effective utilization of the battle rhythm and its included B2C2WGs, adjusted as necessary, enables the staff to support the commander's decision cycle and meet HHQ requirements.

I. Forming the MOC Staff

Though many Navy MOCs are designated by their CCDR as "JTF-capable headquarters," this section deals mainly with the issues associated with expanding a core MOC to a JFMCC role.

Once a MOC has been directed by an appropriate "establishing authority" (typically a combatant commander) to serve as the core for a JFMCC staff, the staff begins the planning process. Planning for how the JFMCC conducts maritime operations to meet the JFC's tasking is typically led by either the N-3 or N-5 through their respective planning B2C2WG (future plans center or future operations cell). Determining how the JFMCC will form, man, and organize also requires analysis by the staff. Optimally, this discussion is led by the chief of staff (COS) or maritime operations center director (MOC-D) and facilitated by the appropriate assistant chief of staff. Next the staff determines how it will be organized to support the commander's decision cycle. Will the JFMCC staff be aligned along the same lines as the core MOC staff? Will the JFMCC staff be organized along traditional staff codes (N-codes) or functionally? What will be the roles and responsibilities associated with each section? These questions depend on the mission assigned to the JFMCC and the commander's desires. Ideally, the MOC already has this construct outlined in existing joint standard operating procedures, which have been used and improved through real-world and exercise experiences.

Once it is determined how the JFMCC staff will organize and what tasks various sections will perform, the staff decides what billets are required to support each section. Billets required for each staff section to perform the tasks should form the basis of the JFMCC joint manning document (JMD). United States Fleet Forces Command N70M has expertise in this area and may be able to provide assistance in developing a JMD. A best practice seen in the fleets is to have standing draft JMDs built and kept on file to support various types of missions, such as major combat operations (MCO), humanitarian assistance/disaster relief, NEOs, etc. The staff then determines which billets they are able to fill from their standing MOC staff and which billets require sourcing outside the command (from other services, reserves, individual augmentees, etc.). Except under extreme situations, the entire MOC staff will not be expected to man the JFMCC staff. It is important to remember, particularly for forward-deployed MOC staffs, that this augmentation may not be available for some time, and there must be a plan to support the commander until then.

In cases when a JFC directs a fleet to form a JTF, the JTF can leverage the United States Joint Forces Command's (USJFCOM's) Joint Enabling Capabilities Command (JECC) for immediate access to essential joint force headquarters capabilities. The JECC combines the capabilities of seven unique organizations to deliver tailored, mission-specific support. These include joint deployable teams with capabilities in operations, plans, and information superiority/knowledge management, as well as a possible joint communications support element, a joint public affairs support element, and an intelligence quick reaction team. The JECC can accelerate the ability of service organizations to transition from a service-specific headquarters to an effective joint force headquarters in response to contingency operations.

When required, some MOCs operate as a distributed (or split) staff. If the staff is split between two or more geographic locations, communications can be more difficult, interactions between staff members can be reduced, and responsibility for some actions can become unclear. As a result, the commander's decision cycle timeline may become lengthy, and some staff personnel may become overloaded while others are under utilized. A MOC SOP should address these issues. In particular, the SOP can delineate specific responsibilities of each part of the staff and list primary and secondary points of contact so that both split staffs and external organizations can easily reach the correct point of contact.

II. Staff Transition From Enduring to Emergent Operations

The Navy's designated MOCs are structured to conduct operations across the ROMO. By design, MOCs are flexible, scalable, and tailorable in order to meet higher headquarters' demand for command and control of forces to achieve operational-level objectives and the tactical execution of tasks that support those objectives.

During normal (enduring) operations, MOC operations are characterized as deliberate in nature. Staff support to planning may take place in the form of periodic OPT meetings to develop plans in support of established operational-level objectives. The number and frequency of decisions by the commander during normal operations can vary in scope and intent. Staff battle rhythm events may meet on a weekly or monthly basis instead of a daily basis. These events may have a tendency to focus more on training (process driven) vice a primary focus on supporting a fast-paced commander's decision cycle (product driven).

Once emergent operations begin, the MOC's degree of transition depends on the span and scope of the assigned mission. Short-duration missions executed under existing command relationships require the least amount of transition. In this case, the commander and MOC staff use the Navy Planning Process to develop the plan, though in a more compressed timeline than under normal operations. The force will most likely be organized from existing assets in theater. The MOC may require limited augmentation to support the operation. Directives to subordinate commanders will most likely be in the form of an EXORD/tasking order or even be contained in the daily intentions message. Adjustments to the current battle rhythm may be minimal.

As the complexity and duration of the crisis increases, requirements to the MOC organization will increase. The MOC may be directed to be the core staff of a JFMCC. In this case, the JMD is activated to satisfy additional manning requirements based on the nature of the mission. If the scope of mission requirements changes, RFFs may need to be generated to fill these additional requirements. The MOC staff may have to support multiple planning efforts of varying levels of complexity. This will require a higher-tempo battle rhythm that is product driven, vice process driven, to support the commander's decision cycle and higher headquarters (HHQ) requirements. All staff sections and B2C2WGs will require access to the commander at some point or another, to receive guidance and obtain decisions on products required by HHQs and required to C2 the fleet. This will require the chief of staff to closely manage the commander's time.

III. Balancing Enduring and Emergent Mission Requirements

A significant challenge to the MOC will be how to employ staff resources to meet mission requirements in support of both enduring and emergent missions. Part of the staff will be susceptible to being too focused on the crisis (emergent) mission, and part will be too focused on the enduring mission. The commander determines how to meet the specific requirements of each and will utilize COS or MOC-D to ensure that decisions are disseminated to the rest of the staff. The MOC can implement various measures to ensure all hands understand where JFMCC responsibilities end and fleet responsibilities begin.

- Keep a portion of the staff off the JFMCC JMD and instead utilize them to address all standing MOC mission requirements.
- Assign responsibility for tracking command relationships for all maritime forces in the AOR. Utilize tools to gain and maintain SA of which forces are under the OPCON/TACON to the JFMCC in support of missions in the JOA and which remain under the OPCON/TACON to the fleet headquarters.
- The MOC staff should also be aware of which component has been designated as the supported component by the JFC and which components are supporting, by phase of the operation. For additional information on these command relationships, see USJFCOM Insights and Best Practices.
- During daily update briefings, maintain separation between issues pertaining to the missions inside and outside the JOA.

IV. Seven-Minute Drills

The seven-minute drill is a valuable tool to validate the utility of a particular B2C2WG and show interrelationships between B2C2WGs. The seven-minute drill explains the purpose of that specific B2C2WG and identifies links to other B2C2WGs. Seven-minute drills by the MOC can be used by the COS or MOC-D to ensure that only necessary B2C2WGs (those producing needed output) are used. Interactions and schedules must be reviewed to determine the correct manning and meeting schedule for B2C2WGs. Meetings should be short and purposeful and should follow an agenda.

A continuing challenge for commanders when defining the MOC B2C2WG structure is balancing the potentially large number of B2C2WG desired for full staff analysis and integration with the limited number of personnel on the staff, time available, and other competing scheduling requirements for the principals and leaders.

One of the means to force discipline on the numbers of B2C2WG is to require the staff proponent of a B2C2WG to defend its need in terms of what it brings to the decision cycle (e.g., specific inputs, outputs, and recipients of that information). The seven-minute drill, outlined below, is a tool to vet B2C2WG. The seven-minute drill is a way the staff proponent summarizes the purpose for the appropriate B2C2WG, its linkage to other B2C2WG, and its support to decision-making requirements.

1. Name of board or cell:
2. Lead J/N-code
3. When/where does it meet in battle rhythm?
4. Purpose:
5. Inputs required from:
6. When?
7. Output/Process/Product:
8. Time of delivery:
9. Membership codes:

See pp. 3-6 to 3-7 for discussion of B2C2WG.

V. Commander's Considerations During Staff Formation

During staff forming, the commander should drive the formation process, much as he or she needs to drive operational planning. In particular, the following are some considerations for the commander before and during the formation of a JFMCC or JTF staff:

- Does the headquarters have a standing requirement (tasked by the combatant commander) to serve as a JTF or JFMCC? Have specific missions (MCO, HA/DR, NEO, etc.) for this standing requirement been identified that can guide training for the staff? What is the expected timeline for joint or Navy augments to arrive to the staff?

- Does the staff have all the requisite subject matter expertise to effectively perform the assigned mission? Does the staff have a plan or understand the process to request required augments (Navy, joint, IA, multinational)? What mitigating actions has the staff taken to fill short-term gaps until required experts arrive?

- If employing a forward command element (afloat or ashore), have specific responsibilities for the split staff been incorporated into the JSOP or other applicable instructions? During split-staff operations, how will the commander's decisions and guidance be shared with the portion of the staff not physically with the commander?

- Who on the staff is responsible for promulgating the battle rhythm? Does the commander know which battle rhythm events require his or her attendance and why?

- When was the last time the headquarters executed operations as a JTF or JFMCC (real world or exercise) and worked with the expected functional component commanders (i.e, JFACC, JFLCC) under crisis conditions? What were the lessons learned? Has the after-action review been briefed to the commander? Is there a process for incorporating lessons learned into the annual joint training plan?

- Does the staff train to and update the command's joint SOP? Who on the staff is responsible for updating the JSOP? Does the JSOP reflect how the staff intends to function as a JTF or JFMCC?

Chap 3
III. Maritime Operational Command

Ref: NWP 3-32, Maritime Operations at the Operational Level of War (Oct '08), chap. 6 and NTTP 3-32.1, Maritime Operations Center (Oct '08), chap. 4.

I. Inherent Relationships to the Levels of War

Operations at the tactical, operational, and strategic levels are conducted concurrently. Tactical commanders fight engagements and battles understanding their relevance to operational objectives and goals. Operational commanders utilize operational art and design to set conditions for battles within a major operation or campaign to achieve military strategic and operational objectives. Combatant commanders (CCDRs) integrate theater strategy and operational art while remaining acutely aware of the impact of tactical events. Because of the inherent interrelationships between the various levels of war, commanders cannot be concerned only with events at their respective echelon, but must also understand how their actions contribute to the military end state.

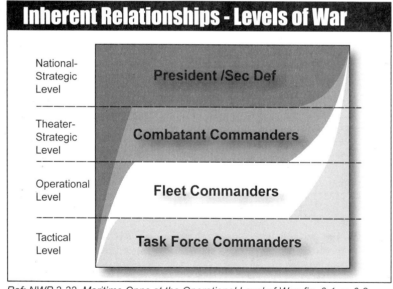

Ref: NWP 3-32, Maritime Opns at the Operational Level of War, fig. 6-1, p. 6-2.

Today's information systems allow for the rapid exchange of information between commanders at the various levels of war. Strategic- and operational-level commanders are now provided insight into the operational environment that was not historically possible. This insight allows these commanders to provide situational awareness (SA) and apply resources not previously available at the tactical level to the tactical problem. Similarly, the enhanced information systems provide the tactical commander means to influence strategic and operational assessments of the operational environment. It is important that while various echelons may be aware of events at other levels, the commanders and their staffs must focus on the appropriate level of war.

II. Exercising Operational Control

Navy commanders at the operational level exercise operational art and design in support of the CCDR's strategic objectives. The operational commander is tasked with interpreting a given CCDR's strategic objectives/end state into a series of major operations as a single campaign. This campaign normally requires the sequenced and synchronized employment of military and nonmilitary sources of national power. The proper sequencing and synchronization of the military forces in the accomplishment of the military mission falls to the operational-level military commander. The strategic commander uses his knowledge of the operational-level commander's intentions to ensure military action is properly synchronized and sequenced with actions of diplomatic, informational, and economic elements of national power. In addition, the strategic commander ensures de-confliction with other theater campaigns. Tactical forces are responsible exclusively for the physical elements of combat operations.

A. Operational-Level Functions

The primary responsibility of the operational commander is to execute those operational-level functions, i.e., **C2, intelligence, fires, movement and maneuver, protection, and sustainment**, which the tactical force requires to succeed in the designated operations area. These functional areas provide the basis from which tactical units derive their freedom of action to engage in physical combat.

See pp. 3-10 to 3-14 for discussion of the operational functions.

These functional areas are mutually supporting, i.e., increased intelligence increases protection, which increases the ability for fires, etc. Conversely, the lapse of any of these functional areas adversely affects the others, and subsequently affects the tactical forces' combat efficiency. The complementary and coordinated application of all the instruments of national power, when synergized consistent with operational art and design, provide the joint force capability required to achieve joint force commander's (JFC's) objectives and the strategic end state. Careful maintenance of an operational perspective ensures operational-level thinking supplants the tactical perspective that has brought the staff and commander previous success and comfort, but is detrimental to success at the operational level.

B. Role of the Operational Commander in C2

Command and control (C2) is one of the six core warfighting functions of operational command. The other core functions are fires, intelligence, movement, maneuver, protection, and sustainment. C2 ties together all the operational functions and tasks and applies to all levels of war and echelons of command across the range of military operations.

C2 is the means by which an operational commander synchronizes and/or integrates force activities in order to achieve unity of effort. Unity of effort over complex operations is made possible through decentralized execution of centralized, overarching plans. Unity of effort is strengthened through adherence to the following C2 tenets: clearly defined authorities, roles, and relationships; information management (IM); implicit communication, timely decision making, coordination mechanisms, battle rhythm discipline; responsive, dependable, interoperable support systems; SA; and mutual trust.

The operational commander's activities are central to operational command C2. The commander directs and assists the staff in planning, i.e., supervises staff and subordinate operational and tactical commanders in their preparation for operations and directs the staff conduct of operations and supervises the conduct of operations by subordinate commanders. In all these tasks, the operational commander leverages experience and judgment to guide the command through the fog and friction of war to accomplish the strategic objectives/goals.

See also p. 2-8. See p. 3-11 for discussion of C2 as an operational-level function.

C. Level of Control

Planning never stops during a campaign/major operation. Once forces start operations in support of the operational commander's synchronized and integrated operational order, the operational commander needs to ensure branch and sequel planning continues. However, once the force movement and maneuver commences the commander also needs to exercise control of the force.

The level of control exercised by the operational commander on the tactical force will depend on the nature of the campaign/major operation, the risk or priority of its success, and the associated comfort level of the commander. Currently, technology and information systems provide the operational commander enhanced SA into the tactical level. The operational commander and staff's ability to exercise operational art and design will be adversely impacted if the C2 tenet of decentralized execution is ignored because of this enhanced SA.

The OPLAN is the operational commander's primary means to control the tactical force. It provides the "what," "where," "when," and "why" information for the various tactical commanders to prioritize, synchronize, and integrate tactical action for achievement of strategic/operational objectives. As the forces move into position and the operational environment's definition is clarified, the operational commander may need to modify the operation plan (OPLAN)/concept of operations (CONOPS). This modification is promulgated to the tactical force using a fragmentary order (FRAGORD). As the OPLAN is executed, the operational commander and staff should guide the operations by minimizing control actions, as they are likely to disrupt the continuity of the plan being executed. Execution control actions may simply be alerting the tactical commander of information not held by the tactical commander or, for an unplanned deviation, determining intent or, if necessary, preempting actions or directing new ones.

The basis for an operational commander exercising control should be better insight into what is required to achieve operational objectives, rather than what is evidenced by the tactical level commander's actions. The commander's sources of insight include knowledge of strategic goals, operational objectives, overall awareness of the tactical situation, knowledge of the tactical plan, familiarity with tactical procedures, awareness of enemy force actions, and status of own forces. Advances in information systems and communications may enhance the operational commander and staff's SA and understanding of the tactical environment.

Decentralized execution exploits the ability of force commanders, warfare commanders, ship commanding officers, and other frontline commanders to make on-scene decisions during complex, rapidly unfolding operations. The level of control used by the operational commander will depend on the nature of the operation or task and the priority of its success and the associated comfort level of the operational commander.

Navy operational-level commanders and staffs monitor tactical actions and utilize measures of performance (MOPs) and measures of effectiveness (MOEs) to assess these actions. During mission analysis the commander and staff develop MOPs and MOEs. These MOPs and MOEs will be used to assess progress toward accomplishing a task, creating an effect, or achieving an objective. Commanders adjust operations based on their assessment to ensure objectives are met and the military end state is achieved. MOEs answer the question "Are we doing the right things?" MOPs answer the question, "Are we doing things right?"

The operational commander's level of knowledge is the basis for operational control (OPCON) actions. A knowledge deficit in any area will cause control actions for that area to be suspect. Maintaining comprehensive knowledge in all six control areas is challenging. This challenge is magnified when the campaign/major OPLAN, with its multiple integrated and synchronized tactical actions, is being executed. Operational commanders must organize their operations centers to prepare and execute campaign/major OPLANs, while exerting appropriate level of control over the tactical force.

II. Maritime Domain Awareness (MDA)

Ref: NWP 3-32.1, Maritime Operations Center (Oct '08), pp. 1-7 to 1-15.

Maritime domain awareness (MDA) is "the effective understanding of anything associated with the global maritime domain that could impact the security, safety, economy, or environment of a nation" (JP 3-32, Command and Control for Joint Maritime Operations, Change 1, 27 May 2008) and is essential in all maritime activities. Maritime domain awareness is one of eight interdependent supporting plans to the National Strategy for Maritime Security and can only be achieved through the combined efforts of Federal, state, and local government agencies, international governments, and commercial and private enterprise. Maritime domain awareness is achieved by collecting, fusing, and analyzing relevant data and disseminating this information to commanders and other Department of Defense (DOD) and interagency decision makers in a timely manner to enhance operational decision-making processes and support of operational plans, including the Maritime Operational Threat Response Plan (MOTR).

See p. 2-17 for discussion of the Maritime Operational Threat Response (MOTR) plan.

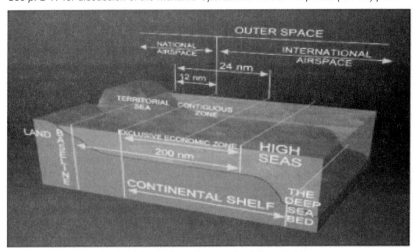

Ref: NWP 3-32.1, Maritime Operations Center, fig. 1-9, p. 1-14.

MOCs persistently monitor the maritime domain to identify potential and actual maritime threats in a timely fashion, to enhance operational-level decision-making processes, and in support of operational plans. The primary method for information sharing, situational awareness, and collaborative planning in the maritime domain is through a global maritime common operational picture (COP). The global maritime COP is developing into a near-real time, dynamically tailorable, network-centric virtual information grid shared by U.S. Federal, state, local agencies and international partners with maritime interests and responsibilities providing the means to monitor activities, identify trends, and differentiate anomalies. Global maritime COP data will be accessible to all users within the limits of security, policy, or regulations.

Maritime operations centers contribute to the active layered defense of U.S. and international interests in a number of vital ways. Key among them is providing input to the global maritime COP concerning matters and events within the commander's assigned operational area. The COP encompasses all oceans, from inland waters to the high seas. As an element of a global MDA network, each MOC contributes information from its operating area that enhances awareness for the collective organizations.

See pp. 2-4 to 2-5 for further discussion of the maritime domain.

MOC staffs must be aware of and plan and position forces to be able to respond to a variety of threats. Although not all-inclusive, threats in the maritime domain include: terrorism, piracy, oil theft, smuggling, drug trafficking, human trafficking, wildlife trafficking, weapons trafficking, environmental degradation, illegal immigration, fisheries violations, and organized crime activities.

Maritime Domain Awareness Goals

Maritime domain awareness serves to unify U.S. government efforts and support international efforts to achieve MDA across the Federal government, with the private sector and civil authorities within the United States, and with allies and partners. It requires close coordination of a broad range of Federal departments and agencies for this lasting endeavor. Maritime domain awareness goals are to:

- Enhance transparency in the maritime domain to detect, deter, and defeat threats as early and as distant from U.S. interests as possible
- Enable accurate, dynamic, and confident decisions and responses to the full spectrum of maritime threats
- Support law enforcement agencies to ensure freedom of navigation and the efficient flow of commerce

Maritime Domain Awareness Process and Functions

The MDA process, in support of Navy and joint operations, is relevant throughout the ROMO. The MDA process can be separated into five functions -- which support MOC planning functions -- all focused to support the commander's decision-making cycle:

- Monitoring
- Collection
- Fusion
- Analysis
- Dissemination

Maritime Domain Awareness Critical Tasks

Fleet MDA tasks have been derived from the Universal Joint Task List (UJTL) and the Universal Naval Task List (UNTL). These tasks apply to commanders at the strategic, operational, and tactical levels. Many of these tasks are intelligence-related and provide relevant information to enable command and control and planning, as well as other tasks shared by intelligence and operations staffs. This is not meant to imply that MDA is solely an intelligence function. The following overarching MDA critical tasks apply to the five functions:

- Direct operational intelligence activities
- Acquire and communicate operational level information and status
- Provide and monitor a common operational picture
- Prepare plans and orders
- Assess the operational situation
- Command subordinate operational forces
- Synchronize employment of forces and functions
- Coordinate and integrate joint/multinational and interagency support
- Collect and share operational information
- Process and exploit collected operational information
- Acquire and communicate operational level information and status
- Produce operational intelligence and prepare intelligence products
- Evaluate intelligence activities in the area of operations (AO)
- Disseminate and integrate operational intelligence

(Maritime Ops Center) III. Maritime Operational Command

D. Control Areas

Ref: NWP 3-32, Maritime Operations at the Operational Level of War (Oct '08), p. 6-7 to 6-8.

Navy operational commanders control staff and subordinate actions. Commander's control actions are binned into the following categories, commonly called control areas. The commander's up-to-date and in-depth knowledge of factors that impact each of these areas is crucial to guiding the operation effectively.

- **Maintain alignment.** The operational commander's task is to ensure that all execution decisions and apportionment requests remain aligned with the operation's mission statement and commander's intent (purpose, sequence, end state, and priorities). There must be a direct correlation between the higher headquarters commander's intent and goals and the operational commander's guidance and the plan formulated to accomplish the mission. All direction during plan development and execution should support the mission statement and commander's intent.

- **Provide SA.** The operational commander must assess the status of plan execution constantly. Using the available common operational picture (COP) and communications and intelligence, the operational commander must determine whether friendly force disposition is in accordance with the plan, whether enemy force disposition is in accordance with expectations, and whether forces are executing according to the plan and procedures.

- **Advance the plan.** The operational commander must monitor all aspects of the plan execution against the timeline. This infers detailed knowledge of all elements of the plan (enemy and own force disposition, branches, and sequels). Rarely are plans executed without deviation. When an unanticipated condition is encountered, the tactical or on-scene commander must adjust the plan correspondingly. The goal is to have every decision and every direction move the plan forward on the time line, toward the desired end state. The operational commander is responsible for attaining this goal.

- **Comply with procedure.** In monitoring execution, the commander oversees compliance with doctrinal tactics, techniques, and procedures (TTP), operation general matter (OPGEN), operation tasks (OPTASKs), special instructions, standard operating procedures (SOPs), operational tasking, and intentions to avoid blue-on-blue engagements and achieve efficiencies in plan execution. As an example of a procedure, the commander and staff must have an in-depth knowledge of the rules of engagement (ROE), and when the need exists for requesting supplemental ROE in order to properly execute the plan, the commander and staff need to know the procedure for making this request.

- **Counter the enemy.** Intelligence preparation of the operational environment (IPOE) and knowledge of enemy capabilities result in assumptions regarding probable enemy objectives and courses of action. The operational commander must be responsive to emerging intelligence, surveillance, and reconnaissance information that differs significantly from expectations and be prepared to adjust the plan in execution. Knowing what the enemy is doing at all times and being quick to countermove on receipt of reliable information is perhaps the number-one goal in C2.

- **Adjust apportionment.** Ground forces; ships; aircraft; air space; command, control, communications; computer infrastructure; and time all are apportioned. Any changes in asset availability, attrition, on-scene requirements, priorities, enemy disposition, or enemy tactics may trigger a need for reapportionment.

The operational commander must monitor for these changes, anticipate requests, and be prepared to adjust, as necessary, to advance the plan. Of all the apportionment factors, the one most frequently adjusted is time.

IV. Operational & Maritime Law

Ref: NWC Maritime Component Commander Handbook (Feb '10), app. G.

Operational law (OPLAW) issues and considerations are pervasive throughout the planning and conduct of operations in the maritime domain. This discussion highlights the fundamental OPLAW considerations of the maritime component commander.

OPLAW is a term used to capture a wide variety of legal and policy considerations that directly impact the employment of military force across the ROMO. OPLAW as it pertains to the maritime domain is frequently subdivided into three major components: Law of the Sea, Law of Armed Conflict, and ROE. Underpinning these three broad categories is a legal and policy framework that includes: domestic, foreign, and international law (which comprises treaty and customary law as well as United Nations Security Council Resolutions), bilateral and multilateral agreements, domestic policy, military policy, Joint and DOD regulations, and service regulations.

A more thorough discussion of these issues can be found in NWP 1-14M/MCWP 5-12.1/COMDTPUB P5800.7A, "The Commander's Handbook on the Law of Naval Operations."

The Relationship Between Law and Policy

Law and policy are two aspects of the OE that a commander must consider when making decisions. However, unlike other aspects of the OE, law and policy often prescribe limits on military actions and thus restrict the commander.

The nature of the relationship between the law and policy is frequently misunderstood. Simply stated, "the law" is a compilation of binding customs, practices, or rules of conduct prescribed or formally recognized as binding and enforced by a controlling authority. "Policy" refers to a definite method of action selected from among alternatives, and in light of given conditions, is intended to guide and determine future decisions. Maritime component commanders must be cognizant of both applicable law and controlling policies promulgated by higher civilian and military authority when making decisions.

It is important to understand that international law is oftentimes less restrictive than U.S. domestic law. Similarly, domestic policy is frequently more restrictive than domestic and international law. Finally, foreign law might also impact the commander's decisions. As such, it is the role of the staff judge advocate (SJA) to provide the commander and the operational planners with all relevant legal authorities and policy considerations, to outline the pros and cons of presented courses of action from a legal perspective, and to highlight legal risk inherent in a particular course of action. Commanders should demand this information as required throughout the decision making cycle with the understanding that their legal advisers should not make decisions. Rather, the role of the SJA is to provide advice and counsel so that the commander may make a decision in light of those legal and policy considerations and risks.

I. Rules of Engagement (ROE) & Rules for the Use of Force (RUF)

ROE and rules for the use of force (RUF) serve three main purposes: political, military, and legal. Politically, ROE/RUF ensure national policy and objectives are reflected in the action of operational and tactical commanders and forces and ensure that U.S. actions do not trigger undesired escalation. Militarily, ROE/RUF provide parameters within which the commander must operate in order to accomplish his assigned mission. For example, ROE may regulate a commander's capability to influence military action by granting or withholding the authority to use particular weapons systems or tactics. Legally, ROE/RUF ensure military actions conform to domestic and international law. Commanders may also issue ROE to reinforce law of armed conflict principles, such as prohibitions on the destruction of religious or cultural property, and minimization of injury to civilians and civilian property.

CJCSI 3121.01B, "Standing Rules of Engagement/Standing Rules for the Use of Force for U.S. Forces (U)," establishes fundamental policies and procedures governing the actions to be taken by U.S. commanders during military operations, contingencies, and routine military department functions including antiterrorism/force protection (AT/FP). Because rules of engagement reflect operational and national policy factors, they often restrict combat operations more than do the requirements of international law.

Standing Rules of Engagement (SROE)

The standing rules of engagement (SROE) establish fundamental policies and procedures governing the actions to be taken by U.S. commanders and their forces during all military operations and contingencies and routine military department functions occurring outside U.S. territory. The SROE also apply to air and maritime homeland defense missions conducted within U.S. territory and territorial seas, unless otherwise directed by the Secretary of Defense.

Standing Rules for the Use of Force (SRUF)

The standing rules for the use of force (SRUF) establish fundamental policies and procedures governing the actions to be taken by U.S. commanders and their forces during all DOD civil support (e.g., military assistance to civil authorities) and routine military department functions (including AT/FP duties) occurring within U.S. territory or U.S. territorial waters.

Self-Defense

CJCSI 3121.01B provides implementation guidance on the inherent right and obligation of self-defense and the application of force for mission accomplishment. A principal tenet of U.S. SROE/SRUF is that commanders always retain the inherent right and obligation to exercise unit self-defense in response to a hostile act or demonstrated hostile intent. Unit self-defense includes defense of other U.S. military forces in the vicinity. Individual self-defense is a subset of unit self-defense and can be limited by the unit commander.

Crafting and Promulgating Supplemental ROE and RUF

Navy component commanders and their staff's command and control tactical assets to accomplish operational objectives. One method of ensuring effective command and control over military forces is through ROE or RUF as appropriate. Circumstances may dictate that combined/joint force maritime component commander (C/JFMCC) forces may require ROE or RUF beyond self-defense in order to safely and effectively accomplish an assigned mission. Similarly, a commander may decide that existing mission-specific ROE are unclear, too restrictive, or otherwise unsuitable. It is incumbent upon operational-level commanders to ensure that tactical forces have

II. Legal Regimes of Oceans and Airspace

Ref: NWC Maritime Component Commander Handbook (Feb '10), pp. G-1 to G-5.

The legal classifications ("regimes") of ocean and airspace areas directly affect maritime operations by determining the degree of control that a coastal nation may exercise over the conduct of foreign merchant ships, warships, and aircraft operating within these areas. The territorial sea and all other zones are measured from baselines.

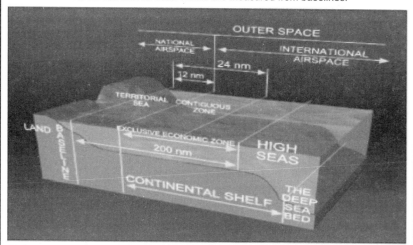

Ref: NWP 3-32.1, Maritime Operations Center, fig. 1-9, p. 1-14.

Internal Waters. Internal waters are those waters landward of the baseline. With limited exceptions, ships and aircraft have no legal right to enter another nation's internal waters without that nation's permission.

Territorial Sea. The territorial sea is a belt of ocean that is measured seaward up to 12 nautical miles from the baseline of the coastal nation and subject to its sovereignty. All ships, including warships, enjoy the right of innocent passage in the territorial sea. Innocent passage involves continuous, expeditious transit in a non-threatening manner. All airspace above the land and territorial sea of a coastal nation is "national airspace," and as a general rule aircraft have no right to enter another nation's national airspace without that nation's permission.

Contiguous Zone. A contiguous zone is an area extending seaward from the baseline up to 24 nautical miles in which the coastal nation may exercise the control necessary to prevent or punish infringement of its customs, fiscal, immigration, and sanitary laws and regulations that occur within its territory or territorial sea. Ships and aircraft may conduct all normal operations in the contiguous zone, to include flight operations and ISR, as long as they exercise due regard for the safety of other ships and aircraft and the coastal nation's limited law enforcement–related rights in the zone.

Exclusive Economic Zone (EEZ). An exclusive economic zone (EEZ) is a resource-related zone adjacent to the territorial sea that may not extend beyond 200 nautical miles from the baseline, where a State has certain sovereign resource–related rights (but not sovereignty). Ships and aircraft may conduct all normal operations in the EEZ, to include flight operations and ISR, as long as they exercise due regard for the safety of other ships and aircraft and the limited resource–related rights of the coastal nation.

High Seas. The high seas include all parts of the ocean seaward of the EEZ. Ships and aircraft may conduct all normal operations on the high seas, to include flight operations and ISR, as long as they exercise due regard for the safety of other ships and aircraft.

the ROE or RUF necessary to successfully accomplish their assigned tasks and missions within the level of risk deemed acceptable by the operational commander or established by higher authority. Commanders must carefully consider what level in the chain of command should possess the authority to use force given the nature of the mission and the operating environment. Additionally, when promulgating ROE or RUF, maritime component commanders should consider any and all means necessary to ensure a clarification of intentions up and down the chain of command exists regarding when tactical units are expected to use force. When operating in a coalition environment, forces will often operate within their own domestic ROE structure. Coalition component commanders should attempt to understand the various multinational partners' ROE constraints and attempt to employ forces within those constraints in order to maximize the effectiveness of the operation.

Maritime Warning Zones

In conjunction with crafting ROE, during COA development commanders and their staffs might consider the utility of establishing a "maritime warning zone" if forces are operating in a geographic area where symmetric and asymmetric land, air, surface, and subsurface threats are believed to exist. In such areas, tactical commanders are often faced with ascertaining the intent of entities (e.g., small boats, low slow flyers, jet skis, swimmers) proceeding toward their units. Oftentimes ascertaining intent is a very difficult problem, especially when operating in the littorals where air and surface traffic is heavy.

Given an uncertain operating environment, operational and tactical commanders may be inclined to establish some type of assessment, threat, or warning zone around their units in an effort to help sort the COP and gain time and battle space to ascertain the intent of inbound entities.

This objective may be accomplished during peacetime while adhering to international law as long as the navigational rights of other ships, submarines, and aircraft are respected. Specifically, when operating outside a nation's territorial seas, commanders may assert notice via notice to airmen and notice to mariners or other similar means that within a certain geographic area, for a certain period of time, dangerous military activities will be taking place. Entities traversing the area may be directed to communicate with tactical commanders in order to state their intentions. Moreover, such notice may include reference to the fact that if ships and aircraft traversing the area are deemed to represent an imminent threat to U.S./coalition naval forces they may be subject to proportionate measures in self-defense.

Of note, ships and aircraft are not required to remain outside such zones and force may not be used against such entities merely because they entered the zone. Commanders may use force against such entities only to defend against a hostile act or demonstrated hostile intent, including interference with declared military activities.

During an armed conflict, within the immediate area of naval operations a belligerent may establish special restrictions upon the activities of neutral vessels and aircraft and may prohibit altogether such vessels and aircraft from entering the area. The geographic context of an "immediate area" is that area within which hostilities are taking place or belligerent forces are actually operating.

Further information regarding the use of belligerent control of the immediate area of operations, to include targeting considerations, can be found in NWP 1-14M, "The Commander's Handbook on the Law of Naval Operations."

III. Law of Armed Conflict Fundamentals
Ref: NWC Maritime Component Commander Handbook (Feb '10), pp. G-6 to G-7.

Legal Issues Involved in Targeting
The legal principles underlying the law of armed conflict — military necessity, distinction, proportionality, and unnecessary suffering — are the basis for the rules governing targeting decisions. The law requires that only military objectives be attacked, but permits the use of sufficient force to destroy those objectives. At the same time, excessive collateral damage must be avoided to the extent possible and, consistent with mission accomplishment and the security of the force, unnecessary human suffering prevented. The law of targeting, therefore, requires that all reasonable precautions must be taken to ensure that only military objectives are targeted so that noncombatants, civilians, and civilian objects are spared as much as possible from the ravages of war.

What May Be Targeted?
Only military objectives may be attacked. Military objectives are combatants, military equipment and facilities (except medical and religious equipment and facilities), and those objects which, by their nature, location, purpose, or use, effectively contribute to the enemy's war fighting or war-sustaining capability and whose total or partial destruction, capture, or neutralization would constitute a definite military advantage to the attacker under the circumstances at the time of the attack. Military advantage may involve a variety of considerations, including the security of the attacking force.

Who May Be Targeted?
- **Lawful Combatants.** Lawful combatants are subject to attack at any time during hostilities unless they are hors de combat; that is, they cease to participate in hostilities due to wounds, sickness, shipwreck, surrender, or capture. Lawful enemy combatants include members of the regular armed forces of a State party to the conflict; civilians who take part in a levee en masse; militia, volunteer corps, and organized resistance movements belonging to a State party to the conflict, which are under responsible command, wear a fixed distinctive sign recognizable at a distance, carry their arms openly, and abide by the laws of war; and members of regular armed forces who profess allegiance to a government or an authority not recognized by the detaining power. Lawful combatants are entitled to combatant immunity — that is, they cannot be prosecuted for their lawful military actions prior to capture.

- **Unprivileged Belligerents.** Unprivileged belligerents are persons engaged in hostilities against the United States during an armed conflict who are not entitled to combatant immunity (e.g., terrorists, civilians directly participating in hostilities, etc.). Unprivileged belligerents who are members of forces or parties declared hostile by competent authority are subject to attack at any time during hostilities. Unprivileged belligerents who are not members of forces or parties declared hostile but who are taking a direct part in hostilities may be attacked only while they are taking a direct part in hostilities, unless they are hors de combat.

Collateral Damage
It is not unlawful to cause incidental injury to civilians, or collateral damage to civilian objects, during an attack upon a legitimate military objective. The principle of proportionality requires that the anticipated incidental injury or collateral damage must not, however, be excessive in light of the military advantage expected to be gained. Naval commanders must take all reasonable precautions, taking into account military and humanitarian considerations, to keep civilian casualties and damage to the minimum consistent with mission accomplishment and the security of the force. For strategic reasons, commanders may elect to issue policies to minimize collateral damage beyond what the law requires.

IV. Legal Bases to Stop, Board, Search, and Seize Vessels

Ref: NWC Maritime Component Commander Handbook (Feb '10), pp. G-3 to G-6.

Nations may desire to intercept vessels at sea in order to protect their national security interests. The act of "intercepting" ships at sea may range from querying the master of the vessel to stopping, boarding, inspecting, searching, and potentially even seizing the cargo or the vessel.

As a general principle, vessels operating outside of any territorial sea are subject to the exclusive jurisdiction of their flag state. Moreover, interference with a vessel seaward of the territorial sea violates the sovereign rights of the flag state unless that interference is authorized by the flag state or otherwise permitted by international law. Finally, inside a coastal nation's internal waters and territorial sea, the coastal nation exercises sovereignty, subject to the right of innocent passage and other international law. Given these basic tenets of international law, commanders should be aware of the legal bases underlying the authorization of MIO when ordered by competent authority to conduct such operations.

There are several legal bases available to conduct MIO, none of which are mutually exclusive. Depending on the circumstances, one or a combination of these bases can be used to justify permissive and non-permissive interference with suspect vessels.

1. United Nations Security Council Resolution

The U.N. Security Council has broad powers to maintain international peace and security. Pursuant to Article 39 of the Charter of the United Nations and Statute of the International Court of Justice, the Security Council is charged with determining the existence of any threat to the peace, breach of the peace, or act of aggression, and shall decide what measures shall be taken in accordance with Articles 41 and 42 to maintain or restore international peace and security, to include Maritime Interception Operations.

2. Condition of Port Entry

A coastal state has broad authority to impose conditions on ships entering its ports or internal waters, to include a requirement that all ships entering port will be subject to boarding and inspection. Such boardings and inspections can be conducted without flag-state consent before or after the ship enters port, provided the port state has pre-notified such a measure as a condition of port entry. The right to board and inspect does not apply to sovereign-immune vessels.

3. Right of Visit

International law allows non-permissive interference with ships where there are reasonable grounds to suspect that the ship is engaged in piracy, slave trade, or unauthorized broadcasting. If a warship encounters a foreign-flagged vessel seaward of the territorial sea, it may board the ship without the flag or master's consent if there are reasonable grounds to suspect that the ship is engaged in one of these universal crimes.

4. Stateless Vessel

Vessels that are not legitimately registered in any one state are without nationality and are referred to as stateless vessels. Such vessels are not entitled to fly the flag of any State and, because they are not entitled to the protection of any state, they are subject to the jurisdiction of all states. Additionally, a ship that sails under more than one flag, using them according to convenience, may not claim any of the nationalities in question and may be assimilated to a ship without nationality. If a warship encounters a stateless vessel or a vessel that has been assimilated to a ship without nationality on the high seas it may board and search the vessel without the consent of the master.

5. Flag State/Master Consent
Seaward of the territorial sea, ships are generally subject to the exclusive jurisdiction of the flag state. Unless another legal basis applies, a warship may not stop and board a vessel seaward of the territorial sea without consent of the flag state or the vessel's master. The master has plenary authority over all activities of the vessel, including the authority to allow anyone to come aboard the vessel. However, the master's consent to allow one to board and inspect the vessel does not allow the assertion of law enforcement authority such as arrest or seizure. Flag state consent is still required to take law enforcement measures against the vessel. Of note, some coalition nations do not hold that the master may grant consent to board the vessel.

6. Bilateral/Multilateral Agreements
International agreements greatly expedite the process by which officials from one state can board suspect vessels of another state. Such agreements can include provisions for advance authority for boarding and search of suspect vessels.

7. Self-Defense
Customary international law, as reflected in Article 51 of the Charter of the United Nations and Statute of the International Court of Justice, authorizes nations to use armed force to protect their national interests against unlawful or otherwise hostile actions and includes a right of anticipatory self-defense.

8. Belligerent Right of Visit and Search
A belligerent is entitled under international law to stop and search ostensibly neutral vessels to ensure such vessels are not transporting contraband (i.e., war materials) to an opposing belligerent or otherwise facilitating an opponent's war effort. Visit and search may not be exercised in neutral waters.

V. Piracy
International law has long recognized a general duty of all nations to cooperate in the repression of piracy. Piracy is an international crime consisting of illegal acts of violence, detention, or depredation committed for private ends by the crew or passengers of a private ship or aircraft beyond the territorial sea of another nation against another ship or aircraft or persons and property on board. (Depredation is the act of plundering, robbing, or pillaging.) In international law, piracy is a crime that can be committed only on or over the high seas, exclusive economic zone, and contiguous zone, and in other places beyond the territorial jurisdiction of any nation. The same acts committed in the internal waters, territorial sea, archipelagic waters, or national airspace of a nation do not constitute piracy in international law but are, instead, crimes within the jurisdiction and sovereignty of the littoral nation.

Only warships, military aircraft, or other ships or aircraft clearly marked and identifiable as being on governmental service and authorized to that effect, may seize a pirate ship or aircraft. A pirate vessel or aircraft, and all persons on board, seized and detained by a U.S. vessel or aircraft should be taken, sent, or directed to the nearest port or airfield and delivered to appropriate law enforcement authorities for disposition, as directed by higher authority.

If a pirate vessel or aircraft fleeing from pursuit by a warship or military aircraft proceeds from the contiguous zone, EEZ or high seas, or international airspace, into the territorial sea, archipelagic waters, or national airspace of another country, every effort should be made to obtain the consent of the nation having sovereignty over the territorial sea, archipelagic waters, or superjacent airspace to continue pursuit. The inviolability of the territorial integrity of sovereign nations makes the decision of a warship or military aircraft to continue pursuit into these areas without such consent a serious matter. However, in extraordinary circumstances where life and limb is imperiled and contact cannot be established in a timely manner with the coastal nation, or the coastal nation is unable or unwilling to act, pursuit may continue into the territorial sea, archipelagic waters, or national airspace.

VI. Other Fundamental Legal Issues Relevant to the C/JFMCC

Ref: NWC Maritime Component Commander Handbook (Feb '10), pp. G-8 to G-9.

1. Humanitarian Assistance/Disaster Relief (HA/DR)

HA/DR is a combination of U.S. military programs with the primary purpose of offering assistance to foreign populations. HA/DR can be funded out of a "stand-alone" appropriation called the Overseas Humanitarian, Disaster and Civic Aid account. The Office of Humanitarian Assistance, Disaster Relief, and Mine Action provides supervision and oversight of DOD HA/DR programs for the Director for Programs, DSCA. That agency coordinates HA efforts within DOD and with other agencies such as Department of State, including the United States Agency for International Development. The U.S. military role in these activities has expanded over the years and now includes military participation to support the goal of mutual security cooperation. These efforts include military assistance for foreign disaster relief and training foreign governments to cope with natural-disaster emergencies before they occur.

2. Humanitarian and Civic Assistance (HCA)

Humanitarian and civic assistance (HCA) is fundamentally an auxiliary activity that U.S. military forces are permitted to carry out in foreign countries during approved deployments such as fleet operations, exercises, or training. HCA activities are directly linked to annual (fiscal year) training programs or operational deployments. Generally, HCA projects cover four types of activities: medical, dental, and veterinary care in rural areas; construction of rudimentary surface transportation systems; well drilling and construction of basic sanitation systems; and rudimentary construction and repair of public facilities.

Congress authorizes the military services to support HCA projects and to act as the HCA executive agent for appropriate combatant commands. CFMCC/JFMCCs should seek legal input when planning for disaster relief or humanitarian and civic assistance since there are many operational-law issues that must be considered, including fiscal law considerations, use of force considerations, and sovereign immunity issues.

3. Environmental Law Considerations

As a matter of customary international law, as reflected in the United Nations Convention on the Law of the Sea, provisions regarding the protection and preservation of the marine environment do not apply to any warship, naval auxiliary, other vessels or aircraft owned or operated by a State. Government vessels and aircraft do have an obligation to operate with due regard for the environment and to adopt appropriate measures which do not impair operations or operational capabilities of such vessels or aircraft.

As for environmental considerations in targeting, it is not unlawful to cause collateral damage to the natural environment during an attack on a legitimate military objective. However, the commander has an affirmative obligation to avoid unnecessary damage to the environment to the extent it is practicable to do so consistent with mission accomplishment. As far as military requirements permit, methods or means of warfare should be employed with due regard for protection and preservation of the natural environment. A commander should consider the environmental damage that will result from an attack on a legitimate military objective as one of the factors during the targeting analysis.

Therefore, when conducting operational planning, the CFMCC/JFMCC must analyze his COAs, and take precautions for the marine environment, but only to the extent reasonable and practical, and not impairing the operation or the force's operational capability.

Chap 4

The Maritime Operations Process

Ref: NWC Maritime Component Commander Handbook (Feb '10), chap. 1.

The operations process consists of the major command and control activities performed during operations: planning, preparing, executing, and continuously assessing the operation. The activities of the operations process may be sequential (especially at the start of an operation). But once operations have begun, a headquarters often conducts parts of each activity simultaneously. Planning and preparation are continuous. While preparing for or executing one operation, commanders and staffs refine base plans or plan for branches and sequels. Preparation begins when a unit receives a mission and accelerates when planning details are developed. Assessing is continuous and influences the other three activities. During execution, the actual command and control of forces is added to the other activities. Subordinate units of the same command may be in different stages of the operations process.

The Operations Process

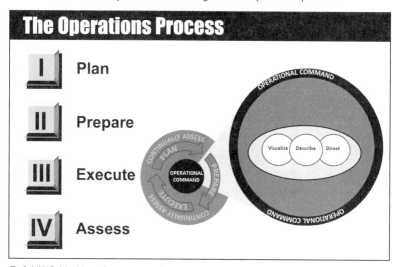

I. Plan
II. Prepare
III. Execute
IV. Assess

Ref: NWC Maritime Component Commander Handbook, fig. 1-1, p. 1-2.

Editor's note: Though the operations process as a concept originated with the U.S. Army, it has many elements that lend themselves to the C2 of maritime forces by the MCC. Although not termed the operations process, this concept is used by the majority of the U.S. Navy's fleet commanders.

The operations process concept is similarly discussed in NWP 3-32, *Maritime Operations at the Operational Level of War*, as the "commander's decision cycle." See pp. 4-14 to 4-15 for further discussion.

(Operations Process) Overview 4-1

I. Operations Process Activities

Ref: NWC Maritime Component Commander Handbook (Feb '10), p. 1-3.

Although operations generally follow the operations process of planning, preparing, executing, and continuous assessment, each receives different emphasis during an operation. After a mission is received, priority is on planning; after an OPORD is issued, planning continues but preparation activities increase. After the operation commences, execution focus increases significantly.

Operations Process Activities

Plan	Plan	Plan	Plan
Prepare	Prepare	Prepare	Prepare
Execute	Execute	Execute	Execute
Assess	Assess	Assess	Assess
Mission Received	OPORD Issued	H-hour	WARNO For New Mission / Current Mission Accomplished

Ref: NWC Maritime Component Commander Handbook, fig. 1-3, p. 1-6.

I. Plan

Planning is the art and science of understanding a situation, envisioning a desired future, and laying out effective ways of bringing about that future. Planning is both conceptual and detailed. Conceptual planning includes framing the problem, defining a desired end state, and developing an operational approach to achieve the desired end state. Conceptual planning generally corresponds to the art of operations and is commander led. In contrast, detailed planning translates the broad concept into a complete and practicable plan. Detailed planning generally corresponds to the science of operations and encompasses the specifics of implementation. Detailed planning works out the scheduling, coordination, or technical issues involved with moving, sustaining, administering, and directing forces.

Planning results in a plan or order that communicates the commander's understanding, visualization, and intent to subordinates, focusing on the desired end state. While planning may start an iteration of the operations process, planning is continuous as commanders and staffs revise plans and develop branches and sequels throughout the conduct of operations.

See pp. 4-5 to 4-6 for discussion of planning.

II. Prepare

Preparation is activities that commands perform to improve their ability to execute an operation. Preparation includes, but is not limited to, plan refinement; rehearsals; intelligence, surveillance, and reconnaissance; organizing, coordination; inspections; and movement. Activities of preparation help the force improve their ability to execute an operation. Preparation creates conditions that improve friendly forces' opportunities for success. Preparation requires staff and subordinate actions to transition the force from planning to execution. Activities of preparation help develop a common understanding of the situation and what is required. They are not solely pre-execution activities but continue into operations. These activities — such as backbriefs, rehearsals, and inspections — help commanders and staffs to better understand their roles in upcoming operations, practice complicated tasks, and ensure equipment and weapons function properly.

See pp. 4-7 to 4-10 for discussion of preparation.

III. Execute

Execution puts a plan into action by applying combat power to accomplish the mission and using situational understanding to assess progress and make execution and adjustment decisions. Navy forces generate combat power by converting potential combat power into effective action. Combat power can be constructive as well as destructive. In peacetime operations, such as a disaster relief operation, combat power is applied mainly for constructive purposes.

Execution focuses on concerted action to seize and retain the initiative, build and maintain momentum, and exploit success. Commanders create conditions for seizing the initiative by acting. Successful operations maintain momentum generated by initiative and exploit success within the commander's intent.

See pp. 4-11 to 4-18 for discussion of execution.

IV. Assess

Assessment is continuously monitoring and evaluating the current situation and the progress of an operation. Assessment involves continuously analyzing the OE to help the commander and staff understand the current situation and how it is evolving during operations. Based on this understanding, commanders and staffs evaluate relevant information to help them judge how operations are progressing toward achieving objectives and the desired end state. Assessment is a primary feedback mechanism that enables the command as a whole to learn and adapt.

Assessment precedes and guides the other operations process activities and concludes each operation or phase of an operation. However, the focus of assessment differs during planning, preparation, and execution. During planning, assessment focuses on developing and maintaining an understanding of the current situation and developing the assessment plan. During preparation and execution, assessment focuses on monitoring the current situation and evaluating the operation's progress toward stated objectives.

During operations, commanders and staffs also assess the underlying framework of the plan itself. This involves reexamining the original design concept and determining if it is still relevant to the situation. Collaboration and dialog with higher, subordinate, and adjacent commanders and staffs, backed up by qualitative and quantitative assessments, contribute to this learning. Commanders also seek expertise outside the military, such as civilian academics, to help them with their assessments. Based on this reexamination, commanders may conduct reframing activities that lead to a new design concept and eventually an entirely new plan, to adapt the force to better accomplish the mission.

See pp. 4-19 to 4-24 for discussion of assessment.

II. Operations Process Supporting Topics

Ref: NWC Maritime Component Commander Handbook (Feb '10), p. 1-5 to 1-6.

The operations process, while simple in concept (plan, prepare, execute, and assess), is dynamic in implementation. Commanders and staffs use the operations process to integrate numerous processes and activities consisting of hundreds of tasks executed throughout the headquarters. Commanders must organize and train their staff to plan, prepare, and execute operations simultaneously while continually assessing.

The operations process is not entirely a new concept. Commanders have been planning, preparing for, and executing operations while assessing their progress for as long as navies have been putting to sea. What is new is defining how each of those functions is interrelated and identifying the various activities that are included in them. When commanding at the tactical level, it was less important to think about how to devote staff resources utilizing the functions of the operations process. Due to the size, duration, and scope of operations at the fleet or JFMCC level it is imperative that the commander is aware of and drives each of these functions to ensure each feeds the next and furthers progress towards the commander's desired end state.

Supporting Topics

Plan	Prepare	Execute

• Navy Planning Process
• Orders and plans
	• Reconnaissance
	• Security
	• Force protection
	• Revise and refine the plan
	• Coordination and liaison
	• Rehearsals
	• Task organization
	• Train
	• Movement
	• Precombat checks and inspections
	• Logistic preparations
	• Integration of new units
		• Decide
		  Execution
		  Adjustment
		• Direct
		  Apply combat power
		  Synchronize
		  Maintain continuity

Assessment During Planning • Monitor the situation • Monitor criteria of success • Evaluate COAs	**Assessment During Preparation** • Monitor preparations • Evaluate preparations	**Assessment During Execution** • Monitor operations • Evaluate progress
Continuous Assessment		
• Situational understanding — sources, solutions • Monitoring — situation/operations, criteria of success • Evaluating — forecasting; seize, retain, and exploit the initiative; variances		

Ref: NWC Maritime Component Commander Handbook, fig. 1-4, p. 1-6.

Fleet and maritime component commanders drive the operations process through operational command. Staffs perform essential functions that amplify the effectiveness of operations; however, commanders play the central role in the operations process through operational command. Operational command is the art and science of understanding, visualizing, describing, and directing in operations against a hostile, thinking adversary. Future plans, future operations (FOPS), current operations (COPS), and the Maritime Assessment Group (MAG) have responsibilities throughout the operations process to conduct conceptual and then detailed planning, then conduct actions to prepare for execution, and then support the tactical execution. The commander makes decisions through each step of the operations process in the commander's decision cycle. The battle rhythm is established to align staff actions with the commander's decisions.

Chap 4
The Operations Process: I. Planning

Ref: NWC Maritime Component Commander Handbook (Feb '10), chap. 2.

The Navy Planning Process (NPP)

Commanders are required to make decisions constantly. Every day, they and their staffs resolve simple, routine, and/or complex problems. To help them think through their options when faced with a force employment decision while applying their knowledge, experience, and judgment, Navy staffs use a decision making tool called the Navy Planning Process (NPP). The NPP, as detailed in NWP 5-01, "Navy Planning," provides maritime planners with the procedures requisite for high tactical/low operational-level planning requirements and applies to contingency planning and crisis action planning situations.

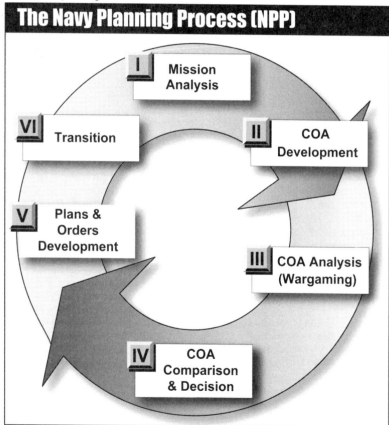

The Navy Planning Process (NPP)

- I. Mission Analysis
- II. COA Development
- III. COA Analysis (Wargaming)
- IV. COA Comparison & Decision
- V. Plans & Orders Development
- VI. Transition

Ref: NWP 5-01, Navy Planning, pp. 1-1 to 1-9.

NWP 5-01 describes the planning process used by tactical and operational-level Navy organizations. NWP 5-01 mirrors the Marine Corps planning process and closely resembles other military planning processes.

See chap. 5, Naval Planning, for discussion of the Navy Planning Process (NPP).

(Operations Process) I. Planning 4-5

Naval Planning and the Levels of War

Ref: NWC Maritime Component Commander Handbook (Feb '10), app. I.

The levels of war -- strategic, operational, and tactical -- help clarify the links between strategic objectives and tactical actions. Among the levels, the planning horizons differ greatly.

Strategic-Level Planning

Joint strategic planning provides strategic guidance and direction for security cooperation planning, joint operations planning, and force planning. Joint strategic planning occurs primarily at the national and theater strategic levels. This planning helps the President, Secretary of Defense, and other members of the National Security Council: formulate political-military assessments, define political and military objectives and end states, develop strategic concepts and options, and allocate resources.

Combatant commanders prepare strategic estimates, strategies, and plans to accomplish their assigned mission. Commanders base these estimates, strategies, and plans on strategic guidance and direction from the President, Secretary of Defense, and Chairman of the Joint Chiefs of Staff.

Operational-Level Planning

Typically, operational-level planning focuses on developing plans for campaigns and major operations and is conducted by joint force commanders (combatant commanders and their subordinate joint task force commanders), and their component commanders (service and functional) conduct operational-level planning. Planning at the operational level focuses on operational art, the application of creative imagination by commanders and staffs — supported by their skill, knowledge, and experience — to design strategies, campaigns, and major operations and to organize and employ military forces. Operational art integrates ends, ways, and means across the levels of war. Operational-level planners use the Joint Operation Planning and Execution System (JOPES), the joint operation planning process, and elements of operational design to develop campaign plans, joint operation plans and orders, and supporting plans.

While Navy components of a joint force assist joint force commanders in developing a campaign plan, Navy forces do not develop independent campaign plans. A campaign plan is a joint operation plan for a series of related major operations aimed at achieving strategic or operational objectives within a given time and space. Navy forces develop supporting plans (operation plans and orders) which nest with the joint force commander's campaign plan. Therefore, where campaigns may follow the six phases, Navy and maritime operational planning usually focuses on major or minor maritime operations within specific portions (phases or multiple phases) of a campaign.

Operational-level planning and tactical-level planning complement each other but have different aims. Operational-level planning involves broader dimensions of time and space than tactical-level planning and is intended to achieve operational objectives. It is often more complex and less defined. Operational-level planners need to define an operational area, estimate forces required, evaluate operation requirements, and assign broad taskings. In contrast, tactical-level planning -- along with specific taskings and actions executed at the tactical level -- proceeds from an existing operational design.

Tactical-Level Planning

Tactical-level planning revolves around how best to achieve objectives and accomplish missions assigned by higher headquarters. Planning horizons for tactical planning are relatively shorter than those of operational-level planning. While tactical-level planning works within the framework of an operational- level plan, tactical planning includes developing long-range plans for solving complex, ill-structured problems. These plans combine offensive, defensive, and stability or civil support operations to achieve objectives and accomplish the mission over extended periods.

See also p. 5-4.

The Operations Process: II. Preparation

Ref: NWC Maritime Component Commander Handbook (Feb '10), chap. 3.

Preparation helps the JFMCC transition between planning and execution. Preparation consists of activities performed by units to improve their ability to execute an operation. Preparation includes, but is not limited to, plan refinement; rehearsals; intelligence, surveillance, and reconnaissance; coordination; inspections; and movement. Preparation creates conditions that improve friendly forces' opportunities for success. Preparation requires commander, staff, and subordinate force actions.

Preparation Functions

 A Improving Situational Awareness

 B Developing a Common Understanding of the plan

 C Practicing and Becoming Procient in Critical Tasks

 D Integrating, Organizing, and Configuring the Force

 E Conducting Operational Functions to Shape the Operational Environment

 F Ensuring Forces and Resources are Ready and Positioned

Ref: NWC Maritime Component Commander Handbook, chap. 3.

The **Future Operations (FOPS) cell** is responsible for most of the preparation functions. Because FOPS integrates subordinate tactical planning into the less detailed operational plan provided by future plans, it needs to be composed of experts from the various warfare areas and possess control capabilities. Preparation activities increase as execution approaches. In a MCC, FOPS has the primary responsibility for ensuring subordinate commands are ready for execution.

See p. 5-75 for discussion of the future operations (FOPS) cell.

II. Integrating Processes and Continuing Activities During Preparation

Ref: NWC Maritime Component Commander Handbook (Feb '10), pp. 3-6 to 3-7.

During preparation, the integrating process and continuing activities are coordinated among the CTF staffs and through meetings, working groups, and boards established in the JFMCC battle rhythm.

Intelligence Preparation of the Operational Environment (IPOE)

During preparation, IPOE continues to provide products to assist commanders in maintaining their situational awareness. IPOE products are updated based on new information collected through ISR and friendly force reporting. CCIRs are updated as they are answered and assumptions are confirmed or invalidated. As IPOE products are updated, ISR synchronization is used to assess ISR against requirements, analyze and identify new IR, and update or change future ISR to continue to answer the CCIRs. The intelligence cell disseminates these products to others on the staff and to subordinate units. During preparation, IPOE continues to support planning (branch and sequel development) and the targeting process. See pp. 5-8 to 5-9.

Targeting

Depending on the situation, targeting during preparation generally focuses on finding and fixing targets. This function includes locating high-payoff targets accurately enough to track, target, and engage them. Finding and fixing of targets is tied directly to ISR operations. At the operational level, the JFMCC needs to consider the employment of lethal/non-lethal and kinetic/non-kinetic fires in the targeting process to meet objectives. See pp. 2-58 to 2-59.

Intelligence, Surveillance, and Reconnaissance (ISR) Synchronization

ISR synchronization continues to identify new IR, including those of subordinate units. It also continues to determine the best means of answering those requirements. As requirements are satisfied, IR and available ISR assets will need to be reevaluated. Intelligence and Operations personnel will have to coordinate recommended re-taskings. COPS personnel will have to ensure new requirements and taskings are promulgated via fragmentary order. Normally surveillance and reconnaissance missions begin early in planning and are the focus of COPS during preparation. Commanders direct surveillance and reconnaissance through warning orders and the ISR annex to the operation order. Commanders consider requesting assistance from sources beyond their control, including theater and national assets. They synchronize reconnaissance operations, as well as the intelligence collection and analysis, with their own organic forces to continuously update and improve their situational understanding.

Security

The force as a whole is often most vulnerable to surprise and enemy attack during preparation. Ships may be concentrated in assembly areas. Parts of the force could be moving to task organize. Required supplies may be unavailable or being repositioned. The security of the force is essential during preparation.

Operational Protection

Operational protection, which includes force protection, is both a warfighting function and a continuing activity. Commanders and staffs continuously plan and execute operational protection functions to defend and preserve the force. This includes protecting personnel (combatants and noncombatants), physical assets, and information of the United States and multinational military and civilian partners. While all protection tasks are important, during preparation commanders particularly emphasize the operational protection subtasks of force protection and operations security.

Force protection comprises preventive measures taken to mitigate hostile actions against personnel, resources, facilities, and critical information. It is distinct from operational protection in that it does not include actions to defeat the enemy or protect against accidents, weather, or disease. It includes protective structures (e.g., reinforcement) and systems (e.g., gas masks, body armor) to reduce the effectiveness of enemy weapon systems. Other methods can range from employing camouflage, concealment, and deception to hardening facilities, command and control nodes, and critical infrastructure.

OPSEC identifies and implements measures to protect essential elements of friendly information. During preparation, forces implement measures that eliminate or reduce the vulnerability of friendly forces to exploitation. These measures include concealing rehearsals, positioning of forces, and other indicators of unit intentions that enemy intelligence may exploit.

Battlespace Management

Water space management is the allocation of water space in terms of antisubmarine warfare attack procedures to permit the rapid and effective engagement of hostile submarines while preventing inadvertent attacks on friendly submarines. It is an important activity during preparation as ships and submarines begin to maneuver against enemy forces. See also p. 2-53.

Airspace management is the coordination, integration, and regulation of the use of airspace of defined dimensions. During preparation the JFMCC needs to consider how he will integrate his aviation assets, manned and unmanned, into the overall joint airspace construct. He also needs to understand the command relationship between his headquarters and the JFACC. Is there a JFACC assigned to the JTF, or is there a "Theater JFACC" coordinating air operations throughout the CCDR's AOR; and how will coordination occur? JFMCC aviation forces will need to practice this construct during preparation in order to enhance mission accomplishment during execution.

Terrain management is the process of allocating terrain by establishing areas of operation, designating assembly areas, and specifying locations for units and activities to de-conflict activities that might interfere with each other. It is an important activity during preparation as units reposition and stage prior to execution. Commanders assigned an area of operations manage terrain within their boundaries. Through terrain management, commanders identify and locate units in the area. Staffs can then de-conflict operations, control movements, and deter fratricide as units are positioned to execute planned missions. Commanders also consider the civilians and civilian organizations located in their area of operations. Though not normally seen as a JFMCC issue, since the maritime domain includes the littorals, the JFMCC can be responsible for battlespace on the land.

III. Preparation Activities

Mission success depends as much on preparation as on planning. Higher headquarters may develop the best of plans; however, plans serve little purpose if subordinates do not receive them in time. Subordinates need enough time to understand plans well enough to execute them. Subordinates develop their own plans and prepare for the operation. After they fully comprehend the plan, subordinate leaders practice key portions of it and ensure their forces are positioned and ready to execute the operation. To help ensure the force is protected and prepared for execution, commanders, staffs, and subordinate units:

- Coordinate and Conduct Liaison
- Build Partnerships
- Conduct Confirmation Briefs
- Conduct Rehearsals
- Conduct Plans-to-Operations Transition
- Revise and Refine the Plan
- Initiate Task Organization
- Integrate Units to Include Multinational Forces
- Train
- Initiate Operational Maneuver and Tactical Force Movements
- Conduct Sustainment Preparation
- Initiate Deception Operations

IV. Rehearsals

Ref: NWP 5-01, Navy Planning (Jan '07), app. N.

Rehearsing is the process of practicing a plan in the time available before actual execution. Rehearsing key combat and logistics actions allows participants to become familiar with the operation and to visualize the plan. This process assists them in orienting themselves to their surroundings and to other units during execution. Rehearsals also provide a forum for subordinate leaders to analyze the plan; however, caution must be exercised in adjusting the plan in order to prevent errors in synchronization. Rehearsals should always be performed before the execution of an operation.

Conducting rehearsals at both the operational and tactical level of war yields a much broader perspective than rehearsing only at the tactical level. The operational level of war focuses on the deployment and employment of component forces, commitment and withdrawal from battle, and the arrangement of battles and major operations in the JOA.

The operational level rehearsal helps the commander weave the series of component tactical actions over days and weeks into a campaign or set of major operations that ultimately addresses the combatant commander's requirements for an end state. The operational-level planning horizon has expanded, and consequently the vision of the future is more important. At the operational level, the questions that involve future vision are:

- What conditions must be produced in the JOA to achieve the objective(s) (ends)?
- What sequence of actions is most likely to produce those condition (ways)?
- What resources are required to accomplish that sequence of actions (means)?
- What is the likely cost or risk to the force in performing that sequence of actions?
- Are the right forces in the right place at the right time?
- Where is the campaign in light of my operational end state?
- What should be done now to influence events three to five days from now?

In the complex world of joint operations, rehearsals are vital for a naval force to the successfully execute OPORDs. Rehearsals allow the staff to practice the OPORD before its actual execution. Through rehearsals, a JFMCC or NCC commander and staff gain an understanding of the CONOPS in its entirety. These rehearsals afford a comprehensive view of the operation, orient the subordinate naval commands and units to one another and, more importantly, give each component a thorough understanding of the JFMCC or NCC's intent, priorities, and guidance.

Joint Operations Rehearsals

There is limited information concerning joint operations rehearsals in joint publications, and the normal alternative of reverting to service publications for guidance can cause considerable confusion.

Types
- Staff Only Rehearsal
- Commanders and Staff Rehearsal
- Partial Force Rehearsal
- Full Force Rehearsal

Techniques
- Map/Chart Technique
- Area Board Technique
- Simulation Supported Technique
- Similar Area Technique
- Actual Area Technique

Navy Rehearsals

Types
- Amphibious Rehearsal
- Assault Rehearsal
- Sweep Rehearsal
- Unit Rehearsal

Techniques
- Complete Rehearsal
- Limited Rehearsal

Chap 4

The Operations Process: III. Execution

Ref: NWC Maritime Component Commander Handbook (Feb '10), chap. 4 and NWP 3-32, Maritime Operations at the Operational Level of War (Oct.'08), chap. 6.

This section provides information on the execution stage of the operations process. Execution combines continued planning, preparation, and assessment with the challenges of a dynamic adversary and the fog of war. Staff organizations and processes organize the chaos of execution to provide orderly, timely, and effective decision-making by the MCC. The MCC combines the art of command with the science of control. Commanders use the MOC to better visualize the operational environment, describe their visualization to subordinates, and direct actions to achieve results. The MCC must not let the science of control (the processes) distract him from understanding the essence of the situation. Understanding the mechanics of the MOC process while focusing on the extent and implications of his decisions can help ensure that the art of command is not diluted by the bureaucracy of the process.

I. Execution Fundamentals

Planning and preparation accomplish nothing if the command does not execute effectively. Execution is putting a plan into action by applying combat power to accomplish the mission and using situational understanding to assess progress and make execution and adjustment decisions.

During execution, many factors conspire against the MCC to complicate the execution of his plan. Not least of these factors are the competing demands for his time. Another factor complicating execution for the commander is building a clear understanding of the operating environment. One key enabler that bridges the situational understanding of the commander and decision-making is the development and monitoring of the commander's critical information requirements (CCIRs). Commanders fight the enemy, not the plan, because the enemy rarely acts as predicted. This is the principal cause of fog; and commanders modify their plans to counter enemy reactions.

See following pages (pp. 4-12 to 4-13) for discussion of execution fundamentals.

The Maritime Operations Center (MOC)

As an operation enters execution, the commander and staff need to modify their tempo and emphasis in order to make timely and informed decisions. The headquarters normally transitions to a functional B2C2WG organization as execution nears. This in combination with an aligned battle rhythm provides a process for the staff to receive, prepare, and present information to the commander for decision-making. The Maritime Operations Center (MOC) is organized to provide the commander support to conduct this process of understanding, assessing, planning, deciding, and directing. The challenges of execution are tied to each of the steps above: difficulty in gaining understanding, assessing progress, developing alternatives, making correct and timely decisions, and providing clear and synchronized direction. The MOC manages time demands and horizons, information flow and management, staff activities, and decision authorities to provide for organized and synchronized decision-making.

The commander's decision cycle is an important way that the MOC develops and distributes MCC decisions. The commander's decision cycle does not compete with the operations process. It is a distinct part of execution that is part of the operations process.

See chap. 3 for discussion of MOC organization and function.

Execution Fundamentals

Ref: NWC Maritime Component Commander Handbook (Feb '10), p. 4-2 to 4-4.

Planning and preparation accomplish nothing if the command does not execute effectively. Execution is putting a plan into action by applying combat power to accomplish the mission and using situational understanding to assess progress and make execution and adjustment decisions.

Competing Demands

During execution, many factors conspire against the MCC to complicate the execution of his plan. Not least of these factors are the competing demands for his time. In addition to monitoring the progress of the operation, the MCC will need to allocate time to planning and preparing for follow-on action in the operation. So his focus will be divided between the actions his forces are currently involved in (current operations) and actions his forces will be executing at a later date (future operations). Included in this will be the various battle rhythm events he will need to attend for the purpose of making decisions and providing guidance to his staff. Lastly, the commander will need to allocate time to meet with the joint force commander, subordinate maritime commanders, and his adjacent commanders aimed at ensuring alignment of current and future operations across the joint and maritime force.

Situational Understanding

Another factor complicating execution for the commander is building a clear understanding of the operating environment. The characteristics of the execution environment are different from the more sterile planning and preparation "laboratories." Information and data quantity increase by an order of magnitude; and the uncertainty of the environment (fog of war) increases the risk associated with action and increases the potential for misalignment of the staff and subordinate tactical forces. Reports from the various headquarters involved in the operation will stream into the MOC, potentially bogging down the ability of the staff to process it. Execution is further challenging because an intelligent enemy is also making decisions to achieve enemy objectives. From planning and preparation's relative stability, the tempo of actions by both friendly and enemy forces increases, causing the situation to become less clear. Attempting to obtain perfect understanding may not only paralyze the current operation but also hamper accomplishment of other essential headquarter functions. The challenge for the staff is to analyze the available information quickly enough to build and maintain situational awareness for the commander, enabling timely and effective decision-making.

Commander's Critical Information Requirements (CCIR)

One key enabler that bridges the situational understanding of the commander and decisionmaking is the development and monitoring of the commander's critical information requirements (CCIRs). During planning CCIRs are geared toward providing the commander information required to inform plan development. During preparation, CCIRs focus on monitoring the friendly and enemy situations in order to enable plan execution. During execution, CCIRs are utilized to inform the commander of progression along the plan. They identify when predetermined decisions such as branch or sequel implementation are required of the commander. CCIRs also inform the commander of changes in the environment that were not anticipated and are thus not supported by the plan developed. The monitoring of CCIRs is a staff-wide, and potentially maritime force–wide, responsibility. Due to the importance of CCIRs, all personnel need to clearly understand what their responsibilities are should a CCIR be triggered. *See p. 5-30.*

Decisionmaking During Execution

Commanders fight the enemy, not the plan, because the enemy rarely acts as predicted. This is the principal cause of fog, and commanders modify their plans to counter enemy reactions. A clear commander's intent does much to allow prompt and effective exercise of subordinates' initiative to react effectively and quickly to changes in the situation.

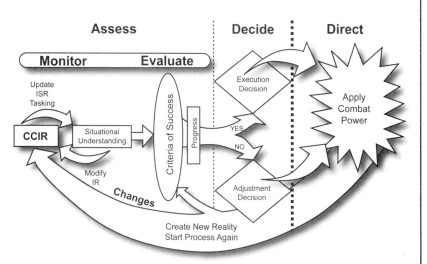

Ref: NWC Maritime Component Commander Handbook, fig. 4-2, p. 4-3.

Ultimately, the commander needs to make decisions that guide the conduct to accomplish the objectives of the operation. To be prepared to make a decision, the commander needs:

- **Clearer understanding of the current operational environment.** This understanding is developed by the focused collection based on CCIR and input from higher, adjacent, and subordinate commands.

- **Evaluation of the operation's progress.** From this clearer understanding, commanders and staffs evaluate the operation's progress according to the plan but more importantly in achieving the operation's objectives. This assessment is based on decision points that indicate potential different paths the operation may take or be taking. These can be identified in the plan or result from unanticipated enemy actions.

- **Consider valid alternatives.** A commander's visualization based on accurate, current situational awareness allows commanders to rapidly and effectively adjust the plan to adapt to changing situations — whether precipitated by the enemy or by changes in friendly force status. The staff provides these alternatives in branch or sequel plans or modifications in force disposition, task assignment, and priorities of support.

- **Decide.** From this evaluation, commanders can make a wide range of decisions — continuing according to the plan, selecting a developed alternative, or developing a new plan. Normally, decisions will be in adjusting off the existing plan. Commanders adjust the disposition of their forces, the tasks assigned to subordinates, and the priorities for support, to achieve the greatest effect at minimum cost.

- **Direct.** Commanders can direct subordinate tactical actions, staff actions, or employment of operational functions. Subordinate tactical actions are usually directed in the form of FRAGOs. Staff actions may include planning (either rapid or unconstrained), assessing (more detailed evaluation of the situation), and monitoring (additional collection of information). Employment of operational functions includes protection, fires, intelligence, maneuver, logistics, and command and control actions to alter the environment and gain advantage over the adversary.

(Operations Process) III. Execution 4-13

II. Commander's Decision Cycle

Ref: NWP 3-32, Maritime Operations at the Operational Level of War (Oct.'08), pp. 6-8 to 6-17.

The commander's decision cycle provides a mechanism for focusing the operational staff to support critical decisions and actions as the operational commander controls plan preparation and execution. The cycle is discussed in NWP 3-32. The commander's decision cycle depicts how command and staff elements determine required actions, codify them in directives, execute them, and monitor their results. The commander's decision cycle has four core phases.

The decision cycle assists the commander in understanding the operational environment and executing operational design during both preparation and execution. Commanders communicate throughout this cycle, both within the headquarters and with higher, adjacent, and subordinate commands. The battle rhythm is constructed to be aligned with MCC decision meetings to ensure information is presented at the proper time and forum for effective and efficient decision-making. Staff centers meet in cross-functional working groups to develop options and recommendations for MCC decisions.

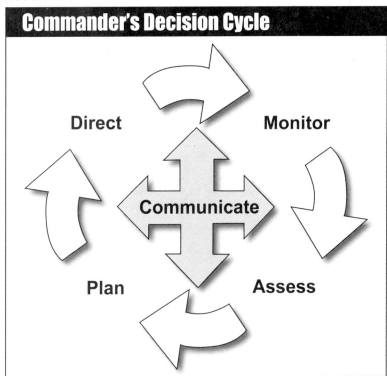

Ref: NWP 3-32, fig. 6-3, p. 6-8.

Event Horizons

This decision cycle nests with other echelon headquarters' decision cycles across all three event horizons. It is continually interfacing with the higher headquarters' decision cycle (which is normally more deliberate and slower-moving), with adjacent units, and with subordinate unit decision cycles (which will likely be moving more rapidly).

See p. 4-18 for a discussion of the three event horizons.

4-14 (Operations Process) III. Execution

The decision cycle assists the commander in understanding the operational environment and executing operational design during campaign preparation and execution. Operational commands assess how they are doing, conduct planning based on this assessment, direct forces as needed to execute the plan, and monitor force execution and its impacts on the adversary's PMESII systems. Outputs of monitoring provide the inputs for the next round of assessment. They communicate throughout this cycle, within the headquarters and with higher, adjacent, and subordinate commands. Operational commanders establish far-, mid-, and near-time horizons to focus staff preparation and execution of the campaign/major operation plan.

A. Assess

Within the commander's decision cycle, assessment is the determination of the effect of operations as they relate to overall mission accomplishment. Fundamental to assessment are judgments about progress in designated mission areas as measured against the expected progress in those same mission areas. These judgments allow the commander and the staff to determine where adjustments must be made to operations and serve as a catalyst for planning. Ultimately, assessment allows the commander and staff to keep pace with a constantly evolving situation while staying focused on mission accomplishment.

B. Plan

Planning is based on assessment and resultant commander's guidance. During campaign preparation the operational commander and staff develop plans whose focus is successful completion of the mission. During campaign/major operation plan execution the commander and staff make adjustments to the current plan or develop new plans to successfully achieve the strategic commander's goals/objectives.

See chap. 5 for discussion of Navy planning process (NPP).

C. Direct

Directing operations begins with receipt of a mission order from higher headquarters to initiate the planning process and continues through completion of the campaign; when directing in a multinational environment, the commander and staff must remain aware of the individual national government ROEs advantages and limitations.

D. Monitor

Monitoring involves observing ongoing activities that may impact the operational commander's operational area (OA) or impact the forces' ongoing or future operations. The baseline for this observation of the situation is the current plan or plans. This baseline allows the staff to observe the current situation against the one envisioned in the plan. This allows the commander and staff to identify where the current situation deviates from the one envisioned in the plan. Staff sections monitor their individual staff functions to maintain current staff estimates.

E. Communications

Communications, within the operational command staff, externally with other commands and DIME elements of national power, are key to the commander's decision cycle. Command-level attention to communication "pipes," systems, tools, and processes are fundamental to successful implementation of the commander's decision cycle.

As it relates to the commander's decision cycle, communications encompasses not only the exchange of information, but also the management of that information. Communications are the catalyst that facilitates the commander's decision cycle. The internal and external operational communications system must support an overall command climate and organizational design.

III. Risk Assessment
Ref: NWP 5-01, Navy Planning (Jan '07), app. E.

Risk is inherent in any use of military force or routine military activity. There are several types of risk; however, the risk discussed in relation to the NPP is associated with the dangers that exist due to the presence of the enemy, the uncertainty of the enemy intentions, and the potential rewards or dangers of friendly force action in relation to mission accomplishment.

Where resources are scarce, the commander may accept risk by applying the principle of economy of force in one area (supporting effort) in order to generate "massed effects" of combat power elsewhere (main effort). In an effort to affect surprise or maintain tempo, the commander may begin action prior to the closure of all units or sustainment. To maneuver or move the force for further actions, FP may be sacrificed somewhat by transiting a part of the force through a contested area. It is the rare situation where forces are so mismatched that the commander is not concerned with risk to the mission, and even in these situations the commander still desires to minimize the individual risk to own forces. The commander alone determines how and where to accept risk.

While risk cannot be totally eliminated, it can be managed by a systematic approach that weighs the costs—time, personnel, and resources—against the benefits of mission accomplishment. Commanders have always risk managed their actions: intuitively, by their past experiences, or otherwise. Risk management won't prevent losses but, properly applied, it allows the commander to take necessary and prudent risks without arbitrary restrictions and while maximizing combat capabilities.

Accepting Risk

Accepting risk is a function of both risk assessment and risk management. The approach to accepting risk entails an identification and assessment of threats, addressing risk, defining indicators, and observing and evaluating.

- **Identity of Threats.** Consider all aspects of mission, enemy, terrain, and weather, time, troops available and civilian (METT-TC) for current and future situations.

- **Assess Threats.** Assess each threat to determine the risk of potential loss based on probability and severity of the threat. Probability may be ranked as frequent: occurs often, continuously experienced; likely: occurs several times; occasional: occurs sporadically; seldom: unlikely, but could occur at some time; or unlikely: can assume it will not occur. Severity may be catastrophic: mission is made impossible; critical: severe mission impact; marginal: mission possible using alternate options; or negligible: minor disruptions to mission. Determining the risk is more an art than a science. Use historical data, intuitive analysis, and judgment to estimate the risk of each threat.

- **Address Risk, Determine Residual Risk, and Make Risk Decision.** For each threat, develop one or more options that will eliminate or reduce the risk of the threat. Specify who, what, where, when, and how. Determine any residual risk and revise the evaluation of the level of risk remaining. The commander alone then decides whether or not to accept the level of residual risk. If the commander determines that the risk is too great to continue the mission or a COA, then the commander directs the development of additional measures to account for the risk, or the COA is modified or rejected.

- **Define Indicators.** Think through the threat—what information will provide indication that the risk is no longer acceptable? Ensure that subordinates and staff are informed of the importance of communicating the status of those indicators.

- **Observe and Evaluate.** In execution, monitor the status of the indicators and enact further options as warranted. After the operation, evaluate the effectiveness of each option in reducing or eliminating risk. For options that were not effective, determine why and what to do the next time the threat is identified.

Applying Risk Management
Risk management requires a clear understanding of what constitutes unnecessary risk, when the benefits actually do outweigh costs, and guidance as to the command level to make those decisions. When a commander decides to accept risk, the decision must be coordinated with the affected units and detailed in each COA.

Risk Assessment for a Joint Force Maritime Component Commander or Navy Component Commander
Risk at the operational/operational-tactical level has characteristics that separate it from risk at higher and lower levels. Since functional and service component commands plan and operate as part of a joint force, JFMCC or NAVFOR operational risk is not independently established. JFMCC or NCC risk is composed of three distinct yet interrelated areas:
- Risk established at the higher headquarters and delegated to the JFMCC or NCC
- Risk shared between the various functional and Service component commanders
- Risk established by the JFMCC or NCC and distributed to tactical level components

Each of these risk areas needs to be analyzed as it relates to operational and strategic objectives. As part of the higher headquarters' mission analysis and COA decision, the higher headquarters determines where risk will be accepted. This accepted risk is communicated to the functional and Service component commander in the higher headquarters intent and CONOPs. This CONOPs determines how the joint force will achieve its objectives. As friendly and enemy factors of space, time, and force are balanced, accepted risk is delegated to the various functional or service components. The supported/supporting functional and Service component commanders may be directed to mitigate or accept risk in order to comply with the commander's method to attack the enemy.

The seize the initiative phase of the JTF operation requires JFMCC to maneuver amphibious forces and escorts within the range of Redland CDCMs and possibly mines in order to support the JFLCC airborne forcible entry. The JFMCC may be forced to accept risk in establishing sea control within CDCM range and a possible minefield. If the JFMCC believes that the risk incurred is unacceptable to mission accomplishment, the JFMCC should communicate this to the JTF while presenting some risk mitigation measures. Such measures could be strikes against CDCM sites, an IO effort to deny Redland targeting information for their CDCMs, or a supporting MCM effort in the AOA.

Risk therefore is a factor that is present in nearly every aspect of military actions whether at the tactical or operational level. Operational commanders need to consider where risk is located and where risk needs to be accepted. The following is a checklist of JFMCC, fleet commander, Navy component commander, numbered fleet commander, or designated task force commander risk areas that should be considered.

- What risk has been accepted by higher headquarters that directly impacts the JFMCC, fleet commander, Navy component commander, numbered fleet commander, or designated task force commander?
- During parallel planning, what additional risks are implied by COA development, analysis, comparison, and the developed CONOPS?
- As a supporting commander, what risk has been incurred to support a different functional component commander? Can these risks be mitigated?
- As a supported commander, what risk is accepted in the JFMCC, fleet commander, Navy component commander, numbered fleet commander, or designated task force commander CONOPs that impacts the other functional component commander? Is there a mitigation method available? What risk mitigation are the other functional component commanders providing to the JFMCC, fleet commander, Navy component commander, numbered fleet commander, or designated task force commander?
- As a higher commander, what risks are accepted that are passed to subordinate tactical commanders? How will these risks be mitigated?

IV. Event Horizons

Ref: NWP 3-32, Maritime Operations at the Operational Level of War (Oct.'08), pp. 6-8 to 6-10 and NWC Maritime Component Commander Handbook (Feb '10), p. 4-6.

Since the MCC is considering multiple operations simultaneously, decisions can be made for the current operation either in the near term, mid term or long term, or for follow-on operations. The far, mid and near time horizons are commonly linked to the future plans, future operations, and current operations elements of the staff, respectively. Each time horizon has a unique commander's decision cycle associated with it. The speed at which decisions must be made is unique to each time horizon. As a rule, decisions in the more distant future can be made more deliberately and in a measured manner.

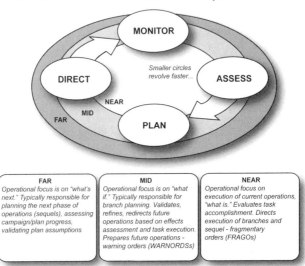

FAR
Operational focus is on "what's next." Typically responsible for planning the next phase of operations (sequels), assessing campaign/plan progress, validating plan assumptions

MID
Operational focus is on "what if." Typically responsible for branch planning. Validates, refines, redirects future operations based on effects assessment and task execution. Prepares future operations - warning orders (WARNORDSs)

NEAR
Operational focus on execution of current operations, "what is." Evaluates task accomplishment. Directs execution of branches and sequel - fragmentary orders (FRAGOs)

Near
Commonly associated with the cross-functional boards, bureaus, cells, centers, elements, groups, offices, planning teams, and working groups (B2C2WG) current operations elements of the staff, the near time horizon focuses on the "what is," and can rapidly progress through the decision cycle — sometimes in minutes for quick-breaking events. Current operations elements of the staff produce a large volume of orders including administrative and tactical FRAGORDs (e.g., change in priorities). These kinds of activities generally do not require full staff integration.

Mid
Commonly associated with the cross-functional B2C2WG future operations elements of the staff, the mid-time horizon focuses on the "what if," and normally moves slower with more deliberate assessment and planning activities, resulting in such things as major FRAGORDs directing major tactical actions (e.g., named operations) and TF movements within theater (e.g., movement of a carrier TF from one carrier operations area to another). It generally requires full staff integration.

Far
Commonly associated with the cross-functional B2C2WG future plans elements of the staff, the far-time horizon is focused on the "what's next," interacts heavily with higher headquarters planning efforts, and moves very deliberately through the decision cycle. It focuses on activities such as development of OPLANs and FRAGORDs to campaign/major OPLAN and policy directives or major force rotations. These kinds of activities normally require full staff integration.

The Operations Process: IV. Assessment

Ref: NWC Maritime Component Commander Handbook (Feb '10), chap. 5.

Assessment is a key portion of the commander's decision cycle. Assessment is the continuous monitoring and evaluation of the current situation and the progress of an operation. Commanders, assisted by their staffs and subordinate commanders, continuously assess the OE and the progress of the operation toward the desired end state in the time frame desired. Based on their assessment, commanders direct adjustments, thus ensuring the operation remains focused on accomplishing the mission. The Navy's and joint operational assessment processes are currently very immature, and assessment processes have little doctrinal basis. As a result of this, as well as differing views on the role of assessment at the operational level, each JFMCC conducts assessment in a different manner. One of the largest criticisms of operational assessment has been that it does not provide a usable product for the commander to use in making decisions. Without a commander's involvement and guidance, operational assessment can suffer from one of two extremes: either it will receive little attention or resources, resulting in the commander's not receiving key data necessary to make critical day-to-day and long-term decisions, or it will consume a staff and tactical force's effort with little return.

Refer to TACMEMO 3-32.2-09, "Operational Assessment" for further discussion.

I. The Assessment Process

The assessment process is continuous and directly tied to the commander's decisions throughout planning, preparation, and execution of operations. The assessment process begins during mission analysis when the commander and staff consider what to measure and how to measure it to determine progress (MOPs/MOEs) toward accomplishing a task, creating an effect, or achieving an objective. During planning and preparation for an operation, for example, the staff assesses the force's ability to execute the plan based on available resources and changing conditions in the operational environment.

A. Measure of Performance (MOP)

A MOP is a criterion used to assess friendly actions that is tied to measuring task accomplishment. MOPs measure task performance. They are generally quantitative, but also can apply qualitative attributes to task accomplishment. MOPs are used in most aspects of combat assessment, since it typically seeks specific, quantitative data or a direct observation of an event to determine accomplishment of tactical tasks. MOPs have relevance for noncombat operations as well and can also be used to measure operational and strategic tasks, but the type of measurement may not be as precise or as easy to observe.

B. Measure of Effectiveness (MOE)

A MOE is a criterion used to assess changes in system behavior, capability, or operational environment that is tied to measuring the attainment of an end state, achievement of an objective, or creation of an effect. MOEs assess changes in system behavior, capability, or operational environment. They measure the attainment of an end state, achievement of an objective, or creation of an effect; they do not measure task performance. These measures typically are more subjective than MOPs, and can be crafted as either qualitative or quantitative. MOEs can be based on quantitative measures to reflect a trend and show progress toward a measurable threshold.

Refer to NWP 5-01, Navy Planning, app. F for selected examples of measures of effectiveness for use in naval warfare.

The Assessment Process

Ref: NWP 3-32, Maritime Operations at the Operational Level of War (Oct.'08), p. 6-11 to 6-13.

The assessment process begins during mission analysis when the commander and staff develop MOPs and MOEs. These MOPs and MOEs will be used to assess progress toward accomplishing a task, creating a result avoiding creation of an undesired result, or achieving an objective. Commanders adjust operations based on their assessment to ensure objectives are met and the military end state is achieved. The assessment process is continuous and is tied directly to the commander's decisions throughout planning, preparation, and execution of operations. Staffs help the commander by monitoring the numerous aspects that can influence the outcome of operations and by providing the commander timely information needed for decisions. MOPs and MOEs should be relevant, measurable, responsive, and resourced.

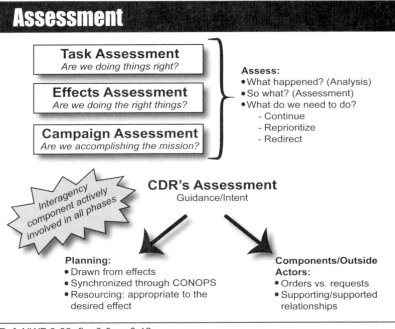

Ref: NWP 3-32, fig. 6-6, p. 6-12.

Assessment occurs at all levels and across the entire range of military operations. Outputs of the **tactical force assessment** are inputs to operational command assessment. Similarly the outputs of the operational level assessment are inputs to strategic command assessment. Assessment at the operational and strategic levels typically is broader than at the tactical level (e.g., combat assessment) and uses MOEs that support strategic and operational mission accomplishment. Strategic and operational level assessment efforts concentrate on broader tasks, effects, objectives, and progress toward the end state. Continuous assessment helps operational commanders determine if the joint force is "doing the right things" to achieve objectives, not just "doing things right."

Operational assessment measures how the campaign is doing in terms of tasks (CONOPS execution), effects, and campaign (end state); and then provides recommendations to adjust the plan or actions by components, tactical force, or outside actors (following the commander's guidance and intent) to achieve desired results.

Assessment is:
- A continuous process that measures the overall effectiveness of employing joint force capabilities during military operations.
- Determination of the progress toward accomplishing a task, creating an effect, or achieving an objective.
- Analysis of the security, effectiveness, and potential of an existing or planned intelligence activity.
- Judgment of the motives, qualifications, and characteristics of present or prospective employees or "agents."

Assessment best practices include:
- Avoid overengineering assessment. A balance is needed between a quantitative and qualitative approach to assessment. Assessment, especially task and campaign objective assessment, is difficult and, in many cases, subjective.
- The operational commander is a critical component of assessment. Commanders counter the staff tendency toward "science of war" solutions by limiting the amount of time and effort their staffs put into quantifying assessments, and by recognizing that proper assessment can only occur by applying their experience, intuition, and own observations in an "art of war" approach to assessment.
- Developing and making recommendations to the commander on "what needs to be done" based on assessments. Staffs cannot expend all efforts on developing the "what happened" and the "so what" of assessment and then not have time or energy to recommend "what needs to be done."
- Always provide recommendations at all levels of assessment — task, effects, and campaign.
- Match frequency of formal assessments to the pace of campaign/major operation planning and execution.
  - **Task assessments** occur fairly frequently and are a focus area within the current operations staff area.
  - **Effects assessments** are a focus area within the future operations staff area and are conducted at times to meet the commander's needs. They are often conducted less frequently than task assessments.
  - **Campaign assessments** during execution are a focus area within the future operations staff area and are conducted to meet the commander's needs. Campaign assessments of plans under review are a focus area within the future plans staff area.
- Conduct task (OPLAN execution) assessments answering "Are we doing things right?"
- Conduct effects assessments answering "Are we doing the right things" to achieve our desired effects?
- Conduct campaign assessment answering "Are we accomplishing the mission?" (achieving end state/objectives).
- Diplomatic, information, military, and economic (DIME) elements of national power perspectives enrich the assessments.
- Periodically revalidate developed objectives, effects, MOE, and MOP. Review basis for operations, assumptions, and systems perspective. Revalidation of the objectives (end state) occurs at the level at which they were developed — normally, the theater-strategic or above level. Review of the desired and undesired effects primarily occurs at the operational level, while review of MOE and MOP to determine if the commander and staff are measuring the correct trends and actions and using the correct metrics occur at the operational and tactical level. These reviews/revalidations keep the units on course by taking into account higher level direction, adversary actions, and other changes in the operational environment.

II. Combat Assessment

Ref: Adapted from The Battle Staff SMARTbook and joint doctrine references.

Combat assessment is the determination of the effectiveness of force employment during military operations. Combat assessment is composed of three elements:

Combat Assessment

 Battle Damage Assessment (BDA)

 Munitions Effects Assessment (MEA)

 Reattack Recommendation

A. Battle Damage Assessment (BDA)

In combination, BDA and MEA inform the commander of effects against targets and target sets. Based on this, the threat's ability to make and sustain war and centers of gravity are continuously estimated. During the review of the effects of the campaign, restrike recommendations are proposed or executed. BDA pertains to the results of attacks on targets designated by the commander. Producing BDA is primarily an intelligence responsibility, but requires coordination with operational elements. BDA requirements must be translated into PIR.

B. Munitions Effectiveness Assessment (MEA)

The targeting team conducts MEA concurrently and interactively with BDA as a function of combat assessment. MEA is used as the basis for recommending changes to increase effectiveness in:

- Methodology
- Tactics
- Weapon systems
- Munitions
- Weapon delivery parameters

MEA is developed by determining the effectiveness of tactics, weapons systems, and munitions. Munitions effect on targets can be calculated by obtaining rounds fired on specific targets by artillery assets. The targeting team may generate modified commander's guidance concerning:

- Unit Basic Load (UBL)
- Required Supply Rate (RSR)
- Controlled Supply Rate (CSR)

Refer to The Sustainment & Multifunctional Logistician's SMARTbook for discussion of UBL, RSR, and CSR.

The need for BDA for specific HPTs is determined during the decide function. Record BDA on the AGM and intelligence collection plan. The resources used for BDA are the same resources used for target development and TA. An asset used for BDA may not be available for target development and TA. The ACE receives, processes, and disseminates the results of attack (in terms of desired effects).

Each BDA has three assessment components.

1. Physical Damage Assessment
Physical damage assessment estimates the quantitative extent of physical damage through munitions blast, fragmentation, and/or fire damage effects to a target. This assessment is based on observed or interpreted damage.

2. Functional Damage Assessment
Functional damage assessment estimates the effect of attack on the target to perform its intended mission compared to the operational objective established against the target. This assessment is inferred on the basis of all-source intelligence and includes an estimate of the time needed to replace the target function. A functional damage assessment is a temporary assessment (compared to target system assessment) used for specific missions.

3. Target System Assessment
Target system assessment is a broad assessment of the overall impact and effectiveness of all types of attack against an entire target systems capability; for example, enemy ADA systems. It may also be applied against enemy unit combat effectiveness. A target system assessment may also look at subdivisions of the system compared to the commander's stated operational objectives. It is a relatively permanent assessment (compared to a functional damage assessment) that will be used for more than one mission.

BDA is more than determining the number of casualties or the amount of equipment destroyed. The targeting team can use other information, such as:

- Whether the targets are moving or hardening in response to the attack
- Changes in deception efforts and techniques
- Increased communication efforts as the result of jamming
- Whether the damage achieved is affecting the enemy's combat effectiveness as expected

BDA may also be passive by compiling information regarding a particular target or area (e.g., the cessation of fires from an area). If BDA is to be made, the targeting team must give intelligence acquisition systems adequate warning for sensors to be directed at the target at the proper time. BDA results may change plans and earlier decisions. The targeting team must periodically update the decisions made during the decide function concerning:

- IPB products
- HPTLs
- TSS
- AGMs
- Intelligence collection plans
- OPLANs

C. Reattack Recommendation

Based on BDA and MEA, the G2/G3 consider the level to which operational objectives have been achieved and make recommendations to the commander. Reattack and other recommendations should address operational objectives relative to the:

- Target
- Target critical elements
- Target systems
- Enemy combat force strengths

III. Assessment Estimate Development and Integration into the Navy Planning Process

Ref: NWC Maritime Component Commander Handbook (Feb '10), fig. 5-8, p. 5-13.

The commander's role is central to the assessment process. Commanders establish priorities for assessment and focus the staff to meet the requirements of time, simplicity, and level of detail based on the situation and the commander's personal desires. This is accomplished by establishing commander's critical information requirements, setting priorities for assessment in the form of measures of effectiveness (MOEs), determining the periodicity and complexity with which he desires to receive assessment briefings, and explicitly stating assumptions, as well as commander's intent, throughout the conduct of operations. While the staff does the detailed work, to include collecting and analyzing information, the commander ultimately assesses the operation.

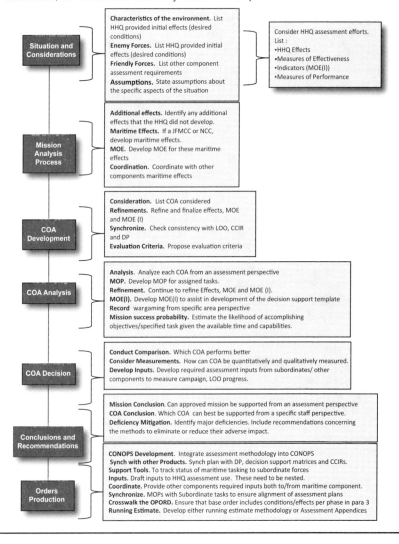

4-24 (Operations Process) IV. Assessment

Chap 5
Navy Planning Overview

Ref: NWP 5-01, Navy Planning (Jan '07), chap. 1.

Military planning is a comprehensive process that enables commanders and staffs at all levels and in all services to make informed decisions, solve complex problems, and ultimately accomplish assigned missions. Military planning is critical at every level of warfare—strategic, operational, and tactical—and in any situation regardless whether the threat is posed by a conventional military, an asymmetric unconventional adversary, or a combination of both. Military planning can be applied whether conditions permit a lengthy, deliberate process or if the situation forces a compressed timeline.

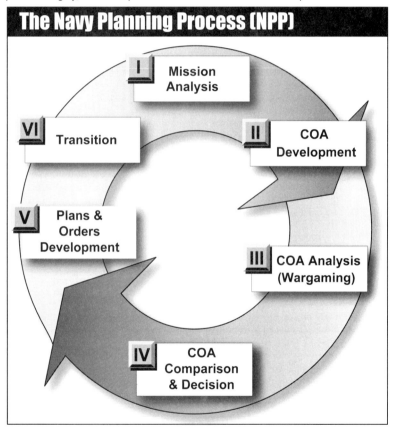

The Navy Planning Process (NPP)
- I. Mission Analysis
- II. COA Development
- III. COA Analysis (Wargaming)
- IV. COA Comparison & Decision
- V. Plans & Orders Development
- VI. Transition

Ref: NWP 5-01, Navy Planning, pp. 1-1 to 1-9.

Military planning, and by extension Navy planning, is the process by which a commander visualizes an end state and then determines the most effective ways by which to reach the end state. Specifically, planning helps the commander direct and coordinate the actions of a force, generate a common situational awareness, develop expectations as to how the dynamic interaction of forces will affect the outcome of an operation, and shape the thinking of the planning team.

(Naval Planning) Navy Planning Overview 5-1

I. The Navy Planning Process (NPP)

Through the Navy Planning Process (NPP), a commander can effectively plan for and execute operations, ensure that the employment of forces is linked to objectives, and integrate naval operations seamlessly with the actions of a joint force. Accordingly, the terminology, products, and concepts in the NPP follow the joint planning process, adhere to joint doctrine, and are compatible with other services.

The NPP assists commanders and their staffs in analyzing operational environment effects and distilling a multitude of planning information in order to provide the commander with a coherent framework to support decisions. The process is thorough and helps apply clarity, sound judgment, logic, and professional expertise. While the full NPP may appear time-consuming, through training, experience, and frequent use, commanders and their staffs can become more proficient, and the NPP has the ability to become a more fluid and adaptable process. Therefore, in the event that experienced planners are faced with a short timeline, the NPP can be easily flexed to support crisis action planning.

The NPP organizes these procedures into six steps that provide commanders and their staffs a means to organize planning activities, transmit plans to subordinates, and share a critical common understanding of the mission. Interactions among various planning steps allow a concurrent, coordinated effort that ensures flexibility, makes efficient use of available time, and facilitates continuous information sharing.

The result of the NPP is a military decision that can be translated into a directive such as an operation plan (OPLAN) or OPORD. The NPP is also the foundation on which planning in a time-constrained environment (often called crisis action planning) is based.

A. Non Time-Sensitive Planning

A final consideration is the type of planning to be conducted in the NPP. Generally, the amount of time available to the organization between receipt of mission and execution determines the type of planning to be used. When time is not a critical factor, planners use a process that reflects the non time-sensitive nature of the planning. At the tactical level of warfare, this means that there is sufficient time available for each echelon in the planning architecture to complete its initial planning before information or orders are passed to subordinates.

Non time-sensitive planning has four phases: strategic guidance, concept development, plan development, and supporting plan development. The Joint Strategic Capabilities Plan (JSCP) initiates planning using the Joint Operation Planning and Execution System (JOPES), and the end product is a plan (operations plan or concept plan). Non time-sensitive planning at the tactical level is normally initiated through the higher joint force headquarters and is frequently in the form of supporting plans or orders.

B. Time-Sensitive Planning

When time becomes a critical factor, planners at the tactical level use a process that reflects the time-sensitive nature of the planning. In the JOPES, crisis action planning is initiated in response to a specific event. Time sensitive planning at the tactical level is also initiated through the joint force headquarters and uses the JOPES crisis planning process. The overall process of time-sensitive planning parallels that of non time-sensitive planning but is better able to respond to changing events. Time-sensitive planning promotes a logical, rapid flow of information and timely preparation of courses of action. Time-sensitive planning also creates situations where planning and execution occur simultaneously.

Refer to NWP 5-01, Navy Planning, app. M for discussion of the Navy Planning Process in a time-constrained environment.

The Navy Planning Process
Ref: NWP 5-01, Navy Planning (Jan '07), pp. 1-4 to 1-5.

The NPP establishes procedures to progressively analyze a mission, develop and wargame courses of action (COAs) against projected enemy courses of action (ECOAs), compare friendly COAs against the commander's criteria and each other, select a COA, prepare an operation order (OPORD) for execution, and transition the plan or order to subordinates tasked with its execution.

Step I. Mission Analysis
Mission analysis drives the NPP. As the first step of the process, its purpose is to review and analyze orders, guidance, intelligence, and other information in order for the commander, planning team, and staff to gain an understanding of the situation and to produce a mission statement.

See pp. 5-17 to 5-34.

Step II. Course of Action Development
Planners use the mission statement, commander's intent, and planning guidance to develop multiple COAs. Then they examine each prospective COA for validity by ensuring suitability, feasibility, acceptability, distinguishability, and completeness with respect to the current and anticipated situation, the mission, and the commander's intent.

See pp. 5-35 to 5-46.

Step III. Course of Action Analysis (Wargaming)
Course of action analysis involves a detailed assessment of each COA as it pertains to the enemy and the operational environment. Each friendly COA is war gamed against selected ECOAs. This step assists planners in identifying strengths, weaknesses, and associated risks, and in assessing shortfalls for each prospective friendly COA. War gaming also identifies branches and potential sequels that may require additional planning. Short of execution, COA war gaming provides the most reliable basis for understanding and improving each COA. This step also allows the staff to refine its initial estimates based on additional understanding that is gained.

See pp. 5-47 to 5-58.

Step IV. Course of Action Comparison and Decision
All retained friendly COAs are evaluated against established criteria and against each other, ultimately leading to a decision by the commander.

See pp. 5-59 to 5-66.

Step V. Plans and Orders Development
The staff uses the commander's COA decision, mission statement, commander's intent, and guidance to develop plans and/or orders that direct subordinate actions. Plans and orders serve as the principal means by which the commander expresses his decision, intent, and guidance.

See pp. 5-67 to 5-72.

Step VI. Transition
Transition is the orderly handover of a plan or order to those tasked with execution of the operation. It provides staffs with the situational awareness and rationale for key decisions necessary to ensure that there is a coherent transition from planning to execution. The process, however, does not end here. The process is continuous. Staffs maintain running estimates that allow for plans and orders refinement. The planning staff continues to examine branches and sequels to plans and orders.

See pp. 5-73 to 5-76.

II. Nesting of the Navy Planning Process Within Other Planning Processes

Ref: NWP 5-01, Navy Planning (Jan '07), pp. 1-5 to 1-6.

Navy forces seldom operate independently without integration and coordination with other services. Navy staffs must be well versed in joint doctrine, particularly **Joint Operations, Joint Operation Planning, Joint Operation Planning and Execution System (JOPES)**, approved joint terminology, and the amphibious planning process contained in Joint Doctrine for Amphibious Operations.

The NPP must link commanders, as well as their staffs, with higher headquarters, laterally to other service and functional component commands, and to subordinate commanders and their staffs. In the case of a Navy component commander (NCC) or joint force maritime component commander (JFMCC), for instance, a similar concurrent planning technique also should be used with subordinate forces such as carrier strike groups (CSGs), expeditionary strike groups (ESGs), surface strike groups (SSGs), missile defense surface action groups (MDSAGs), as well as task forces and task groups organized for specific missions. Likewise, in the case of a CSG or ESG, planning efforts among the warfare commanders, ships, squadron, and other forces attached must be linked.

NWP 5-01, Navy Planning, app. I provides a detailed illustration of the nested functions from a joint task force (JTF) to tactical component commands.

Navy Planning and the Levels of War

Military planning occurs at all levels—from the strategic level and the Chairman of the Joint Chiefs of Staff (CJCS) down to the individual ship, submarine, or aircraft. The levels of planning differ in their complexity, scope, and purpose; however, they are all linked. In particular, the operational level, where the NPP in conjunction with the joint planning process can be most effectively implemented, is where the crucial link from strategic military objectives to tactical war fighting is established.

See p. 4-6 for further discussion of naval planning and the levels of war.

Navy Planning Process and Effects

An essential aspect of planning is for commanders and their staffs to understand how their activities are related to the operational constructs of their superior's plans or operations. In the joint planning process, effects (more properly "desired effects") are nested between objectives and tasks. Objectives are generally broadly stated and in isolation are insufficient for planning and assigning tasks to subordinates. Desired effects are essentially detailed "sub-objectives." When and if desired effects are realized, the aggregate result should be achievement of the objective. At the theater strategic and operational levels of war, commanders designate objectives and supporting desired effects; this enables the assignment of tasks to component commanders.

At the tactical level of war, which is the focus of the NPP, effects typically are associated with direct results of offensive and defensive tactical actions, often involving weapons employment. At the tactical level, an effect typically is the proximate, first-order consequence of an action; for example, the destruction of a target by precision-guided munitions or the evacuation of noncombatants during a noncombatant evacuation operation (NEO), both of which usually are immediate and easily recognizable. As such, at the tactical level, there is little need for using effects language in subordinate unit taskings. A clear task associated with a purpose and an equally clear commander's intent convey sufficient information for tactical units to understand their nesting with their superior's plans or operations. It is important, however, for organizations operating at the operational level of war to understand how effects are integrated into the joint operation planning process.

Joint Operation Planning and Execution System (JOPES)

Planning for the employment of military forces is an inherent responsibility of command. It is performed at every echelon of command and across the range of military operations. Joint planning integrates military actions with those of other instruments of national power and our multinational partners in time, space, and purpose to achieve a specified end state. The military's contribution to national strategic planning consists of joint strategic planning with its three subsets: security cooperation planning, joint operation planning, and force planning.

JOPES formally integrates the planning activities of the entire Joint Planning and Execution Community (JPEC) during the initial planning and plan refinement that occurs both in peacetime and when faced with an imminent crisis. While JOPES activities span many organizational levels, the focus is on the interaction which ultimately helps the President and SecDef decide when, where, and how to commit US military capabilities in response to a foreseen contingency or an unforeseen crisis. The majority of JOPES activities and products occur prior to the point when the CJCS approves and issues the execute order, which initiates the employment of military capabilities to accomplish a specific mission.

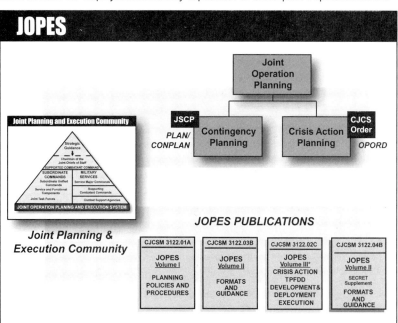

JOPES consists of a multivolume set of CJCS unclassified and classified manuals. JOPES Volume I provides the foundation for joint operation planning. JOPES is used in the development and implementation of OPLANs and OPORDs prepared in response to requirements from the President, the SecDef, or the CJCS.

Refer to The Joint Forces Operations & Doctrine SMARTbook (Guide to Joint, Multinational & Interagency Operations), chap. 3 for complete discussion of joint strategic planning, to include the Joint Operations Planning and Execution System (JOPES).

(Naval Planning) Navy Planning Overview 5-5

III. Roles in the Navy Planning Process

Ref: NWP 5-01, Navy Planning (Jan '07), pp. 1-6 to 1-8.

The NPP is a dynamic process that requires close cooperation and involvement between the commander, staff, and a planning team to ensure that time is used efficiently and that the most effective plan to meet the commander's intent is developed. The synergy among the commander, staff, and planning team is critical; however, the relationships with the joint force commander (JFC), Service component command, functional component command, and subordinate commands are equally important in the NPP.

Time for the NPP will often be insufficient. Effective interaction and a flexible approach to the process are required when time becomes critical. Commanders must understand how they can shorten the process time through more specific guidance and increased participation. In effect, the less time to plan, the more involved the commander must become.

The Commander's Role

The commander is more than simply the decision maker in this process. It is the commander's experience, knowledge, and judgment that provide the bearing and direction under which the staff must operate. The commander provides the necessary focus and guidance to the planning team and staff. While unable to devote all of one's time to the NPP, the commander must be keenly aware of the current status of the planning effort and make sound decisions based upon the detailed work of the planning team.

- **Personal Experience**: The commander is usually the most experienced and seasoned member of the team. The commander will have had wide experiences that will almost always help guide the process to an effective solution.

- **Personal Judgment**: Age and experience produce an ability to judge what works and what doesn't often by just knowing. The commander is the honest broker, keeping competing agendas and preferences out of the deliberations.

- **Personal Relationships**: Commanders share a special relationship with other commanders. It is through these relationships that a commander often gains a greater understanding of the intent and desire of a superior commander while having a greater understanding of the status and capabilities of subordinates.

Commanders must also understand the old axiom that "you often get what you inspect, not what you expect." While the commander's participation is not "inspecting" the NPP, his presence lends credibility to the process and keeps other less essential requirements from distracting the staff and planning team. The commander's role becomes more critical when time and requirements start to compete with one another, and only the commander can adequately shorten the process. This is done through sound, precise guidance and up-front decisions on courses of action, war-gaming, and formality of the process. If the planning team is left trying to guess what the commander desires in these areas, it will most often default to either a slow, deliberate process or it may try to eliminate instead of compressing parts of the process.

The Planning Team and Staff's Role

The commander's staff must be properly organized to support the NPP. However, staffs are functional organizations and are often torn between current and future operations planning and day-to-day sustainment of the naval force. Therefore, the commander must draw personnel who are embedded within the various staff codes into a planning team that is capable of providing a holistic and dedicated effort to the NPP. The staff agencies within the command that normally direct the planning effort for the commander are often divided between current plans or long-term plans within a Navy component or functional component (JFMCC) command. Based upon their unique organization and staff resources, smaller units may direct another staff to guide planning.

Staff Estimates
Additionally, each functional staff code should concurrently conduct **staff estimates** in support of the planning team to ensure that proposed COAs are suitable, acceptable, and feasible from each functional perspective. Staff estimates are essential throughout planning. They form the basis for supporting annexes and appendixes of an OPLAN or OPORD and are continuously updated as the situation or conditions within the operational environment change. At all times, though, it is imperative that the planning team and staff communicate and synchronize their efforts to ensure that planning for an operation is conducted to achieve a common goal.

See also p. 5-18. Refer to NWP 5-01, Navy Planning, app. K for staff estimate formats.

As the plan is briefed and discussed, it establishes a common purpose and clearly understood objectives within the organization and chain of command. Planning is the link that binds the members and activities of an organization together. The more effectively that staffs plan and exercise the plan, the more efficiently they can react to changing circumstances. Ultimately, planning enhances operational success by enabling the command to react faster and more effectively than the enemy. Due to the unique nature of Navy forces, the constraints imposed by operating in a naval environment, and the necessity to affect the transition to a joint or multinational arena, it is imperative that Navy forces be efficiently organized and properly staffed. This facilitates planning future operations and anticipating changes as they arise in a dynamic environment. The planning team and staff have several critical functions in relation to the commander. While planning team members and staff officers have specific functions related to their jobs, they must all fully understand the importance of the commander's guidance in moving the process toward completion.

Planning team and staff personnel must expect and react to the commander's guidance. They should not allow themselves to move forward in the process without a clear understanding of the commander's desires. As time and requirements begin to compete with each other, making decisions and assumptions at the planning team level becomes imperative. However, if the team does not have a clear understanding of the commander's intent and guidance, those decisions and assumptions may move them farther away from the final product rather than closer together and may cause the planning to diverge from the naval force's objectives. The commander's intent gives the force direction in the absence of specific orders, and the best commander's intent is written by the commander.

It is understood that echelons below a Navy component commander or JFMCC do not always have the manpower, resources, or time to follow the entire NPP; however, it is imperative that these lower echelon commands adapt, combine, and then incorporate the functions and responsibilities of the NPP into their organization as limitations permit. Normally, lower level commands find that the bulk of the planning effort resides with the commander and a select number of staff or command personnel. However, even smaller naval organizations will find the NPP to be a useful framework for the analysis of an operational task and the development of COAs to address the issue.

Liaison
Another critical element of the staff and planning team organization is liaisons. In order to ensure that planning efforts remain in concert with those of senior, subordinate, and lateral commands, continuous and reliable communications are important. Often due to time and space considerations, it is impossible to be in the same location as other commands. Liaison can take many forms. It can and often is established informally among corresponding functional staff elements (such as operations directorate of a joint staff (J-3), Navy component command operations officer (N-3), Army or Marine Corps component command operations officer (G-3)). However, to ensure true unity of purpose and to avoid contradictory planning efforts, a formal liaison relationship is best.

IV. Intelligence Preparation of the the Operational Environment (IPOE)

Ref: NWP 5-01, Navy Planning (Jan '07), app. B.

All planners need a basic familiarity with the IPOE process in order to become critical consumers of the products produced by the intelligence community. Some steps in the IPOE are conducted in parallel with the mission analysis and require input from other members of the maritime planning group. Although the specifics of the process vary depending on the situation and force involved, there is general agreement on the four major steps of IPOE.

For a more detailed discussion of the IPOE process, refer to JP 2-01.3, Joint Tactics, Techniques, and Procedures for Joint Intelligence Preparation of the Battlespace.

Step One: Define the Operational Environment

This first step is an initial survey of the geographic and non-geographic dimensions of the operational environment. It is used to bound the problem and to identify areas for further analysis. There are generally three tasks that must be accomplished.

- Identify the AO and the area of interest
- Determine the significant characteristics of the operational environment. This sub-step is an initial review of the factors of space, time, and forces and their interaction with one another.
- Evaluate existing databases and identify intelligence gaps and priorities. In this sub-step, intelligence personnel review the information found in various automated databases, Intelink sites (the classified version of the Internet), and other intelligence sources, both classified and unclassified. Intelligence requests and requirements may take the form of priority intelligence requirements (PIRs), requests for information (RFIs), production requests (PRs), and collection requirements.

Area of Operations: Defined by LAT/LONG or displayed on a map/chart for clarity and reference. The higher headquarters normally assigns this.

Area of Interest: Adjacent geographic area where political, military, economic, or other developments have an effect within a given theater; it might also extend to the areas enemy forces occupy that may endanger the accomplishment of one's mission; in practical terms, the area of interest determines the maximum scope of intelligence-gathering activities for the geographic combatant command; any theater (of war) also encompasses the pertinent parts of the cyberspace.

Step Two: Describe the Operational Environment Effects

The purpose of this step is to determine how the operational environment affects both friendly and enemy operations. It begins with an identification and analysis of all militarily significant environmental characteristics of each operational environment dimension. These factors are then analyzed to determine their effects on the capabilities and broad COAs of both enemy and friendly forces. Sub-steps include:

- Analyze the factor of space of the operational environment
- Analyze the factor of time of the operational environment
- Determine the operational environment effects on enemy and friendly capabilities and broad COAs

Summarize the Key Elements of the Factor of Space: military geography (area, position, distances, land use, environment, topography, vegetation, hydrography, oceanography, climate, and weather), politics, diplomacy, national resources, maritime infrastructure and positioning, economy, agriculture, transportation, telecommunications, culture, ideology, nationalism, sociology, science and technology.

Summarize the Key Elements of the Factor of Time: preparation, duration, warning, decision cycle, planning, mobilization, reaction, deployment, transit, concentration, maneuver, accomplish mission, rate of advance, reinforcements, commit reserves, regenerate combat power, redeployment, reconstruction

Step Three: Evaluate the Enemy (Factor of Forces)

The third step is to identify and evaluate the enemy's forces and its capabilities; limitations; doctrine; and tactics, techniques, and procedures to be employed. In this step, analysts develop models that portray how the enemy normally operates and identifies capabilities in terms of broad ECOAs that the enemy might take. Analysts must take care not to evaluate enemy doctrine and concepts by mirror imaging U.S. doctrine. Sub-steps include:

- Identify enemy force capabilities
- Consider and describe general ECOAs in terms of DRAW-D (Defend, Reinforce, Attack, Withdraw, or Delay)
- Determine the current enemy situation (situation template)
- Identify broad COAs that would allow the enemy to achieve objectives

Summarize the Key Elements of the Factor of Forces (Enemy): defense system, armed forces, relative combat power of opposing forces (composition, reserves, reinforcements, location and disposition, strengths), logistics, combat efficiency (morale, leadership, doctrine, training, etc.)

Step Four: Develop Enemy Courses of Action

Accurate identification of the full set of ECOAs requires the commander and his staff to think as the enemy thinks. From that perspective, it is necessary first to postulate possible enemy objectives and then to visualize specific actions within the capabilities of enemy forces that can be directed at these objectives and their impact upon potential friendly operations. From the enemy's perspective, appropriate physical objectives might include own-forces or their elements, own or friendly forces being supported or protected, facilities or lines of communication, and geographic areas or positions of tactical, operational, or strategic importance. The commander should not consider ECOAs based solely on factual or supposed knowledge of the enemy intentions.

The real COA by the enemy commander cannot be known with any confidence without knowing the enemy's mission and objective, and that information is rarely known. Even if such information were available, the enemy could change or feign the COA. Therefore, considering all the options the enemy could physically carry out is more prudent.

To develop an ECOA, one should ask the following three questions: Can the enemy do it? Will the enemy accomplish his objective? Would it materially affect the accomplishment of my mission? Each identified ECOA is examined to determine whether it meets the tests for suitability, feasibility, acceptability, uniqueness, and consistency with doctrine.

No ECOA should be dismissed or overlooked because it is considered as unlikely or uncommon, only if impossible. Once all ECOAs have been identified, the commander should eliminate any duplication and combine them when appropriate. Each ECOA is evaluated, prioritized, and ranked according to the probability of adoption. This final step in the IPOE process is designed to produce, at a minimum, two ECOAs: the enemy's most likely COA and most dangerous COA, giving the commander a best estimate and a worst-case scenario for planning. However, if time allows, other ECOAs are also developed. Each ECOA usually includes a description of expected enemy activities, the associated time and phase lines expected in executing the COA, expected force dispositions, associated COGs, a list of assumptions made about the enemy when projecting the COA, a list of refined high-value targets, and a list of named areas of interest, which are geographical areas where intelligence collection will be focused.

V. Operational Art & Design

Achievement of objectives does not lend itself to mechanistic, deterministic, scientific models or simple linear processes — developing a solution requires study of the interplay of literally hundreds, if not thousands, of independent variables. In other words, developing a solution for strategic objectives is more of an art than a science.

A. Operational Art

Operational art serves as a bridge and as an interface between maritime strategy and naval tactics. It is the application of creative imagination by commanders and staffs — supported by their skill, knowledge, and experience — to design strategies, campaigns, and major operations and to organize and employ military forces. Operational art is the thought process commanders use to visualize how best to efficiently and effectively employ military capabilities to accomplish their mission.

In applying operational art, the operational commander draws on judgment, perception, experience, education, intelligence, boldness, and character to visualize the conditions necessary for success before committing forces.

Operational art requires broad vision, the ability to anticipate, and the skill to monitor, assess, plan, and direct. It helps commanders and their staffs order their thoughts and understand the conditions for victory. Without operational art, campaigns and operations would be a set of disconnected tactical actions.

The operational commander uses operational art to consider not only the employment of military forces, but also their sustainment and the arrangement of their efforts in time, space, and purpose. This includes fundamental methods associated with synchronizing and integrating military forces and capabilities.

Operational art helps the operational commanders overcome the ambiguity and uncertainty of a complex operational environment. It governs the deployment of forces, their commitment to or withdrawal from a joint operation, and the arrangement of battles and major operations to achieve operational and strategic military objectives. Among the many considerations, it requires commanders to answer the following:

- **ENDS** - What conditions are required to achieve the objectives?
- **WAYS** - What sequence of actions is most likely to create those conditions?
- **MEANS** - What resources are required to accomplish that sequence of actions?
- **RISK** - What is the likely cost or risk in performing that sequence of actions?

Strategic objectives/goals and the 12 principles of joint operations bound the commander's operational art, which guides war fighting at the strategic, operational, and tactical levels of war and is derived from experience across the range of military operations.

In generic terms, operational art at sea is that component of military art concerned with the theory and practice of planning, preparing, and conducting major operations and maritime campaigns aimed at accomplishing operational or strategic objectives in a given part of a maritime operations area. Only by applying tenets of operational art is it possible to accomplish objectives determined by national strategy and policy in the most decisive manner and with the fewest losses in personnel and material by friendly forces. The main role of operational art is to properly prioritize, sequence, and synchronize or orchestrate the use of all available military and nonmilitary sources of one's power.

See following pages (pp. 5-12 to 5-13) for discussion of the elements of operational art and design.

B. Operational Design

Operational design provides a deep understanding of the problem to be solved and sets the conditions for tactical success. Operational planning — particularly for extensive operations that require a campaign — uses various elements of operational design to help commanders and staffs visualize the arrangement of force capabilities in time, space, and purpose to accomplish the mission. Operational design is the conception and construction of the framework that underpins an operational plan and its subsequent execution. While operational art is the manifestation of informed vision and creativity, operational design is the practical extension of the creative process. Together they synthesize the intuition and creativity of the commander with the analytical and logical process of design. Operational design must include:

- Understanding the strategic guidance
- Identifying the enemy's critical strengths and weaknesses
- Developing an operational concept that will achieve strategic and operational objectives

These three key considerations become the framework for planning a campaign or major operation, identifying the enemy's centers of gravity and critical factors, and developing an operational concept to achieve strategic objectives. It is intrinsic in the Navy planning process (NPP). The NPP is aligned with joint planning guidelines and with planning procedures of JP 5-0, Joint Operation Planning. Thus, Navy commands following the process can effectively integrate into and operate as part of a joint force. NPP, described in NWP 5-01, Navy Planning, provides a logical set of planning steps through which the commander and staff interact. It supports operational design by establishing design elements to assist the commander and staff in visualization and shaping of the operation to accomplish strategic objectives. These elements of operational design comprise a tool that is particularly helpful in visualizing what the campaign should look like, shaping the commander's intent, and determining courses of action (COA). These provide a basis for selecting a COA and developing the detailed concept of operations (CONOPS).

Success for the operational commander and staff is directly related to execution of operational art and design. Throughout these efforts, the following five principles are highlighted as keys for successful operational design:

- Determine what needs to be done and why, but not the tactical specifics of how
- Organize subordinate commanders to take best advantage of all of the military force capabilities
- Articulate the geometry of the operational environment to provide sufficient control measures in terms of boundaries and fire control measures without over-controlling the fight
- Establish command relationships that promote interdependence among the components, instill a "one team, one fight" mentality, provide command authority commensurate with responsibilities, and build trust and confidence
- Decentralize authorities to empower subordinates to operate within the commander's intent and take advantage of unforeseen opportunities

Inputs to Operational Design

Operational commanders and their staffs apply operational art — supported by their skill, knowledge, and expertise — to design campaigns and operations, and organize and employ military forces. The NPP discussed in NWP 5-01 provides the foundation for Navy planning at the operational level. In addition to doctrine and tactical expertise, operational commanders and their staffs need to have a clear understanding of the strategic end state and a systems perspective of the operational environment.

See following pages (pp. 5-12 to 5-13) for discussion of the elements of operational art and design.

Elements of Operational Art & Design

Ref: JP 5-0, Joint Operation Planning (Dec '06), chap. IV.

Operational art encompasses operational design — the process of developing the intellectual framework that will underpin all plans and their subsequent execution. The elements of operational design are tools to help supported JFCs and their staffs visualize what the joint operation should look like and to shape the commander's intent. The emphasis applied to an operational design's elements varies with the theater's strategic objectives. The strategic environment is not the only factor that affects operational design. Other factors such as the availability of HNS, diplomatic permission to overfly nations and access en route air bases, the allocation of strategic mobility assets, the state of the theater infrastructure, and forces and resources made available for planning all have an impact on the operational design. In the final analysis, the goals of a sound operational design are to ensure a clear focus on the ultimate strategic objective and corresponding strategic COGs, and provide for sound sequencing, synchronization, and integration of all available military and nonmilitary instruments of power to that end.

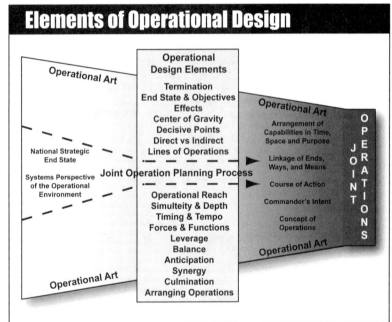

Ref: FM 5-0, Joint Operation Planning, fig. IV-1, p. IV-5.

The commander and staff use the tenets of operational art and design to define the mission (what the commander has been told to do and the reason for it) and assemble/examine information relating to the mission. This information constitutes the initial estimate of the situation. Using this information the staff conducts a mission analysis and the commander formulates an operational idea. The operational idea is provided to the staff as the commander's planning guidance for the development of courses of action. Also critical for follow-on plan development, the initial estimate of the situation provides a loose collection of diverse references that the operational commander and staff can consider in the development of the basic plan and continuous estimate-of the situation refinement.

Refer to The Joint Forces Operations & Doctrine SMARTbook, chap. 3 for complete discussion of the elements of operational art and design.

An Estimate of the Situation

The commander and staff use the tenets of operational art and design to define the mission (what the commander has been told to do and the reason for it) and assemble/examine information relating to the mission. This information constitutes the initial estimate of the situation. Using this information the staff conducts a mission analysis and the commander formulates an operational idea. The operational idea is provided to the staff as the commander's planning guidance for the development of courses of action. Also critical for follow-on plan development, the initial estimate of the situation provides a loose collection of diverse references that the operational commander and staff can consider in the development of the basic plan and continuous estimate-of the situation refinement. Information collected from exercising operational art and design is cataloged as follows:

- **Determination of specified, implied, and essential tasks.** Specified tasks are specifically assigned to a unit by higher headquarters. Implied tasks emerge from analysis of the order, the commander's guidance, and the enemy. Those tasks that most contribute to mission success are deemed essential, and they become the central focus for operations planning. Essential tasks are those that define mission success and apply to the force as a whole.
- **Ultimate and intermediate objectives.** The first and the most important step in designing a major operation is to properly determine and articulate its ultimate and intermediate objectives. The ultimate objective defines operational success. Intermediate objectives support the accomplishment of the ultimate objective.
- **Force requirements.** An important element of operational design is calculation of the overall size and force mix for the entire campaign or major operation. The principal factors in this process are the type of operation, combat potential of friendly and enemy forces, the number and scale of intermediate objectives and their sequencing, the distances between the base of operations and the prospective operating area, and meteorological conditions.
- **Balancing of operational factors against the objectives.** The operational commander and the planners must also properly balance friendly factors of space, time, and force, first against the ultimate objective and then against each intermediate objective.
- **Identification of the enemy and friendly operational centers of gravity, critical vulnerabilities, and decisive points (DPs).** After determining the appropriate objective of a major operation, the operational commander and his planners must determine corresponding enemy and friendly operational centers of gravity, the source of strength — physical or moral — required to achieve the objective.
- **Initial lines of operations (LOOs).** Physical LOO's may be either interior or exterior. Interior LOOs originate from a central position. A force operates on exterior lines when its operations converge on the adversary.
- **Direction (axis).** An important element of a design for a maritime offensive campaign or major offensive operation aimed to seize control of a part of the coast, large island, or archipelago is a direction or axis, a swath of land, and the sea/ocean area and airspace above it extending from one's base of operations to the ultimate physical objective via selected intermediate objectives.
- **The operational idea.** The operational idea (or scheme) is the very core of a design for a major operation. It is the framework upon which a CONOPS is built.
- **Operational sustainment.** How forces will be sustained and how the military effort will be maintained is an integral part of operational design.
- **Decisive Points.** Identification of decisive points (DPs) remains an important feature of the COG analysis and its subsequent defeat or neutralization. Joint doctrine defines decisive point as "a geographic place, specific key event, critical factor, or function that, when acted upon, allows commanders to gain a marked advantage over an adversary or contribute materially to achieving success."

(Naval Planning) Navy Planning Overview 5-13

VI. Design in Military Operations

Ref: JWFC Pamphlet 10, Design in Military Operations (Sept '10).

Design is a methodology for applying critical and creative thinking to understand, visualize, and describe complex, ill-structured problems and develop approaches to solve them. It is a repeatable methodology of reasoning that helps commanders understand how to change a complex-adaptive system from "what is now" to "what is feasible and better"— from the conditions in the operational environment when operations begin (the observed system) to the conditions intended when operations end (the desired system). The difference, or gap, between the current and desired system states is the problem commanders and staffs must solve — how to bridge this gap. In its purest form, design is creative and critical thinking that builds a current and coherent understanding of the relevant relationships in the target environment. Thinking about the environment in terms of complex adaptive systems can help commanders and planners understand the operational environment and key relationships.

Once they understand the environment and the true nature of the problem, commanders consider how to solve the problem. This involves determining which factors in the broad environment are relevant to the current operation and the problem at hand. "Framing" is a term sometimes used to include these relevant factors and exclude others. Commanders identify those actors, tensions, and forces that might support, oppose, or are otherwise affected by potential solutions, and then visualize a broad operational approach to achieve the best solution. They capture their understanding and visualization in their planning guidance, which subordinate commanders, staff, and others will use in subsequent detailed planning.

Because commanders and staffs cannot predict with certainty how their actions will change the environment, particularly when the opposition begins to react, they maintain a posture of skepticism toward the finality of any solution, and remain prepared to reframe their understanding of the environment, the problem, and the broad operational approach as evidence accumulates that the system is not responding as expected. We must try to understand how the actions in one part of the system can affect the system as a whole.

1. Frame the Environment

Framing the environment establishes context for describing the problem and developing an operational approach. The term "framing" is used to indicate the process of identifying the relevant aspects of the environment and distinguishing them from the aspects that are not relevant to the operations at hand. In framing the environment, commanders and staffs review relevant directives, documents, data, guidance and any assigned tasks. If required, commanders and staffs inform their higher authority of new information or the basis for differing perspectives of the environment. In particular, commanders and staffs collaborate with their superiors to resolve differences of interpretation of higher-level objectives and the suitability of available ways and means to accomplish them based on an understanding of the directing authority's motivations and intentions underlying the tasks or assigned missions. Combatant commanders and staffs, and national leaders, may have a clear strategic perspective of the problem, while operational level and tactical commanders and staffs often have a better understanding of local circumstances.

The environmental frame depicts the **observed system** (the current state of the environment), identifying the tendencies and potentials of relevant actors and operational variables that define current system behavior and possibilities for change. Based on higher guidance, the environmental frame also defines the set of conditions that constitute the **desired system** (the desired future state of the environment) that would meet the intentions of the directing authorities. A **condition** is a reflection of an aspect of observed and desired systems. In other words, the observed system is typically comprised of a number of existing conditions, while the desired system is comprised of a set of potential desired conditions.

Framing the environment is continuous. Over time it can reveal the dynamic nature of human interactions and the importance of identifying contributing factors. Useful items to consider during environmental framing include (but are not limited to):
- **Higher Headquarters Design Products**
- **Systems Perspective.** A systems perspective of the operational environment strives to provide an understanding of interrelated systems (political, military, economic, social, information, infrastructure, culture, language, religion, and others) relevant to a specific joint operation without regard to fixed geographic boundaries.
- **Key Actors, System Tendencies, Potentials, and Relationships.** Commanders and staffs focus on understanding key aspects of the system including groupings, relationships, or interactions among **relevant actors**.
- **Tendencies and Potentials.** In developing understanding of the interactions and relationships of **relevant actors** in the operational environment, commanders and staffs consider natural tendencies and potentials in their analyses. A tendency is the inclination to make decisions or behave in a certain manner. A potential is the inherent ability or capacity for growth or development of a specific interaction or relationship.

2. Frame the Problem

Once armed with an understanding of the environment, the design effort shifts to problem framing. The essential activities in framing the problem continue to be thinking critically and conducting open, frank discussion with stakeholders, considering their diverse perspectives, and thereby discovering and understanding the underlying nature of the problem. In framing the problem, commanders and staff address as a minimum:
- What systems (related conditions, actors, or relationships) may oppose us? Commanders and staffs refer back to their understanding of the environment to identify all the actors and influences (friendly, neutral, and hostile) that may impede movement from the existing state to the desired state.
- What systems may help us?
- What organizational challenges and requirements must we accommodate?
- What resources we can draw upon to achieve our goals?
- How much time is allocated by higher authority for solving the military problem?

Refining the commander's understanding extends beyond analyzing interactions and relationships in the environment. It also identifies areas of tension and competition—as well as opportunities and challenges—that commanders and staffs must address to transform current conditions toward the desired system. Tension is the resistance or friction among and between actors (tension may also be positive).

The **problem frame** is a refined component of the environmental frame that helps the commander define, in text and graphics, the broad areas for action that will transform existing conditions toward desired conditions that should comprise the desired system. These areas for action can provide the basis for the eventual functional lines of effort and geographic lines of operations that can be the centerpiece of the operational approach.

A concise **problem statement** clearly defines the problem that must be overcome to achieve the desired transformation. The problem statement broadly describes the requirements for transformation, changes in the operational environment, and critical transitions. The problem statement accounts for the key time and space relationships inherent in the problem frame.

3. Develop the Operational Approach

As the commander and staff gain an understanding of the problem within the context of the operational environment, potential solutions should become evident. The configuration of tensions, competition, opportunities, and challenges may reveal ways to interact with various aspects of the environment in order to transform it to the desired system. Analyzing these options often requires coupling potential actions to a problem by quickly wargaming their possible outcomes.

Design in Military Operations (Continued)

Ref: JWFC Pamphlet 10, Design in Military Operations (Sept '10).

Considering broad approaches to help solve the problem provides focus and sets boundaries for the selection of possible actions that lead collectively to achieving the desired system. The staff converges on the types and patterns of actions necessary to achieve the desired system by creating a conceptual framework that links desired conditions to potential actions. Likewise, the entire staff considers how to orchestrate actions to solve the problem in accordance with the broad approach.

The operational approach is a description of the broad actions that will create the conditions that define the desired system. This operational approach provides the logic that enables commanders and staffs to begin visualizing and describing possible combinations of activities and tasks to reach the desired system, given the factors, actors, relationships, and tensions identified in the environmental and problem frames.

- **Initiative**. In developing an operational approach, commanders and staffs consider how potential actions will enable the force to maintain the initiative. The staff evaluates what combination of actions might derail opposing actors from achieving their goals while moving the observed system toward the desired system. This entails evaluating an action's potential risks and the relevant actors' freedom of action. Likewise, identifying the possible emergence of unintended consequences or threats, commanders and staffs may discover exploitable opportunities.
- **Resources and Risks**. The staff provides an initial estimate of the resources required to support the operational approach. The commander and staff identify and consider risk throughout the iterative application of design.

4. Document the Results

Commanders and staffs link the design effort to subsequent planning by issuing guidance to the joint force staff and components and to other interested stakeholders. The commander's planning guidance can be as detailed or as broad as the commander desires, and could vary significantly according to the nature of the operation. It often includes at least the mission statement, commander's intent, assumptions, operational limitations (constraints and restraints), and results of the design effort. The primary design product could be in the form of a **design concept** or similarly named product, since it describes the commander's visualization of a broad operational approach for moving from the observed system to the desired system.

5. Reframe (as Required)

Reframing is a process of revisiting earlier design hypotheses, conclusions, and decisions that underpin the current operational approach. In essence, reframing reviews what the commander and staff believe they understand about the operational environment, the problem, and the desired system toward which the force is moving (hypothesis). During execution, commanders and staff use **reframing indicators** as they continuously monitor and evaluate their design, plans, and actions against this baseline to detect significant unanticipated changes. If required, they adjust the operational approach to ensure alignment with the desired direction and determine whether that direction itself remains relevant to the environment and the higher commander's desires and expectations. Generally, the decision to reframe can be triggered by factors such as:

- An assessment challenges the commander's and staff's understanding of the operational environment, existing problem, or relevance of the operational approach
- A scheduled periodic review shows a problem
- Failure to make required progress
- Key assumptions or hypotheses prove invalid
- Unanticipated success
- A major event causes "catastrophic change" in the environment

I. Mission Analysis

Ref: NWP 5-01, Navy Planning (Jan '07), chap. 2.

As the first step, mission analysis drives the entire planning process. Its purpose is to give the naval commander, staff, and planning team an overall assessment of the situation. Mission analysis begins with a review of orders, plans, intelligence products, and guidance provided by higher headquarters in order to produce an operations mission statement and to identify tasks necessary to accomplish the operational mission. Following the mission analysis briefing, this step in the planning process ends when the commander issues planning guidance and a warning order (WARNORD), initiating the COA development process.

I. Mission Analysis

Key Inputs	Key Outputs
▪ Higher Headquarters 　▪ Plan, orders and guidance 　▪ Intelligence products 　▪ Staff estimates ▪ Navy Commander 　▪ Initial guidance	▪ Approved mission statement ▪ Commander's planning guidance ▪ Commander's intent ▪ Commander's critical information requirements ▪ Warning order

1. Identify Source(s) of the Mission
2. Determine Support Relationships
3. Analyze the Higher Commander's Mission
4. Determine Specified, Supplied and Essential Tasks
5. State the Purpose
6. Identify Externally Imposed Limitations
7. Analyze Available Forces and Assets
8. Determine Critical Factors, Centers of Gravity and Decisive Points
9. Develop Planning Assumptions
10. Conduct Initial Risk Assessment
11. Develop Proposed Mission Statement
12. Conduct Mission Analysis Briefing
13. Develop Initial Commander's Intent
14. Develop Commander's Critical Info. Requirements (CCIR)
15. Develop Commander's Planning Guidance
16. Develop Warning Order

Ref: NWP 5-01, Navy Planning, fig. 2-1, p. 2-1.

Before the commander and planning team can begin mission analysis, they must understand the possible area of operations (AO), probable mission, available forces, and political, military, and cultural characteristics of the area. The planning team acquires information from higher headquarters, national-level intelligence sources,

(Naval Planning) I. Mission Analysis 5-17

other military and interagency governmental organizations, nongovernmental organizations, and their innate knowledge of the operational environment. Acquiring higher headquarters intelligence preparation of the operational environment (IPOE) products is essential during the early stages of mission analysis. If operations are already underway, the commander and the planning team will have to rapidly acquire essential information to gain understanding and situational awareness.

A thorough mission analysis focuses the activities of the commander, staff, and planning team, thereby saving time and effort. To plan effectively, planners should have access to all documents relative to the mission and AO, such as on-the-shelf OPLANs, standing rules of engagement (ROE), standard operating procedures (SOPs), and existing IPOE documents. Additionally, the staff should begin developing preliminary staff estimates early in the mission analysis phase.

Mission analysis should provide answers to the following questions:

- What tasks must the command do for the mission to be accomplished?
- What is the purpose of the mission received?
- What limitations have been placed on our own forces' actions?
- What forces/assets are available to support the operation?
- What additional assets are needed?

Inputs

Gathering specific direction and guidance from higher headquarters, as well as the commander's guidance, initiates the mission analysis process.

A. Higher Headquarters Plans, Orders, and Guidance

A verbal or written directive from higher headquarters provides the necessary initial information needed for mission analysis. This information is normally contained in OPLANs, concept plans (CONPLANs), WARNORDs, and/or OPORDs. If higher headquarters directives or guidance is unclear, the commander, staff, or planning team should immediately seek clarification. Liaison officers can provide valuable information and should actively facilitate the NPP.

B. Higher Headquarters Intelligence Products

Higher headquarters intelligence products form the basis of the naval commander's own intelligence support. Higher headquarters intelligence products include joint intelligence preparation of the operational environment (JIPOE) materials, which are derived from the intelligence estimate, intelligence summaries, and Annex B of an OPORD. At a minimum, higher headquarters JIPOE should provide identification and analysis of the enemy's objectives, critical strengths and critical weaknesses, centers of gravity (COGs), critical capabilities, critical requirements (CRs), and critical vulnerabilities (CVs). It also should estimate the ECOAs that are most likely to be encountered based on the current situation. Joint intelligence preparation of the operational environment products from higher headquarters and those from the commander's intelligence staff may include the modified combined obstacle overlay (MCOO) and threat situation templates.

See pp. 5-8 to 5-9 for an overview of Intelligence Preparation of the Operational Environment (IPOE) process.

C. Higher Headquarters Staff Estimates

A higher headquarters staff also assembles and continuously modifies estimates that pertain to the individual functions of each staff code. Based on the stated mission, operational environment situation, and enemy and friendly forces, these estimates provide a current status and an assessment of the ability to meet the requirements of the mission assigned, identify shortfalls and potential issues, and weigh the various COAs as they pertain to the specific staff code functions. For instance, the logistics staff at a higher headquarters creates an estimate of the force's logistics capabilities. This estimate then assists the subordinate in developing a staff estimate specifically focused on the logistics of the subordinate force's logistics capabilities. See Appendix K for selected staff estimate formats. Depending on the level of command and the time available, the staff estimate may be a formal detailed written document or an informal oral presentation. No matter what form it takes, its value to the planning process cannot be overemphasized.

Refer to NWP 5-01, Navy Planning, app. K for sample staff estimate formats.

D. Commander's Initial Guidance

The Navy commander not only receives guidance and products from higher authority, but also is responsible for developing his own initial guidance to support mission analysis. In order to generate this initial guidance, the commander assesses the situation based on higher headquarters direction and his view of the operational environment, understanding of the enemy, and personal experience. The commander also must consider the capabilities of the forces assigned/attached—combat readiness, material support, unit morale, and other factors—to accomplish the mission. This mental exercise can consist simply of the commander's thoughts or can be a detailed analytical effort. Regardless, the commander develops a vision of the operational environment that then helps to produce guidance to the naval force. *The Marine Corps Planning Process (MCWP 5-1) has a similar step that is referred to as the commander's battle space area evaluation (CBAE).*

Once the analysis is completed, the commander should issue the commander's initial guidance. Ideally, the commander personally issues this guidance to the staff, planning team, and subordinate commanders and seeks immediate feedback to ensure that the vision of the operational environment and guidance is understood. Depending on the time available, the commander may provide either general or specific guidance for the planning team, staff, and subordinate commanders to consider (e.g., an enemy COG, a certain command and control relationship, etc.). It is critical to establish a realistic timeline and to adhere to it. Also consider that time will be needed for subordinate-level planning (1/3-2/3 rule). Generally, the commander's initial guidance should include specific force employment considerations, any additional planning limitations, a general assessment of the potential mission and related tasks, a general assessment of the operational environment, and an initial commander's intent.

Process

With inputs from higher headquarters and an understanding of the commander's guidance, the planning team formally conducts the mission analysis process. This is a flexible process, normally consisting of several non-sequential activities tailored to the situation, time available, and the commander's guidance. The process is neither rigid nor static; it is continuous, evolving, iterative, and dynamic. The commander's planning team gathers information that is continuously refined throughout the planning process, and the staff uses this information to prepare functional staff estimates, which provide a logical and orderly examination of all factors that affect mission accomplishment. At the conclusion of mission analysis, the planning team should develop proposed commander's planning guidance and a mission analysis briefing for the commander and staff. The initial step in mission analysis is determining and refining known facts. Much of this can be derived from higher headquarters directives, estimates, and intelligence products. The remaining known facts reside within the commander's own staff products (i.e., staff estimates). Together, these facts highlight the potential threat and forces currently available and provide a general idea of the time available for planning before operations commence.

1. Identify Source(s) of the Mission

The source of the mission is normally found in a higher headquarters directive (i.e., OPLAN, OPORD, or WARNORD). Depending on the scope of the operation, consider also reviewing applicable United Nations Security Council Resolutions (UNSCRs), alliance directives, National Security Presidential Decision Directives, and other authoritative sources for additional information. For instance, Operation ALLIED FORCE in Kosovo and Operation IRAQI FREEDOM both involved key UNSCRs that shaped the mission of naval forces in each conflict.

2. Determine Support Relationships

It is critical for the staff and planning team to be clear in their understanding of support relationships. Support relationships typically exist at the operational level between service and/or functional component commanders (i.e., JFMCC and joint force air component commander (JFACC)) but also may be established at the tactical level (i.e., between a CSG commander and an ESG commander). By ascertaining the proper support relationship, the naval commander, staff, and planning team can determine what organization or command is the main effort for an operation. This information provides the necessary chain-of-command information and normally is found in the source of mission document(s), such as an OPORD. The support relationship may change throughout the different phases of an operation.

For example, the second phase of an operation may place the main effort on an amphibious demonstration by a Marine expeditionary unit (MEU) within an ESG. The JFC would identify the JFMCC as the supported commander, and the other functional component commanders would be supporting commanders. However, the third phase of the operation may focus the main effort on an airborne forced entry into an enemy airfield. In the third phase of the JFC's establishing directive, the supported commander is now assigned to the joint force land component commander (JFLCC). The JFMCC would be a supporting commander and may be tasked to provide supporting fires or maintain maritime superiority in the AO to prevent the enemy from bringing in arms and supplies to reinforce its ground forces and conduct a counterattack at the airfield.

3. Analyze the Higher Commander's Mission

The higher commander's mission statement, which is normally contained in Paragraph 2 (mission) of the higher commander's directive, and the capabilities and limitations of the naval force must be studied. The commander must draw broad conclusions as to the character of the forthcoming military action. However, the commander should not make assumptions about issues not addressed in the higher headquarters' directive. If the higher headquarters' directive is unclear, ambiguous, or confusing, the commander must seek clarification. The following is an example of a mission statement from a higher joint force headquarters.

The higher commander's intent normally is found in Paragraph 3 (execution) of the higher commander's directive, although its location in the text may vary. Sometimes the higher commander's intent may not be transmitted at all. When this occurs, the subordinate commander and staff should derive an intent statement and confirm it with the higher headquarters. The intent statement of the higher commander should then be repeated in Paragraph 1 (situation) of the naval commander's own OPORD to ensure that the staff, supporting commanders, and subordinates understand it. Each subordinate commander's intent must be framed and embedded within the context of the higher commander's intent, and they must be nested both vertically and horizontally to achieve a common military end state throughout the command.

Mission Statement & Commander's Intent

Ref: NWP 5-01, Navy Planning (Jan '07), pp. 2-4 to 2-5.

Example Mission Statement

The following is an example of a mission statement from a higher joint force headquarters:

When directed, commander, joint task force (CJTF) BLUE SWORD conducts multinational operations in the joint operations area (JOA) to defeat the Redland 23rd Guards Division and destroy terrorist forces and their infrastructure in Redland in order to eliminate the terrorist base of operations in the region.

Example Commander's Intent

The following is an example of a commander's intent from a JTF commander, joint force commander, or combatant commander.

GENTEXT/EXECUTION//

(U) COMMANDER'S INTENT. THE PURPOSE OF THE OPERATION IS TO ELIMINATE THE TERRORIST BASE OF OPERATIONS THAT OPERATES FREELY IN REDLAND AND THREATENS PINKLAND SOVEREIGNTY.

(U) METHOD: MY DESIRE IS TO NEUTRALIZE CONVENTIONAL REDLAND MILITARY FORCES WITH PRIMARY FOCUS IN THREE DISTINCT AREAS: ENABLERS SUCH AS REDLAND COMMAND AND CONTROL AND LOGISTICS; REDLAND GROUND, AIR, AND NAVAL FORCES STAGED TO CONDUCT AN OFFENSIVE INTO PINKLAND; AND PARAMILITARY AND TERRORIST GROUPS COLLABORATING WITH REDLAND TO ATTACK PINKLAND AND OTHER FRIENDLY FORCES IN THE REGION.

(U) TASK FORCE OPERATIONS MUST PRESERVE THE SOVEREIGNTY OF NEIGHBORING NEUTRAL COUNTRIES AND TAKE ALL NECESSARY STEPS TO MINIMIZE DAMAGE TO CIVILIAN INFRASTRUCTURE WITHIN REDLAND.

(U) WE WILL EXECUTE OPERATIONS THROUGH A JOINT, MULTINATIONAL COALITION AND WILL INTEGRATE OUR OPERATIONS WITH THE GOVERNMENTAL AND NONGOVERNMENTAL ORGANIZATIONS THAT ARE EXERCISING OTHER MEANS OF OUR NATIONAL POWER TO BRING THIS CRISIS TO AN END. OUR COMMAND STRUCTURE WILL BE CLEAR, AND OUR CONTROL WILL PERMIT FULL AND EFFECTIVE COORDINATION AMONG SUBORDINATE ELEMENTS. WE WILL CONTINUOUSLY LIAISE WITH PINKLAND TO SYNCHRONIZE OUR RESPECTIVE OPERATIONS SINCE IT WILL NOT BE WITHIN THE STRUCTURE OF THE COALITION TASK FORCE.

(U) WE WILL MAXIMIZE OUR ABILITY TO LEVERAGE ALL OF THE TOOLS IN OUR KIT BAG, INCLUDING SUPERIOR, PRECISION FIREPOWER AND UNRIVALED MOBILITY, TO DOMINATE THE OPERATIONAL ENVIRONMENT. SPEED AND TIMING ARE ESSENTIAL— TAKE FULL ADVANTAGE OF EVERY OPPORTUNITY IN ORDER TO GAIN MOMENTUM AGAINST REDLAND. I EXPECT MY SUBORDINATE COMMANDERS TO PROVIDE THOROUGH SOLUTIONS THAT ARE PRACTICAL BUT INNOVATIVE AND THAT KEEP THE ELEMENTS OF SPEED AND TIMING AS FUNDAMENTAL INGREDIENTS.

(U) THE END STATE FOR OUR OPERATION IS THE DEFEAT OF THE 23RD GUARDS DIVISION AND THE DESTRUCTION OF THE TERRORIST FORCES AND THEIR CAMPS IN REDLAND. CONDITIONS SHOULD EXIST FOR A STABLE ENVIRONMENT IN REDLAND IN WHICH GOVERNMENTAL AND NONGOVERNMENTAL ORGANIZATIONS CAN HAVE FREE ACCESS TO REDLAND TO HELP TRANSITION THEIR GOVERNMENT TO A NEW CIVIL AUTHORITY.

4. Determine Specified, Implied, and Essential Tasks

Every mission consists of two elements: the tasks to be accomplished by one's own forces and the purpose of those tasks. Before going further, it is necessary to illustrate how tasks, operations, and missions are related.

If a mission or operation has multiple tasks, then the priority of each task should be clearly expressed. Using information provided by higher headquarters and the commander's initial guidance, the planning team identifies specified and implied tasks. Specified tasks are specifically assigned to a unit by higher headquarters.

Ref: NWP 5-01, Navy Planning, fig. 2-2, p. 2-6.

Specified Tasks
Specified tasks are derived primarily from the execution paragraphs of the OPORD, but they may be found elsewhere, such as in the mission statement, coordinating instructions, or annexes. Implied tasks are not specifically stated in the higher headquarters order but must be performed in order to accomplish specified tasks.

Implied Tasks
Implied tasks emerge from analysis of the order, the commander's guidance, and the enemy. Routine, inherent, or SOP tasks are not included in the list of tasks.

Essential Tasks
Those tasks that most contribute to mission success are deemed essential, and they become the central focus for operations planning. Essential tasks are those that define mission success and apply to the force as a whole. Essential tasks can come from either specified or implied tasks. If a task must be successfully completed for the commander to accomplish his purpose, it is an essential task. Only essential tasks are included in the proposed mission statement. The following example shows the three types of tasks that a commander may experience.

Though not elaborated in this example, the planning team also must determine the follow-on tasks that may be required at a later time due to the effects of the operation, the situation in the operational environment, the enemy's actions, and the dynamic nature of the operational environment. These tasks, commonly seen in a directive or guidance as "be prepared to" or "BPT" shape the planning team's efforts as well as the specified, implied, and essential tasks.

Influence of the Task List Libraries on the NPP

Key references that must be considered when generating tasks for the Navy are the Universal Joint Task List (UJTL) and Universal Naval Task List (UNTL). The UNTL includes the UJTL plus the Navy Tactical Task List (NTTL). The UJTL and UNTL provide a wide range of tasks that a joint and a naval force, respectively, can perform across a broad spectrum of operations. Because both lists are composed using joint terminology, there is an understanding at both the joint and naval command level of exactly the types of tasks that a naval force can accomplish.

Paired down from these extensive lists are the specific, mission-essential tasks that a joint and a naval force must be able to perform. These are termed joint mission-essential tasks (JMETs) and Navy mission-essential tasks (NMETs).

The reason the UJTL, UNTL and, more importantly, the JMET and NMET are critical to the NPP is that they delineate the capabilities within and specify the types of missions that can be performed by a naval force. In order to properly conduct the NPP, the commander, staff, and planning team must know what the force can do and to what standards they have trained. Additionally, when communicating with subordinate commands and other services either directly or through directives such as orders and plans, it is important that the naval commander use the terms and types of tasks that are delineated in the UNTL and core Navy mission-essential task list.

5. State the Purpose

The purpose follows the statement of task(s). To clearly delineate the two, the statement "in order to" should be inserted between the task(s) and purpose. Purpose is normally found at the beginning of the execution section of the superior's directive. If the superior's directive also contains an intent statement, it also should be reviewed to help analyze the purpose of the operation. The purpose always dominates the tasks. A task or tasks can be accomplished or changed due to unforeseen circumstances, but the purpose remains essentially the same if the original mission remains unchanged. Purpose should answer the "why" question.

Example

Purpose: In order to facilitate the defeat of the 23rd Guards Division and the destruction of the terrorist forces and their infrastructure in Redland.

6. Identify Externally Imposed Limitations

Limitations may be specifically stated in higher headquarters direction or implied in ROE. They normally are divided into two categories: restraints and constraints. Both directly impact mission analysis and the planning process. Restraints and constraints collectively comprise limitations on the commander's freedom of action. It is also important to note that these limitations are externally imposed and do not include self-imposed limitations. Restraints and constraints may be included in the ROE, commander's guidance, or instructions from higher headquarters.

Possible Limitations

	Limitations	Implications
RESTRAINTS (CAN'T DO)	Do not use mines to close Redland naval base. Do not violate neutral country territorial waters.	Accept risk of Redland naval forces at sea or use other means to close the port. Restricts operational area.
CONSTRAINTS (MUST DO)	Integrate coalition naval forces. Establish maritime superiority no later than D+1.	Increases command and control (C2) complexity. Limits preparation time.

Ref: NWP 5-01, Navy Planning, table. 2-1, p. 2-7.

Commander/Staff Interaction During/After Mission Analysis

Ref: NWC Maritime Component Commander's Handbook (Feb '10), p. 2-9 to 2-11.

As the OPT returns to give the commander the results of MA, the commander must prepare to provide additional feedback and direction. The commander normally has a strategic- and operational-level awareness and professional interactions that differ significantly from those of OPT members. Additionally, while the OPT/staff is closely coordinating with linked OPT/staffs, the commander is in direct communication with those staffs' commanders. The commander needs to consider his or her thoughts on those intangibles that could dramatically change the progression of the operation — nonmilitary instruments of national power, strategic communications (SCs), IO, morale, and readiness of enemy and friendly forces. The MA brief is one of the first opportunities to collectively analyze these various insights.

1. The Mission Analysis Brief Is a Decision Brief

The commander, major subordinate commanders, staff principals, liaison officers (LNOs), and the OPT should be present at the brief. The commander is normally asked to approve a proposed mission statement and commander's critical information requirements, confirm the initial risk assessment, review assumptions, and provide commander's intent and planning guidance. The challenge for the commander is to dedicate the time to carefully reflect and ensure preparedness to make required decisions and advance the OPT into COA development. The commander needs to personally declare what is approved in the brief; e.g., mission statement.

2. Approved Mission Statement

The OPT proposes a mission statement to the commander that provides the who, what, where, when, and why of the assigned mission. It does not include the how, which is generated in COA development by subordinate commands. The commander should focus on the why; if the purpose is correct the rest of the mission naturally flows. The challenge of the purpose is that it is linked to higher operational and strategic objectives and has to be carefully discerned. Next consider the essential tasks within the mission statement. These tasks, if not accomplished, will result in mission failure.

3. Commander's Critical Information Requirements

Review the staff-proposed commander's critical information requirement (CCIRs). Do they include friendly and enemy perspectives on indications that a decision has to be made to continue planning or execute the plan? These CCIRs will evolve, but resources should be applied to satisfy each information requirement.

4. Initial Risk Assessment

The initial risk assessment is a listing of threats and risks and a first-level analysis of their likelihood and consequences. Normally, risk is divided into risk to mission and risk to force. Future planning steps will address how each of these risks will be mitigated or accepted. Resources (either forces or functions) will be applied to mitigate the risks. Discuss with the staff where and how much risk is acceptable.

5. Assumptions

Assumptions have significant planning implications and can sometimes be deceiving. What appears to be common sense early in the planning process may eventually be proved wrong, which can make the validity of the entire plan questionable. Question the OPT as to why an assumption needs to be made and ask battle staff what resources can be applied to prove/disprove the assumption. HHQ assumptions are only "treated as fact"; they do not become facts unless proven true. The JFMCC staff needs to analyze

HHQ assumptions and provide feedback if those assumptions are unrealistic or could overly constrain JFMCC planning.

6. Commander's Intent
The commander should write a clear, concise statement that is understandable by subordinates. Commander's intent must be crafted to allow subordinate commanders sufficient flexibility in accomplishing their assigned mission(s). It must provide a "vision" of those conditions that the commander wants to see after the military action is accomplished. The commander must define how the "vision" will generally be accomplished by the force and assets, and the conditions/status of friendly and enemy forces with respect to the OE as the end state. The commander, and not his staff, writes the best commander's intent.

7. Purpose
The purpose is the reason for the military action with respect to the mission of the next higher echelon. The purpose explains why the military action is being conducted. This helps the force pursue the mission without further orders, even when actions do not occur as planned. Thus, if an unanticipated situation arises, participating commanders understand the purpose of the forthcoming action well enough to act decisively and within the bounds of the higher commander's intent. Review the validity of why the operation is being conducted. Ensure the "why" of the mission statement matches this purpose.

8. Method
The "key tasks," in doctrinally concise terminology, that explain the offensive form of maneuver, the alternative defense, or other action to be used by the force as a whole. As information becomes available, refine task/event sequence, lines of operation, warfighting function synchronization, an emphasized organic or nonorganic capability, or obvious phasing. Details as to specific subordinate missions are not discussed.

9. End State
The operational end state describes the set of required conditions that indicates the achievement of operational objectives. It should address what the environment, friendly forces, and enemy forces will look like at the end of the operation. A preliminary end state (military end state) describes the conditions when military force is no longer the principal means to achieving the strategic aim. The type of end state the commander focuses on depends on the command's position/responsibilities. As the JFMCC, the end state is normally a military end state and may describe phase-change conditions.

10. Planning Guidance
PG is focused on advancing the OPT into COA development. The PG will direct the staff to develop options that comply with the commander's direction. Due to time constraints the commander may also direct the OPT to develop or avoid specific COAs to avoid wasted staff effort. The commander's PG must focus on the essential military tasks and associated objectives that support the accomplishment of the assigned mission.

The guidance should be published in written form. No format for the planning guidance is prescribed; however, the guidance should be sufficiently detailed to provide clear direction and to avoid unnecessary effort by the staff or subordinate commanders. The more detailed the guidance, the more specific staff activities will be. And the more specific the activities, the more quickly the staff can complete them.

See p. 5-33 for further discussion of commander's planning guidance.

11. Post Mission Analysis Actions
Issue a warning order (WARNORD) to formalize decisions and direct subordinate action. Submit to HHQ request for information (RFI), request for forces (RFF) (force shortfalls), or immediate desired actions (deployments).

7. Analyze Available Forces and Assets

Commanders and their staffs must review the forces that have been provided for planning, identify their locations (if known), and initially determine if there is a need to modify the current task organization and support relationships. This is also the time to determine what and when reserve forces will be available. When accomplishing this, the staff should refer to the specified and implied tasks to identify what broad force structure and capabilities are necessary to accomplish these tasks. This is just the initial analysis in which the staff identifies any potential shortfalls between the tasks and forces available to carry them out. More detailed requirements are addressed in the Course of Action Development.

Examples

Task: Establish maritime superiority in the AO in order to allow the transit of joint forces into the JOA.

Forces: List all forces assigned. Notional CSG: command element, destroyer squadron, aircraft carrier, carrier air wing, 3 CRUDES, 1 submarine, 1 TAO/ TAOE; GBR maritime task group (2 x DD, 1 x SSN) located in AO

Observation: Tactical control (TACON) of GBR SSN needs to be shifted to JFMCC for antisubmarine warfare (ASW) planning and execution; JFACC needs to be supporting commander and provide 2 x KC-10/135, 1 x AWACS, 1 x GLOBAL

During this initial analysis, planners must evaluate the ability of the command to successfully accomplish the mission, given available forces and resources including combat and support. They should identify critical shortfalls in either of the two areas or in the areas of specific subject matter or technical expertise and request support from higher headquarters. An initial assessment of the anticipated C2 structure and relationship is important during mission analysis because it sets the stage for friendly COA execution.

8. Determine Critical Factors, Centers of Gravity, and Decisive Points

The next step in the mission analysis process requires a progressive analysis of three key components: critical factors, COGs, and decisive points (DPs). It is important to remember also that each of these elements is applicable at the various levels of war—strategic, operational, and tactical. It is important for the commander to be cognizant of the strategic and tactical aspects of the critical factors and COGs, but focus must be mainly on the operational aspect.

Determination of critical factors leads to finding the COGs, which in turn leads to assessing DPs. This analysis must be done for both the enemy and friendly naval forces. While the intelligence staff identifies enemy critical strengths and weaknesses and COGs as part of the IPOE process, subordinate commanders and the planning team assist the commander in identifying friendly critical factors and COGs.

Once the initial COG assessment is completed, planners should determine critical capabilities and CRs. By taking the analysis down to this level of detail, the planners can then start to determine the critical vulnerabilities (CVs) that are deficient or vulnerable to attack. To be a critical vulnerability, it must allow for decisive or significant results. While determining CVs for the enemy begins to define the CONOPS, it is also important to note friendly vulnerabilities in order to properly protect the friendly COGs.

Refer to NWP 5-01, Naval Planning, app. C for discussion of COG analysis: objectives, critical factors, critical capabilities, critical requirements, critical vulnerabilities, and decisive points. Refer to NDP 1, Naval Warfare, for a more detailed discussion on the role of the COG analysis in operational design.

9. Develop Planning Assumptions

Assumptions are made for both friendly and enemy situations. They encompass issues over which a commander normally does not have control. A valid assumption should answer the following questions: Is it logical? Is it realistic? Is it essential for planning to continue? Does it avoid assuming away an enemy capability?

Assumptions are used in planning at every level. Subordinate commanders must treat assumptions given by higher headquarters as facts. While commanders can assume the success of friendly supporting forces, they cannot assume success for their own. As planning continues, additional assumptions may be needed, and previous assumptions may be discarded. Keep a record of assumptions in order to track and validate them as they are confirmed or disapproved. If assumptions cannot be validated before execution, they become part of the inherent risk of the operation and may require branches. During COA development, the commander may require the planning team to develop operations branches for all assumptions pertaining to ECOAs.

Examples

Country _____ will remain neutral but will deploy the major part of its naval forces near the AO.

Country _____ will (not) permit overflight for carrier-based tactical aviation and Tomahawk land attack missile (TLAM).

Strait of _____ will remain open during hostilities for all friendly shipping.

Country _____ will (not) allow basing of ships if they do (not) conduct combat missions against country _____.

10. Conduct Initial Risk Assessment

During mission analysis, the commander conducts a personal initial risk assessment. Risk is inherent in any use of military force or routine military activity. Risk falls into two broad categories: risk to mission and risk to forces (i.e., force protection (FP)).

Commanders, their staffs, and planning teams should identify and assess potential risk so that they can take the appropriate steps to mitigate it. This can be accomplished by conducting a vulnerability analysis or vulnerability assessment. This risk may be stated or implied in higher headquarters intent or guidance. Risk also may be determined from individual staffing. While the staff is involved in the risk assessment, it is the commander who ultimately has to determine how and where risk will be accepted.

See pp. 4-16 to 4-17 for an expanded discussion of risk management.

11. Develop Proposed Mission Statement

Based on mission analysis, the planning team drafts a restated mission for the commander to review, edit, and approve in concert with (or following) the mission analysis briefing. The mission statement should be a clear and concise statement of the essential tasks along with the purpose of those tasks. If the mission contains multiple tasks, they should be listed in the sequence that they are to be accomplished. A proper mission statement should contain the following items: who (which forces) will execute the mission, what type of action (e.g., defend) is contemplated (include essential tasks only), when will the action begin, where will the action occur (AO), why (purpose) each force conducts its part of the operation (including objectives).

Example: Carrier Strike Group (CSG)-Level Mission Statement

Mission Statement: On order (when), CTF BLUE SWORD (who) supports Deception Plan X-Ray and establishes maritime superiority (what) in the BLUE SWORD JOA (where) in order to facilitate the defeat of the 23rd Guards Division and the destruction of the terrorist forces and their infrastructure in Redland (why).

(Naval Planning) I. Mission Analysis 5-27

12. Conduct Mission Analysis Briefing

The planning team presents a mission analysis briefing to the commander and staff to obtain approval of the mission statement, intent, and follow-on planning guidance. The mission analysis briefing reviews the specific products developed and refined during mission analysis before proceeding to COA development. Additionally, consider reviewing the following: operational environment situation update; intelligence estimate and IPOE products (including enemy COGs, ECOAs, and DPs); higher headquarters mission and commander's intent; commander's guidance, purpose, and tasks (specified, implied, essential); assumptions, limitations (restraints and/or constraints), and ROE; force structure and shortfalls (combat forces, support resources, subject matter experts); initial staff estimates across functional areas (logistics; transportation; communications system support; intelligence, surveillance and reconnaissance (ISR); personnel; etc.); friendly COG analysis to include DPs, request for information (RFI), and operational information requests; recommended commander's critical information requirements (CCIRs), priority intelligence requirement (PIR), and friendly force information requirement (FFIR); and proposed mission statement.

The mission analysis briefing ensures a common and thorough understanding of the proposed mission and tasks along with the underlying mission analysis. The briefing focuses on relevant conclusions reached throughout the analysis process and creates a common understanding and focus for the follow-on planning. The briefing format reflects some of the typical information often found in a mission analysis briefing. Exact content varies based upon the level of command, type of operation, organization SOP, and the commander's needs.

See facing page (p. 5-29) for sample briefing format.

13. Develop Initial Commander's Intent

A commander's intent is broader than the mission statement; it is a concise, free-form expression of the purpose of the force's activities, the desired results, and how actions will progress toward that end. It is a clear and succinct vision of how to conduct the action. In short, the commander's intent links the mission and the CONOPS. The intent expresses the broader purpose of the action that looks beyond the why of the immediate operation to the broader context of that mission, and it may include how the posture of the force at the end state of the action will transition to or facilitate further operations (sequels).

Commander's intent is not a summary of the CONOPs. It should not tell specifically how the operation is being conducted but should be crafted to allow subordinate commanders sufficient flexibility and freedom to act in accomplishing their assigned mission(s) even in the "fog of war." While there is no specified joint format for commander's intent, a generally accepted construct includes the purpose, method, and end state.

Purpose
The reason for the military action with respect to the mission of the next higher echelon. The purpose explains why the military action is being conducted. This helps the force pursue the mission without further orders, even when actions do not unfold as planned. Thus, if an unanticipated situation arises, participating commanders understand the purpose of the forthcoming action well enough to act decisively and within the bounds of the higher commander's intent.

Method
The "how," in doctrinally concise terminology, explains the offensive form of maneuver, the alternative defense, or other action to be used by the force as a whole. Details as to specific subordinate missions are not discussed.

End State
Describes what the commander wants to see in military terms after the completion of the mission by the friendly forces.

Sample Mission Analysis Briefing

Ref: NWP 5-01, Navy Planning (Jan '07), fig. 2-3, p. 2-10.

BRIEFER	SUBJECT
CofS or N-5/N-3	Purpose and Agenda
	Area of Operations (AO)
J-2/N-2	Initial Intelligence Estimate Brief: terrain analysis, meteorological and oceanographic (METOC) analysis, threat integration with situation templates, enemy's COGs, and enemy ECOAs.
J-5/J-3, N-5/N-3	Higher Headquarters' Mission and Intent
	Facts: Source(s) of the mission, and supporting and supported command relationships
	Assumptions; Limitations: restraints, can't do and/or constraints, must do
	Specified, implied, and essential tasks
	Available forces and assets and noted shortfalls (U.S. and coalition)
	Centers of gravity DPs (friendly)
	Initial force movement control center (FMCC) force structure analysis
	Risk assessment and vulnerability assessment
	End state; Proposed mission statement; Proposed initial CCIR
	Time analysis including projected planning milestones
	Conclusions: shortfalls and war-stoppers, recommendations
J-1/N-1	Current Manning
	Facts: personnel strengths and morale, replacements and medical returned to duty (RTD), critical shortages
	Assumptions: replacements, coalition support, other
	Conclusions: projected strengths on D-day, projected critical Navy enlisted classification (NEC) status on D-day, shortfalls, war-stoppers, recommendations
J-4/N-4	Sustainment
	Facts: Class I, II, III(p), IV, VI, VII, X status, status of supply services, critical shortages
	Assumptions: resupply rates, host nation support, other
	Conclusions: projected supply level status on D-day, shortfalls, warstoppers, projected treatment capability, recommendations, ordnance/weapons
	Facts: Class V status, distribution system, restrictions, critical shortages
	Assumptions: resupply rates, host nation support, other
	Conclusions: projected supply status on D-day, projected distribution system, shortfalls and war-stoppers, and recommendations
	Fueling
	Facts: Class III(b) status, distribution system, restrictions, critical shortages
	Assumptions: resupply rates, host nation support, other
	Conclusions: projected supply status on D-day, projected distribution system, shortfalls and war-stoppers, recommendations
	Fixing
	Facts: maintenance status (equipment readiness); class IX status; repair times, evacuation policy, and assets; critical shortages
	Assumptions: coalition support, other
	Conclusions: projected maintenance status on D-day
J-6/N-6	Communications Architecture and Status
	Facts: operational status of communications circuits and command, control, communications, computers & intelligence (C4I) systems; bandwidth allocation; communications paths for various C2 functions; planned outages and degradations
	Assumptions: bandwidth stability, C4I system reliability Conclusions: projected C4I systems and communications status during operations, impact of loss or degradation of C4I systems or communications
Medical	Facts: MEDEVAC procedures, lay down of medical treatment capabilities and resources, and critical environmental health concerns (prevalent diseases, hazardous animals, pollutants, potability of local water sources)
	Assumptions: aircraft to move injured or sick, shore-based med. facilities w/i flying distance
	Conclusions: critical shortages in supplies or personnel, number of wounded and sick that organic medical services can handle
Others	Others as Appropriate to the Mission
COS or J-3/N-3	Proposed Restated Mission; Commander's Guidance Requested

The commander is responsible for formulating the single unifying concept for a mission. Having developed that concept, the commander then prepares an intent statement from the mission analysis, the intent of the higher commander, and his own vision to ensure that subordinate commanders are focused on a common goal. The task here is to clearly articulate the intent so that it is understandable two echelons below. When possible, the commander delivers it, along with the order (or plan), personally (and/or via video teleconferencing (VTC)).

Face-to-face delivery ensures mutual understanding of what the issuing commander wants by allowing immediate clarification of specific points. While intent is more enduring than the CONOPs, the commander can, and should, revise intent when circumstances dictate. The following offers an example of a Navy component commander's intent (it is associated with the commander's intent).

The commander's intent is essential to focus further planning, and it enables the commander to indirectly control events during the execution of the operation. While brevity and clarity are imperative, the intent must be crafted so that commanders two levels down have the flexibility to accomplish their mission in lieu of further guidance. For an NCC or JFMCC commander, this means that a level down to a task force (e.g., CTF) should be considered when writing the commander's intent. For a CTF, the commander needs to consider down to the unit level.

See facing page (p. 5-31) for sample Navy Component Commander's Intent format.

14. Develop Commander's Critical Information Requirements (CCIR)

The planning team must then generate a list of CCIRs. Generation of CCIRs is a vital step in mission analysis and the NPP as a whole, and it is imperative that the planning team pays close attention to the content and wording of this list. Furthermore, CCIRs are constantly evaluated and updated for their relevance and applicability.

Commander's critical information requirements can have a significant impact on the commander's decisions and actions and can influence the course of events for an operation. The list should not be extensive; it must focus solely on the absolutely critical pieces of information that the commander needs to know. The key question to answer when thinking of a CCIR is, "What does the commander need to know and when does he/she need to know it?" Commander's critical information requirements focus the commander's staff, information collection efforts, efficient processing, and flow of information throughout the command. While the planning team can recommend CCIRs, only the commander can approve them.

Commander's critical information requirements are subdivided into two categories: PIRs and FFIRs.

Priority Intelligence Requirements (PIRs)

A priority intelligence requirement is an intelligence requirement, stated as a priority for intelligence support, that the commander and staff need to understand the adversary or the operational environment (JP 2-0).

Friendly Force Intelligence Requirements (FFIRs)

A friendly force information requirement is information the commander and staff need to understand the status of friendly force and supporting capabilities (JP 3-0).

A danger that commanders and planning teams need to avoid is creating a list of CCIRs that are too extensive, that do not relate to the mission, and that are not measurable or observable. Ultimately, a poorly crafted list causes the commander, staff, and subordinates to dedicate too much time to the list, forces assets to be used to track and report on unnecessary events, and detracts from the mission.

Sample Navy Component Commander's Intent

Ref: NWP 5-01, Navy Planning (Jan '07), pp. 2-11 to 2-12.

A commander's intent is broader than the mission statement; it is a concise, free-form expression of the purpose of the force's activities, the desired results, and how actions will progress toward that end. It is a clear and succinct vision of how to conduct the action. In short, the commander's intent links the mission and the CONOPS. The intent expresses the broader purpose of the action that looks beyond the why of the immediate operation to the broader context of that mission, and it may include how the posture of the force at the end state of the action will transition to or facilitate further operations (sequels).

GENTEXT/EXECUTION//

(U) PURPOSE: NEUTRALIZATION OF THE REDLAND MARITIME CAPABILITY IN ORDER TO SUPPORT OPERATIONS AGAINST THE 23RD GUARDS DIVISION AND THE ELIMINATION OF THE TERRORIST FORCES AND INFRASTRUCTURE IN REDLAND.

(U) METHOD: OUR OPERATION MUST REMAIN FOCUSED ON FOUR KEY REQUIREMENTS. FIRST, WE MUST ASSIST IN SETTING THE CONDITIONS FOR THE JTF'S INTRODUCTION OF FORCES INTO REDLAND—THEY CANNOT BE HAMPERED BY ANY CHALLENGES FROM THE SEA. SECOND, THE ESG MUST BE READY TO IMMEDIATELY EMPLOY THE AMPHIBIOUS READY GROUP (ARG)/MEU INTO EITHER OF THE BLOCKING POSITIONS AS SOON AS THE JFC DIRECTS ITS EXECUTION—WE CANNOT LOSE TIME FOR REPOSITIONING. THIRD, OUR DECEPTION MUST REMAIN CREDIBLE UNTIL THE AIRBORNE BRIGADE IS SECURE IN ITS LODGMENT IF WE ARE TO DRAW PRESSURE OFF OF THE FORCIBLE ENTRY UNITS. FOURTH, AND ABOVE ALL OTHERS, REMEMBER THAT THE TERRORIST ELEMENTS AND THEIR INFRASTRUCTURE IN REDLAND ARE THE PRIMARY OBJECTIVES—REMAIN FLEXIBLE TO EXPLOIT OPPORTUNITIES THAT MIGHT PRESENT THEMSELVES TO ALLOW US TO RENDER A DECISIVE BLOW.

(U) TASK FORCE OPERATIONS MUST RECOGNIZE THE TERRITORIAL WATERS AND AIRSPACE OF NEIGHBORING NEUTRAL COUNTRIES, PREVENT DAMAGE TO NEUTRAL COMMERCIAL SHIPPING, AND TAKE ALL NECESSARY STEPS TO MINIMIZE DAMAGE TO INFRASTRUCTURE WITHIN REDLAND.

(U) THE END STATE FOR OUR OPERATIONS WILL BE THE ESTABLISHMENT OF MARITIME SUPERIORITY AND A NEUTRALIZED REDLAND NAVAL FORCE THAT CAN RECONSTITUTE AND PROVIDE MARITIME SECURITY ONCE A NEW REDLAND REGIME, FREE OF TERRORISTS, IS IN PLACE.

The commander's intent is essential to focus further planning, and it enables the commander to indirectly control events during the execution of the operation. While brevity and clarity are imperative, the intent must be crafted so that commanders two levels down have the flexibility to accomplish their mission in lieu of further guidance. For an NCC or JFMCC commander, this means that a level down to a task force (e.g., CTF) should be considered when writing the commander's intent. For a CTF, the commander needs to consider down to the unit level.

To illustrate how CCIRs can affect a commander's decision making and the course of an operation, take the example of a CSG that is conducting presence as part of the second phase of a joint operation. One of the PIRs is to know if Redland is transporting illegal arms and members of a terrorist organization via commercial shipping.

Example: Possible Navy Component Commander's Critical Information Requirements

Priority intelligence requirement: Are there indications that Redland submarines are preparing to deploy? Has Redland placed uploaded missile launchers at its coastal defence cruise missile (CDCM) sites? Are there indications that Redland is loading mines on ships or civilian vessels?

Friendly force information requirement: Closure of the ESG into the JOA. Significant change in meteorological conditions that will delay or prevent offensive operations against Redland naval forces. Completion of mine clearing operations in the Redland port.

Lesson Learned: A danger that commanders and planning teams need to avoid is creating a list of CCIRs that are too extensive, that do not relate to the mission, and that are not measurable or observable. Ultimately, a poorly crafted list causes the commander, staff, and subordinates to dedicate too much time to the list, forces assets to be used to track and report on unnecessary events, and detracts from the mission.

15. Develop Commander's Planning Guidance

The commander's planning guidance focuses the planning during COA development. It should be sufficiently specific to capture the commander's expectations but not so restrictive that it inhibits meaningful COA development. The content of the planning guidance is dependent on the commander's leadership style; it can be very detailed and specific or general in nature. This guidance may also be expressed in terms of Navy operational functions, given in detail in NWP 3-56, Composite Warfare Commander's Manual, or tactical war fighting functions, such as air defense (AD), surface warfare (SUW), or mine warfare (MIW) types of operations, forms of maneuver, etc.

Planning guidance should include the commander's vision of decisive and shaping actions; this assists the planning team in determining the main effort, phases of the operation, location of critical events, and other aspects of the operation the commander deems pertinent to COA development. Additionally, the planning team may develop governing factors. The planning team can use these factors later to evaluate one COA against another in terms of effectiveness and efficiency. Though the planning team typically drafts the governing factors, they belong to the commander, who may modify the factors at any time and who must ultimately approve them.

Planning guidance may include, but is not limited to, the following: specific COA to consider with associated prioritized governing factors; threat vulnerabilities; risk assessment; waterspace and airspace management; weather and oceanographic condition factors; additional limitations; ROE; territorial sea/airspace considerations; C4I architecture considerations; CCIRs (PIR and FFIR); special circumstances such as combat search and rescue (CSAR); ISR and information operations (IOs) priorities; fires, effects, and targeting direction; security and FP measures; decisive and shaping actions; selection and employment of the main effort; types of operations and phasing arrangements; forms of maneuver; CZ relationships/task organization; timing of the operations; logistics and transportation priorities; and phasing.

Although the planning team may develop draft planning guidance for the commander, the commander must review and approve the guidance before it is disseminated throughout the command and to subordinates.

See facing page (p. 5-33) for further discussion.

Commander's Planning Guidance

Ref: NWP 5-01, Navy Planning (Jan '07), pp. 2-13 to 2-14. See also p. 3-25.

Planning guidance should include the commander's vision of decisive and shaping actions; this assists the planning team in determining the main effort, phases of the operation, location of critical events, and other aspects of the operation the commander deems pertinent to COA development. Additionally, the planning team may develop governing factors. The planning team can use these factors later to evaluate one COA against another in terms of effectiveness and efficiency. Though the planning team typically drafts the governing factors, they belong to the commander, who may modify the factors at any time and who must ultimately approve them.

A staff can take an element of commander's guidance (for example, "ensure the plan offers me flexibility") and employ it as a governing factor for COA development and later comparison ("which COA offers the greatest flexibility?"). Generic governing factors, phrased in this example to support the COA comparison provided later, may include:

- Which is more decisive?
- Which is least complicated by the rules of engagement?
- Which allows the greatest flexibility in selecting the time and place of the action?
- Which is easiest to support from the perspective of command, control, and communication?
- Which offers the greatest flexibility?
- Which offers best logistic/sustainability?
- Which makes the enemy's logistic support most difficult?
- Which offers the least operational risk?
- Which is most dependent on weather and oceanographic conditions?
- Which offers the best use of our transportation links?
- Which has the most effect on the enemy's COGs?
- Which allows the accomplishment of the objective in the shortest time?
- Which will best facilitate the attainment of the next objective?
- Which offers the fewest losses to friendly forces?
- Which inflicts the largest losses on the enemy?
- Which offers the greatest hope of splitting the enemy's coalition?
- Which will most strengthen the cohesion of our coalition?
- Which will reduce the enemy morale the most?
- Which offers the most favorable ratio of relative combat power?
- Which will best facilitate future operations?

Planning guidance may include, but is not limited to, the following: specific COA to consider with associated prioritized governing factors; threat vulnerabilities; risk assessment; waterspace and airspace management; weather and oceanographic condition factors; additional limitations; ROE; territorial sea/airspace considerations; C4I architecture considerations; CCIRs (PIR and FFIR); special circumstances such as combat search and rescue (CSAR); ISR and information operations (IOs) priorities; fires, effects, and targeting direction; security and FP measures; decisive and shaping actions; selection and employment of the main effort; types of operations and phasing arrangements; forms of maneuver; CZ relationships/task organization; timing of the operations; logistics and transportation priorities; and phasing.

(Naval Planning) I. Mission Analysis 5-33

16. Develop Warning Order

Once the naval commander approves or modifies the results of mission analysis, the planning team may draft and issue a WARNORD to subordinate units. The WARNORD should include the approved mission statement, the commander's intent, the commander's planning guidance, and any other information that will assist subordinate units with their planning (e.g., changes in task organization, earliest time of movement, etc.).

The WARNORD should be written in the standard situation, mission, execution, administration and logistics, command and control (SMEAC) five-paragraph format for a military directive. It serves notice to subordinate units of forthcoming military operations. The commander may transmit additional WARNORDs as planning matures and more information becomes available. This process allows parallel planning at multiple levels of command for pending operations.

A sample WARNORD is included in NWP 5-01, Appendix L, Formats for Orders.

Outputs

- Approved Mission Statement
- Commander's Intent
- Commander's Planning Guidance
- WARNORDs

Key Points

- Mission analysis sets the stage for all follow-on planning
- The commander must be involved early in the process
- Establish a realistic timeline and adhere to it
- Staff estimates, IPOE, and commander's assessment/guidance are on-going processes
- The better prepared the planning team is for the mission analysis briefing, the better the outputs of mission analysis will be for COA development, analysis, and selection

II. Course of Action Development

Chap 5

Ref: NWP 5-01, Navy Planning (Jan '07), chap. 3.

A COA is any concept of operation open to a commander that, if adopted, would result in the accomplishment of the mission. For each COA, the commander must visualize the employment of his forces and assets as a whole—normally two levels down—taking into account externally imposed limitations, the factual situation in the AO, and the conclusions previously reached during mission analysis.

II. COA Development

Key Inputs

- HIGHER HEADQUARTERS
  - WARNORD
  - OPORD
- NAVY COMMANDER
  - Mission Statement and Commander's Intent
  - Commander's Planning Guidance
  - Updated IPOE
  - ECOAs

Key Outputs

- Approved COAs
- Refined ECOAs
- Wargaming Guidance
- Evaluation Criteria
- Initial Staff Estimates

1. Analyze Relative Combat Power
2. Generate COA Options
3. Test for Validity
4. Recommend C2 and Support Relationships
5. Prepare COA Sketch & Statement
6. Prepare COA Briefing
7. Develop COA Analysis & Evaluation Guidance

Ref: NWP 5-01, Navy Planning, fig. 3-1, p. 3-1.

After receiving guidance, the planning team develops COAs for analysis and comparison. The commander should involve the entire planning team in COA development. The commander's guidance and intent focus the planning team's creativity to produce a comprehensive, flexible plan within the time constraints. When possible, the commander's direct participation helps the staff gain quick, accurate answers to questions that arise during the process. Course of action development is a deliberate attempt to design unpredictable (difficult for the enemy to deduce) alternatives. A good COA positions the force for future operations and provides flexibility to meet unforeseen events during execution; it also provides the maximum latitude for initiative by subordinates.

During COA development, the commander and staff continue the risk management process, focusing on steps one through three:
- Identify risks
- Assess risks
- Develop controls and make risk management decisions

See pp. 4-16 to 4-17 for further discussion of risk management.

Inputs

Input may vary but, as a minimum, they should include a WARNORD or OPORD, and a mission statement and commander's intent. The naval commander's restated mission should include a statement and initial commander's intent, planning guidance, updated IPOE, available friendly forces, information operations themes, and ECOAs.

Process

1. Analyze Relative Combat Power

Combat power is generated through a combination of maneuver, firepower, protection, leadership, and information in combat directed against the enemy. The naval commander integrates and applies the effects of these elements with other potential combat multipliers against the enemy. The goal is to generate overwhelming combat power to accomplish the mission at minimal cost. By analyzing force ratios and determining and comparing each force's strengths and weaknesses as a function of combat power, planners can gain some insight into the friendly capabilities pertaining to the operation, the type of operations possible from both friendly and enemy perspectives, how and where the enemy may be vulnerable, what additional resources may be required to execute the mission, and how to allocate existing resources.

Planners initially make a rough estimate of force ratios. At the fleet level, relative combat power should be an evaluation of rough ratios of combat units two levels down. Planners must not develop and recommend COAs based solely on mathematical analyses of force ratios. Although some numerical relationships are used in this process, the estimate is largely subjective and should primarily consider the effects the force will generate, rather than the size of the force. Planning requires assessing both tangible and intangible factors, such as friction or enemy will and intentions. Numerical force ratios do not include the human factors of warfare that many times are more important than the number of ships, submarines, or aircraft. The staff must carefully consider and integrate the intangible factors into their comparisons. Planners can compare friendly strengths against enemy weaknesses, and vice versa, for each element of combat power. From these comparisons, they may deduce particular vulnerabilities for each force that may be exploitable or may need to be protected. These comparisons may provide planners insights into effective force employment. By using historical minimum-planning ratios for various combat missions and carefully considering time, space and force factors, and enemy templating assumptions, the planner can generally conclude what types of operations can be conducted successfully. This step provides the planners with what might be possible, not with a specific course of action.

NWP 5-01, Navy Planning, app. D provides methods for analyzing relative combat power.

Commander/Staff Interaction During/After Course of Action Development

Ref: NWC Maritime Component Commander's Handbook (Feb '10), p. 2-13.

COA development should consider all joint force capabilities and focus on contributing to the defeat/neutralization of the enemy's center of gravity (COG) and the protection of the friendly COG. At the completion of COA development, the commander provides additional guidance to the OPT to advance to the COA analysis step. The OPT will have brainstormed and developed a set of friendly COAs that describe different ways to accomplish the objectives.

The commander considers each COA and decides which one(s) to continue to develop it. The commander also decides whether the options developed span the possible ways to attack the problem. Although not a formal brief like the MA brief, this is a reality check for the OPT. Do the options meet the commander's expectations? Now is the time to eliminate some of the COAs or direct the OPT to develop different options. Ensure the COAs conform to previous guidance and adequately present methods for mitigating or assuming risk. Identify if conditions (HHQ guidance, the OE, assumptions) changed that require additional options. The commander should review previous guidance and intent to evaluate whether the options conform. The commander should test each COA for validity:

- **Suitable**. Does the COA accomplish the mission and comply with guidance?
- **Feasible**. Does the COA accomplish the mission with the forces and functions provided and within the time and space constraints?
- **Acceptable**. Do the COA's advantages justify the cost? (Risk)
- **Distinguishable**. Do the COA's differ significantly from each other? Are the COA's broad enough to span the possible? The task organization may define the uniqueness of the COA.
- **Complete**. Is there enough detail to describe actions two command levels down?

Additionally, the commander can provide evaluation criteria and war-gaming guidance. Decision criteria are those governing factors by which a COA will eventually be assessed in the COA comparison step. These criteria can normally be discerned from the commander's intent. For war-game guidance:

- Identify which friendly COA and enemy COA to war-game
- Identify war game methodology
- Identify specific critical events to focus on; e.g., "gain and maintain maritime superiority" might be specifically war-gamed if it is required to occur early in the operation and is a prerequisite to follow-on operations

Once again, the time available will often be a primary consideration for this guidance. If planning time is not compressed, greater breadth and depth can occur during the COA analysis step.

The commander needs to thoroughly review the established preliminary command and control arrangements between forces for each COA. This structure should consider the types of units to be assigned to a headquarters or component. The maritime component commander's span of control and decision authorities need to be considered while making C2 arrangements. C2 arrangements should take into account the entire OE organization. They should also account for the special C2 requirements of operations that have unique needs, such as amphibious landings or special operations.

The commander should also consider CCIRs, COG, intent, and the OE. Are collection resources answering CCIRs? Are planning CCIRs becoming execution CCIRs? Are there any changes/modifications to the evaluation of COG or the OE?

2. Generate Course of Action Options

Based on the commander's guidance and the results of mission analysis, the planning team generates options for COA development. A good COA should be capable of defeating all feasible ECOAs. In a totally unconstrained environment, the goal is to develop several appropriate COAs. Since there is rarely enough time to do this, the commander's guidance may specify a limit to the number of COAs.

Course of action options should focus on ECOAs arranged in order of probable adoption. Brainstorming is the preferred technique for generating options. Brainstorming requires time, imagination, and creativity, but it produces the greatest range of options. The planning team must be unbiased and open-minded in evaluating proposed options. The operational design for a COA can vary by many different characteristics, including but not limited to the form of operational maneuver, the designation of the sector of main effort, use and timing of operational pauses, level of operational deception, or method of insertion of amphibious forces. Team members can quickly identify COAs obviously not feasible based on their particular areas of expertise. They can also quickly decide if they can modify a COA to accomplish the requirement or eliminate it immediately. If one team member identifies information that might affect another's analysis, the member should share it immediately, eliminating wasted time and effort.

When developing possible COAs, the staff may wish to use the DRAW-D (defend, reinforce, attack, withdraw, or delay) concept to consider friendly COAs. Staff members can also quickly decide if they can modify a COA to accomplish the requirement or eliminate it immediately.

In developing COAs, staff and planning team members must determine the doctrinal requirements for each type of operation they are considering, including doctrinal tasks to be assigned to subordinate units. One valuable resource during this stage is the NMETL. By reviewing this list, planning teams and staffs can ensure that tasks are properly stated and designated for naval forces. In addition, COA development must look at possibilities created by attachments, such as combining an SSG with an ESG.

The planning team first determines the decisive point, if not already determined by the commander. This is where the unit masses the effects of overwhelming combat power to achieve a result with respect to space, time, and enemy forces that will accomplish the unit's purpose. This is defined as the main effort. Next, the planning team determines supporting efforts (those tasks other than the main effort) that must be accomplished to enable the main effort to succeed. The planning team then determines the purposes of the main and supporting efforts.

The main effort's purpose is directly related to the mission of the unit; the supporting effort's purpose relates directly to the main effort. Next, the planning team determines the essential tasks for the main and supporting efforts to achieve these purposes. Once staff members have explored each COA's possibilities, they can examine each (changing, adding, or eliminating COAs as appropriate) to determine if it satisfies COA selection criteria.

Example: Tentative Course of Action

Deterrence/show-of-force will consist of CSG positioned to support freedom of navigation operations and strikes, ESG positioned to conduct amphibious operation, and SSG along with coalition ships conducting TBMD, protection of shipping, and maritime interception. Intelligence, surveillance, and reconnaissance and maritime patrol aircraft (MPA) assets support operational environment preparation and MDA. Special operations forces (SOF) prepared to perform liaison and targeting support. Achieve local air superiority and maritime superiority in the AO. Use air/sea interdiction to sever ground lines of communications (GLOCs) and sea lines of communications (SLOCs).

Lesson Learned

The planning team must avoid the common pitfall of presenting one good COA among several throwaway COAs. When presented with a collection of viable COAs, the commander often will find that COAs can be combined or desirable elements moved from one to another.

3. Test for Validity

The staff should review a tentative COA for validity. This test should address suitability, feasibility, acceptability, distinguishability, and completeness.

1. Suitability (Adequacy)

The COA must accomplish the mission and comply with the commander's guidance; however, the commander may modify guidance at any time. When the guidance changes, the staff records and coordinates the new guidance and reevaluates each COA to ensure that it complies with the change.

2. Feasibility

The force must have the capability to accomplish the mission in terms of available time, space, and resources.

3. Acceptability

The tactical or operational advantage gained by executing the COA must justify the cost in resources, especially casualties. This assessment is largely subjective.

4. Distinguishability

Each COA must differ significantly from any others. Significant differences may result from use of reserves, different task organizations, day or night operations, or a different scheme of maneuver. This criterion is also largely subjective.

5. Completeness

Each COA must include the following: major operations and tasks to be performed; major forces required; concepts for deployment, employment, and sustainment; time estimates for achieving objectives; and the desired end state and mission success criteria. The order from higher headquarters normally provides the what, when, and why for the force as a whole. The "who" in the COA does not specify the designation of units; it arrays units by type (for example, generic task/action groups or platform/capability). Designation of specific units occurs later in the NPP.

4. Recommend Command and Control and Support Relationships

Planners next establish preliminary C2 arrangements to groupings of forces for each COA. This structure should consider the types of units to be assigned to a headquarters or component and the span of control. If planners need additional C2 cells, they note the shortage and resolve it later. Command and control arrangements take into account the entire battle space organization. Planners also consider the special C2 requirements of operations that have unique requirements, such as amphibious landings.

See pp. 2-34 to 2-35 for discussion of command and support relationships.

5. Prepare Course of Action Sketch and Statement

Current joint military planning consists of six steps for a campaign or operation; the basic phases are: Phase 0 (shaping), Phase 1 (deterrence), Phase 2 (seize the initiative), Phase 3 (dominance), Phase 4 (Stabilize the environment), Phase 5 (enable civil authority). The details of each phase are explained in JP 5-0.

At a combatant command echelon or a JTF, the COA sketch and statement incorporate the actions by each service or functional component in each phase of the campaign or operation. The JFMCC or NCC may be involved in each of the phases as well. Together, the statement and sketch cover the "who" (generic task organization), "what" (tasks and purposes), "when," "where," "how," and "why" (purpose of the operation) for each subordinate unit/component command, any significant risks, and where they occur for the force as a whole.

The critical role that naval forces have in providing presence, participating in exercises and other events as part of a combatant command's theater security cooperation efforts, and presenting a deterrence force illustrates the impact of naval forces during the initial two phases of a campaign or operation. However, for commands under a JFMCC or NCC, such as a TF, the scope and span of the entire campaign or operation likely will be beyond the planning needs or ability of the staff. Therefore, in most cases these lower echelon commands focus solely on specific phases in the overall operation, but it still is imperative that the commander and staff understand how their force's actions fit into the overall campaign or operation and be prepared to plan for these other phases if necessary.

See pp. 5-42 to 5-43 for an example JTF course of action sketch, statement and narrative. See pp. 5-44 to 5-45 for an example JFMCC/NCC course of action sketch, statement and narrative.

Course of Action Sketch

The COA sketch provides a picture of the force employment concept of the COA and could include an array of generic forces and control measures, such as:

- Unit or command boundaries that establish the AO or operating area (OPAREA)
- Unit deployment/employment
- Control graphics (fire support area, carrier operating area, amphibious operating area, ordnance jettison area, water space management, territorial waters, etc.)
- Sequencing of events (i.e., ISR → maneuver → and staging → mine clearance → strikes → amphibious demonstration)
- Designation of the decisive (main effort) and shaping (supporting effort) operations
- Enemy known or templated locations (i.e., ships and submarines in port, at sea, or unlocated).

Planners can enhance the sketch with identifying features to help orient the commander and staff. The sketch may be on any media; what it portrays is more important than its form.

Course of Action Statement

The COA statement describes how the forces will accomplish the commander's intent. It concisely expresses the commander's concept for operations and governs the design of supporting plans or annexes. Planners develop a concept by refining the initial array of forces and using graphic control measures to coordinate the operation and to show the relationship of friendly forces to one another, the enemy, and the operational environment. During this step, units are converted from generic to specific types of units. The purpose of this step is to clarify the initial intent about the deployment, employment, and support of friendly forces and assets, and to identify major objectives and target dates for their attainment. In drafting the tentative CONOPS, each COA should state by phase, in broad but clear terms, what is to be done, the size of the forces deemed necessary, and the time in which force needs to be brought to bear. A COA statement should be simple, clear, and complete. It should address all the elements of organizing the operational environment.

6. Prepare Course of Action Briefing

At this stage of the process, the planning team may propose (or the commander may require) a briefing on the COAs developed and retained. The purpose of this briefing is to gain the naval commander's approval of the COAs to be further analyzed, to receive guidance on how COAs are to be compared and evaluated, or to receive guidance for revision of briefed COAs or the development of additional COAs.

After the briefing, the commander gives any additional guidance. The commander may direct which ECOAs should be used in war gaming and, based on time available, may also direct that most dangerous, most likely, or both be used during COA analysis. If the commander rejects all COAs, the planning team begins again. If one or more of the COAs is accepted, then staff members develop evaluation criteria in order to begin COA analysis.

See p. 5-46 for an example COA briefing format.

7. Develop Course of Action Analysis and Evaluation Guidance

At this point the planning team recommends additional analysis and evaluation criteria to the commander. These criteria should:

- Reflect the criteria for success established during the mission analysis
- Provide a reasonable basis for comparing the relative merits of the COAs under consideration
- Focus on the force-oriented objectives and DPs identified in the mission analysis
- Be quantifiable and measurable.

The commander's governing factors, developed initially in mission analysis, are a critical input. At this stage in the NPP, the commander should confirm or modify the governing factors that the planning team developed during mission analysis.

Outputs

Prior to COA analysis (wargaming), the planning team requires the following:
- Course of action analysis and evaluation guidance
- Refined commander's intent
- Wargaming guidance
- Approved COAs
- Refined ECOAs
- Initial staff estimates.

Key Points

- A COA is any concept of operation open to a commander that, if adopted, would result in the accomplishment of the mission
- A COA must meet validity tests of suitability, feasibility, acceptability, distinguishability and, most important, be complete with respect to the who, what, where, when, why, and how

Example of Joint Task Force Course of Action Sketch and Statement

Ref: NWP 5-01, Navy Planning (Jan '07), fig. 3-2, p. 3-6.

Joint Task Force Course of Action

[Map sketch titled "JTF BLUE SWORD" showing REDLAND, ORANGELAND, WHITELAND, GREYLAND, and PINKLAND. Features include: Terrorist Camp, Red City, OBJ CAT, OBJ RAT, OBJ DOG, Air Base, Naval Base, NSW locations, CDCM Sites, Airbase (JFACC beddown), Port (ISB), Prepo Shipping, Deception Ops area, AO. Unit annotations: CTF Central (ESG) - Amphib Ops OBJ RAT/CAT - SASO; JTF Sword C2 HQs during Phase 3; CTG South (SSG) - AD of CSG - Deception Ops - MIO and Escort Ops - Strike; CTF Main (JFMCC HQ/CSG) C2 IO/Deception Ops Strike (TACAIR) on Redland; CVOA. Title: COA Airborne Forced Entry (JFLCC Main Effort). COA 1 D Day: TBD]

Ref: NWP 5-01 Navy Planning, fig. 3-2, p. 3-6.

COA STATEMENT
The airborne forced entry COA is an aggressive offensive operation aimed at neutralizing the Redland 23rd Guards Division in order to attack and destroy the terrorist organization in Redland. This COA is conducted in six phases.

Example of Joint Task Force Course of Action Narrative

Ref: NWP 5-01, Navy Planning (Jan '07), fig. 3-3, p. 3-7.

Phase 0 (shaping) begins with elements of the JTF forces participating in exercise Freedom Assurance (FA) with Pinkland and coalition partners in order to demonstrate coalition resolve to Redland and to finalize status of forces agreement (SOFA) and host nation (HN) agreement with Pinkland for use as an intermediate staging base (ISB). During this phase, all aspects of coalition power will be exercised in order to ensure interoperability and to establish SOPs and common tactics, techniques, and procedures (TTPs). JFMCC is the main effort during Phase 0 with all others supporting. Phase 0 ends if or when the Redland situation stabilizes or if events warrant moving to Phase 1.

Phase 1 (deterrence) begins with the JTF conducting deterrence ops in the vicinity of (IVO) Redland in order to (IOT) deter Redland aggression. The JFLCC will stage airborne forces at ISB ALPHA in Pinkland. The JFACC establishes air superiority over ISB ALPHA, protects air lines of communications (ALOCs), supports flow of JFLCC forces into ISB with STRATAIR, and prepares to support Phase 2 airborne forced entry operations. JFMCC conducts a maritime show of force IVO Redland territorial waters, protects SLOCs, and prepares to support amphib ops into objective (OBJ) CAT or RAT. JFSOCC conducts special reconnaissance (SR) in Redland in support of (ISO) the JTF collection plan and prepares to conduct direct action (DA) against terrorist camps. The JTF establishes an operational HQ onboard a JFMCC command ship. JFMCC show of force is the main effort during Phase 1 with all others supporting. Phase 1 ends if Redland resumes aggression or stands down.

Phase 2 (seize the initiative) begins with JTF forces seizing the initiative in preparation for subsequent decisive ops. On order (O/O), JFLCC conducts airborne forced entry into Redland airfield and seizes OBJ DOG; O/O flow in follow-on forces, and BPT to accept TACON of MEU after amphib ops into OBJ CAT and RAT. JFACC establishes air superiority over Redland, supports the JFLCC airborne op, disrupts movement of Redland forces into JFLCC AO in priority of 2nd, 3rd, and 1st Red Guard brigades (BDEs) (RGB), supports JFSOCC DA ops, and BPT to support JFMCC amphib ops. JFMCC establishes sea control in the Redland Sea, BPT to conduct amphib ops to establish blocking positions in either OBJ CAT or RAT, supports Deception Plan X-Ray in southern Redland, and B/BPT to release TACON of MEU to JFLCC. JFSOCC destroys terrorist camp complex, denies Redland force movement along northern portion of Hwy 15, and destroys remnants of the terrorist force. JFLCC is the main effort in Phase 2 with all others supporting. Phase 2 ends when JFACC has gained air superiority over the objective areas, the enemy threat at the AIRFIELD and DOG are neutralized, and JTF force build-up is sufficient for transition to decisive ops.

Phase 3 (dominance) begins with the JFLCC conducting offensive operations IOT destroy Redland ground forces. The JFLCC will transition to stability and support operations (SASO) as Redland forces capitulate. JFACC maintains air superiority over Redland and the ISB and provides close air support (CAS) to the JFLCC. JFMCC maintains maritime sea control in the Redland Sea and BPT to support MEU amphib ops into OBJ CAT or RAT based on disposition of Redland forces. JFLCC continues to be the main effort in this phase with all others supporting. Phase 3 ends when Redland forces have been destroyed or surrender to the JTF.

Phase 4 (stability) begins with the JTF HQ transitioning from the afloat HQ to a land based HQ in Redland. JFLCC conducts SASO throughout Redland and reestablishes critical infrastructure. JFMCC supports SASO and O/O redeploys nonessential maritime assets. JFACC continues to provide CAS to JFLCC and JFSOCC ops, and O/O redeploys nonessential assets. JFSOCC continues to kill or capture fugitive terrorists, conducts sensitive site exploitation (SSE), and O/O redeploys nonessential assets. Joint psychological operations (PSYOP) task force (JPOTF) conducts PSYOP IOT influence the Redland population to cooperate with security forces and directs displaced personnel to coalition aid stations. JFLCC continues to be the main effort in Phase 4 with all others supporting. Phase 4 ends when security conditions are adequate to transfer authority to a legitimate Redland government.

Phase 5 (enable civil authority) begins by enabling a legitimate Redland government to assume control of its sovereign territory. The JTF will provide support to DOS and O/O transition control to an international organization, stand down the JTF, and redeploy. JFLCC forms and trains new Redland security force, conducts joint security patrols and operations with the security forces, and O/O redeploys forces. JFACC continues to provide CAS and supports redeployment via STRATAIR. JFMCC supports SASO, redeploys forces as situation permits, and transitions port security to new Redland maritime/coast guard. JFLCC continues to be the main effort in Phase 5 with all others supporting. Phase 5 ends when a legitimate Redland government has control of its sovereign territory, and an international organization has assumed responsibility for Redland security and stabilization.

Example of JFMCC or NCC Course of Action Sketch and Statement

Ref: NWP 5-01, Navy Planning (Jan '07), fig. 3-4, p. 3-8.

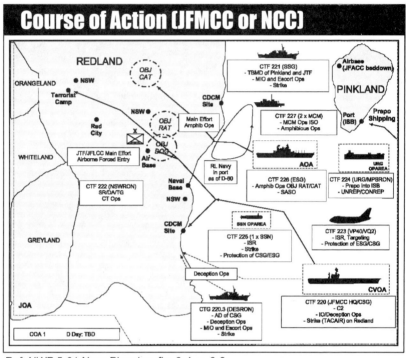

Ref: NWP 5-01 Navy Planning, fig. 3-4, p. 3-8.

COA STATEMENT

The naval force supports the airborne forced entry with an aggressive operation to establish maritime superiority in the AO in order to facilitate the MEU amphibious operation, neutralize the Redland Navy, protect commercial shipping and Pinkland, and interdict the movement of weapons to Redland and terrorists by sea.

Example of JFMCC or NCC Course of Action Narrative

Ref: NWP 5-01 (Jan '07), Navy Planning, fig. 3-5, p. 3-9.

Phase 1 (deter): Supported Commander: JFMCC. Main effort: movement of pre-positioned shipping to intermediate staging base (ISB). Phase 1 begins with the flow of pre-positioned shipping to Pinkland ISB. During Phase 1, JFMCC continues to project power by conducting surface, air, subsurface, and amphibious operations off the coast of Redland and establishes MDA and local sea control in the AO. JFMCC HQ integrates, accepts, and establishes C2 with coalition maritime forces. IO and ISR conducted in conjunction with (ICW) JTF IO themes and collection plan. JFMCC be prepared to move to Phase 2 and conduct amphibious operations and strikes into Redland. Phase 1 ends with the completion of JFMCC achieving local sea control and MDA in the AO, pre-positioned shipping received at ISB, and coalition C2 structure completed.

Phase 2 (seize the initiative): Supported Commander: JFMCC. Main effort: maritime superiority. Phase 2 begins with the insertion of naval special warfare (NSW) into Redland to provide special reconnaissance (SR) and task group (TG) for strike operations. JFMCC maintains MDA and local sea control in the AO and continues to flow pre-positioned shipping into ISB. JFMCC BPT conduct amphibious operations and strikes into Redland in order to support JFLCC airborne forced entry. MCM forces conduct mine clearance and establish Q-routes for amphibious operation. JFMCC BPT destroy Redland naval forces including ships, submarines, aircraft, and coastal defenses and BPT provide TBMD against Redland missiles launch against Pinkland or JTF. JFMCC conducts deception operation in south Redland to support MEU and JFLCC ops. On order, JFMCC surface forces conduct maritime intercept of designated suspicious vessels and provide escort of Pinkland shipping. IO and ISR conducted ICW with JTF IO themes and collection plan. Phase 2 ends with pre-positioned shipping at ISB, NSW inserted into Redland, Q-routes established, and JFMCC forces prepared to transition to Phase 3.

Phase 3 (dominance): Supported Commander: JFLCC. Main effort: Airborne assault into Redland airfield. Phase 3 begins with JFMCC offensive operations to destroy Redland Navy and support JFACC in strikes on Redland land targets to enable JFLCC airborne forced entry. MEU and NSW operational control (OPCON) to JFLCC and joint forces special operations component command(er) (JFSOCC), respectively. JFMCC air defense be prepared to provide TBMD, and surface forces be prepared to conduct maritime intercept and provide escort of Pinkland shipping. Naval construction brigade (NCB) and maritime special operations (MSO) elements deployed to ISB. IO and ISR conducted ICW with JTF IO themes and collection plan. Phase 3 ends with Redland Navy defeated at sea and prevented from offensive operations and MEU and NSW OPCON shift complete.

Example Course of Action Briefing
Ref: NWP 5-01, Navy Planning (Jan '07), p. 3-4.

At this stage of the process, the planning team may propose (or the commander may require) a briefing on the COAs developed and retained. The purpose of this briefing is to gain the naval commander's approval of the COAs to be further analyzed, to receive guidance on how COAs are to be compared and evaluated, or to receive guidance for revision of briefed COAs or the development of additional COAs. This is another place where a collaborative session may facilitate subordinate planning. The COA briefing includes:

- Updated IPOE
- Enemy course of action (event templates)
- The restated mission
- The commander's and the higher commander's intent (two echelons above)
- The COA statement and sketch
- The rationale for each COA, including
    1. Considerations that might affect ECOAs
    2. Deductions resulting from a relative combat power analysis
    3. Reasons units are arrayed as shown on the sketch
    4. Reasons the staff used the selected control measures
    5. Updated facts and assumptions.

After the briefing, the commander gives any additional guidance. The commander may direct which ECOAs should be used in war gaming and, based on time available, may also direct that most dangerous, most likely, or both be used during COA analysis. If the commander rejects all COAs, the planning team begins again. If one or more of the COAs is accepted, then staff members develop evaluation criteria in order to begin COA analysis.

Example Format and Sequence

BRIEFER   **SUBJECT**

METOC   METOC analysis and impact on COAs

J-2/N-2   Updated intelligence estimate and IPOE with ECOAs
- Terrain, oceanographic, and weather analysis (ICW METOC; focus on effects on enemy)
- Situation template(s)

J-3/N-3   Restated mission
- Higher and own commander's intent
- COA statement and sketch as a single entity
    - Statement includes scheme of maneuver and addresses
    - Sketch includes array of forces and control measures
    - Operational environment framework includes the main effort and any significant risk accepted
- COA rationale
    - Considerations affected by possible enemy COA to be wargamed
    - Deductions resulting from relative combat power analysis
    - Why units are arrayed as shown on the sketch
    - Why selected control measures are used

J-1/N-1, J-4/N-4, J-5/N-5   Updated facts and assumptions, if available

J-3/N-3   Request commander's guidance on which COAs to further analyze

5-46 (Naval Planning) II. Course of Action Development

Chap 5
III. Course of Action Analysis (Wargaming)

Ref: NWP 5-01, Navy Planning (Jan '07), chap. 4.

The heart of the NPP is the analysis of opposing courses of action. In the previous steps of the planning process, enemy courses of action (ECOAs) and COAs were examined relative to their basic concepts—ECOAs were developed based on enemy

III. COA Analysis (Wargaming)

Key Inputs	Key Outputs
▪ NAVY COMMANDER 　▪ Refined Commander's intent 　▪ Wargaming guidance 　▪ Approved COAs 　▪ Refined ECOAs 　▪ Initial staff estimates	▪ Wargame results ▪ List of critical events and decision points ▪ Updated IPOE ▪ Subordinate Commander's estimate of supportability ▪ Branches and sequels identified for further planning

1. Organize for Wargaming
2. List all Friendly Forces
3. Review Assumptions
4. List Known Critical Events
5. Determine the Governing Factors
6. Select the Wargaming Method
7. Record and Display Results
8. War Game the Combat Actions and Assess the Results
9. Refine Staff Estimates
10. Update and Refine Intelligence Preparation of the Operational Environment Products

Ref: NWP 5-01, Navy Planning, fig. 4-1, p. 4-1.

capabilities, objectives, and the estimate of the enemy's intent, and COAs were developed based on friendly mission and capabilities. In this step, the planning team conducts an analysis of the probable affect that each ECOA has on the chances of success of each COA. The aim is to develop a sound basis for determining the feasibility and acceptability of the COA. Analysis also provides the planning team with a greatly improved understanding of its COAs and the relationship between them.

Course of action analysis identifies which COA best accomplishes the mission while positioning the force for future operations. It helps the commander and staff to

- Determine how to maximize combat power against the enemy while protecting the friendly forces and minimizing collateral damage
- Have as near an identical visualization of the combat action as possible
- Anticipate operational environment events and potential reaction options
- Determine conditions and resources required for success
- Determine when and where to apply the force's capabilities
- Focus intelligence collection requirements
- Determine the degree of flexibility in each COA

Course of action analysis is conducted using wargaming. The war game is a disciplined process, with rules and steps that attempt to visualize the flow of the operation. Specifically, war gaming

- Considers the results from the friendly COG determination, friendly COAs, enemy COG determination, enemy COAs, as well as the characteristics of the physical environment.
- Relies heavily on joint doctrinal foundation, tactical judgment, and operational experience.
- Focuses the planning team's attention on each phase of the operation in a logical sequence.
- Highlights critical tasks and provides familiarity with operational possibilities otherwise difficult to achieve.
- Is an iterative process of action, reaction, and counteraction. War gaming stimulates ideas and provides insights that might not otherwise be discovered.
- War gaming is a critical portion of the planning process and should be allocated more time than any other step. At a minimum, each retained COA should be war gamed against both the most likely and the most dangerous ECOA.
- While the focus is on the analysis portion of the war gaming process, this stage also allows the various staff components to refine their estimates. Refined estimates help the planning team in determining feasibility and acceptability.

Inputs

- Refined commander's intent
- Commander's wargaming guidance
- Approved COAs
- Refined ECOAs
- Initial staff estimates

Process

During COA analysis, the planning team evaluates the effectiveness of each friendly COA against all of the ECOAs by using the naval commander's evaluation criteria. The planning team makes adjustments to identified problems and weaknesses of the friendly COAs and identifies branches and sequels. Each friendly COA is war gamed independently against each selected ECOA. Course of action analysis helps

Commander/Staff Interaction During/After Course of Action Analysis

Ref: NWC Maritime Component Commander's Handbook (Feb '10), p. 2-15.

The heart of the commander's estimate process is the analysis of different courses of action. Analysis is nothing more than war gaming — either manual or computer assisted. The aim is to develop a sound basis for determining the feasibility and acceptability of the COA's. Analysis also provides the planning staff with a greatly improved understanding of their COA's and the relationship between them.

During war gaming, the staff attempts to capture an operation's dynamics through a series of action/reaction/counteraction sequences. During that process, the staff attempts to capture key elements that collectively define the synchronization of the operation.

The commander may decide to receive an optional back brief after COA analysis, to be updated on planning status and potentially provide additional guidance. War gaming is a "what if" game of friendly versus enemy COA's. The COA analysis identifies which COA best accomplishes the mission while also identifying any gaps and seams in the plan. During COA analysis the commander and staff identify potential:

- Advantages
- Disadvantages
- Risk
- Branches and sequels
- Decision points
- Commander's critical information requirements

The commander should put the appropriate level of emphasis on wargame participation. Since the non-OPT staff has a significant role in the depth of the research, lack of adequate participation may cause substandard results. The commander is not required to analyze each of the wargaming results but could review the synchronization matrix and critical event list. Once again, consider each friendly COA for validity in light of the analysis. Specifically, is the JFMCC scheme of maneuver and assignment of tasks feasible with the forces/capabilities available, or does it rely too heavily on RFFs to execute? Has the OPT recommended that specific COAs be discarded?

The commander should identify if new gaps and seams have been identified. Is there a need for additional forces? Are the assumptions still valid? Were the staff estimates mature enough to provide detail to conduct the COA analysis?

CCIRs should change from planning to execution-type CCIRs. Planning CCIRs are information requirements to continue planning. Execution CCIRs are information requirements during the conduct of the operation to drive a decision. A decision support matrix should be developed to identify the commander's decisions and possible branch plans for deviations from the plan.

- Ensure collections are looking at PIRs
- Ensure the information management (IM) plan reflects CCIR priorities
- Consider and track higher command's CCIRs

War gaming stimulates ideas and provides insights that might not otherwise be discovered. It highlights critical tasks and provides familiarity with operational possibilities otherwise difficult to achieve. War gaming is a critical portion of the planning process and should be allocated more time than any other step.

the commander determine how best to apply his combat power against the enemy's CVs while protecting his own CVs. War gaming pits each friendly COA against every ECOA. This process is repeated until all COAs are wargamed against each of the ECOAs. The COAs are not compared to each other until the next step, COA comparison. During this step, the continued refinement of staff estimates provides the staff and subordinate commanders a view of the COA. These views later assist the commander during COA comparison and decision.

1. Organize for Wargaming

Gather the necessary tools, materials, and data for the war game. Units need to war game on maps, charts, computer simulations, and other tools that accurately reflect the nature of the terrain. For naval forces, due to such limitations as systems and space onboard a ship, wargaming most likely can be performed on a map/chart or a computer system such as the Global Command and Control System (GCCS). The staff then posts the COA on a map/chart (which can be hard copy or electronic) displaying the JOA/AO and other significant control measures. Tools required include but are not limited to:

- A display of critical mission analysis information: higher and own (mission, commander's intent, assumptions, and CCIRs)
- Event template
- Recording method
- Completed COAs, to include control measures and ISR collection plan
- Means to post enemy and friendly unit symbols
- Chart or map of JOA/AO (either paper or digital)
- Updated estimates and common operating picture

2. List All Friendly Forces

The commander, planning team, and staff consider all units that can be committed to the operation, paying special attention to support relationships and limitations. The friendly force list remains constant for all COAs that the staff analyzes. These friendly forces should have been recorded during mission analysis.

3. Review Assumptions

The commander, planning team, and staff review assumptions (as developed in mission analysis) for continued validity and necessity.

4. List Known Critical Events

Critical events are essential tasks, or a series of critical tasks, conducted over a period of time that require detailed analysis. This may be expanded to review component tasks over a phase(s) of an operation or over a period of time (C-day through D-day). The planning staff may wish at this point also to identify decision points (those decisions in time and space that the commander must make to ensure timely execution and synchronization of resources). These decision points are most likely linked to a critical event (e.g., commitment of the reserve force).

Example: Critical Events

Conduct mine countermeasures in support of amphibious operations.

Establish maritime superiority in the AO.

Conduct deception operation in southern Redland.

Execute operational fires in support of ground force forcible entry operations.

5. Determine the Governing Factors

Governing factors are those criteria the staff uses to measure the effectiveness and efficiency of one COA relative to other COAs following the war game. They are those aspects of the situation (or externally imposed factors) that the commander deems critical to the accomplishment of the mission. Potential influencing factors include elements of the commander's guidance and/or commander's intent, selected principles of war, external constraints, and even anticipated future operations for involved forces or against the same objective. Governing factors change from mission to mission. Though these factors are applied in the next step when the COAs are compared, it is helpful during this wargaming step for all participants to be familiar with the factors so that any insights into a given COA that influence a factor are recorded for later comparison. The criteria may include anything the commander desires. If not received directly from the commander, they are often derived from the intent statement.

6. Select the Wargaming Method

There are varieties of wargaming methods that can be used, with the most sophisticated being computer-aided modeling. Though many of the war gaming techniques have been developed primarily for ground force operations, they can be adapted for the purpose of war gaming a naval operation. There are four basic war gaming methods available to the operational commander: the sequence of essential tasks, avenue in depth, belts, and box methods. The sequence of essential tasks method, which focuses on critical events, is probably the most useful war gaming method and is the method illustrated in this publication.

- **Sequence of Essential Tasks Method.** The sequence of essential tasks, also known as the critical events method, highlights the initial shaping actions necessary to establish a sustainment capability and to engage enemy units in the deep battle area. At the same time, it enables the planners to adapt if the enemy executes a reaction that necessitates the reordering of the essential tasks. This technique also allows wargamers to concurrently analyze the essential tasks required to execute the CONOPS.

- **Avenue in Depth Method.** Avenue in depth focuses on one avenue of approach at a time, beginning with the main effort. This technique is good for offensive COAs or for defensive situations when operating space inhibits mutual support.

- **Belts Method.** Belts divide the operating space into areas that span the width of the AO. This technique is based on the sequential analysis of events in each belt; that is, events are expected to occur more or less simultaneously. This type of analysis often is preferred because it focuses on essentially all forces affecting particular events in one time frame. A belt normally includes more than one event.

- **Box Method.** The box technique is a detailed analysis of a critical area, such as a landing beach or strike target. When using it, the planning team isolates the area and focuses on the critical events within that area. The assumption is that the friendly units not engaged in the action can handle the situation in their region of the operational environment and the essential tasks assigned to them.

Time and resources available to support war gaming undoubtedly influence the method selected. However, war gaming also can be as simple as using a detailed narrative in conjunction with a map/chart or situation sketch. Each critical event within a proposed COA should be war gamed based upon time available using the action, reaction, and counteraction method of friendly and enemy interaction.

7. Record and Display Results

Recording the war game's results gives the staff a record from which to build task organizations, synchronize activity, develop decision support templates, confirm and refine event templates, prepare plans or orders, and analyze COAs based on identified strengths and weaknesses. Staff members can use the war game worksheet to record any remarks regarding the strengths and weaknesses they discover. The amount of detail depends on the time available. With more time, the war game worksheet can be expanded to include component actions, operational functions, or other required areas. Details and methods of recording and displaying war game results are best addressed in unit standard operating procedures.

The war game worksheet allows the staff to synchronize the COA across time and space in relation to the enemy COA. It uses a simple format that allows the staff to game each critical event using an action, reaction, and counteraction method with an ability to record the timing of the event, force/assets requirements, and remarks/observations. While individual commands may choose to use war game worksheets that serve their needs, they provide a sample of recording methods that can be easily adapted to a variety of organizations and missions.

See following pages (pp. 5-54 to 5-55) for example wargame sketch and worksheets. See NWP 5.01, Naval Planning, app. A for blank wargame worksheets.

8. Wargame the Combat Actions and Assess the Results

During the war game, the commander and staff try to foresee the dynamics of an operation's action, reaction, and counteraction. The staff normally analyzes each selected event by identifying the tasks the force two echelons below must accomplish. Identifying the COA's strengths and weaknesses allows the staff to make adjustments as necessary.

The staff and planning team considers all possible forces including templated enemy forces outside the AO that could react to influence the operation. They evaluate each friendly move to determine the assets and actions required to defeat the enemy at each turn. The staff and planning team should continually evaluate the need for branches to the plan that promote success against likely enemy moves in response to the friendly COA.

The commander, staff, and planning team look at many areas in detail during the war game, including all enemy capabilities, deployment considerations and timelines, ranges and capabilities of weapons systems, and desired effects of fires. They look at setting the conditions for success, protecting the force, and shaping the operational environment.

Experience, historical data, SOPs, and doctrinal literature provide much of the necessary information. During the war game, staff officers conduct a risk assessment in their area of expertise and responsibility for each COA. The staff continually assesses the risk to friendly forces from catastrophic threats, seeking a balance between mass and dispersion.

The planning team and staff identify the operational functions required to support the scheme of maneuver and the synchronization of the sustaining operation. If requirements exceed available assets, the staff recommends the priority for use to the commander based on the commander's guidance and intent and on the situation. For instance, a JFMCC or NCC may not have enough aircraft, ships, and submarines to locate, track and, if required, attack an enemy submarine force while also supporting other missions such as strikes, maritime intercept, ISR, and TBMD. If the commander's guidance and intent place emphasis on tasks other than submarine prosecution, the staff may recommend that enemy submarines of a certain type or possibly operating in a designated area be the focus. To maintain flexibility, the commander

Wargaming Roles & Responsibilities

Ref: NWP 5-01, Navy Planning (Jan '07), p. 4-13.

General Role of Staff Members
Each of the primary staff officers has a distinct role during the war game. Though the size and composition of the command may differ, the general role of the staff members is as follows:

- **Planning team chief or chief of staff (COS)**: Acts as the unbiased controller for the war game and ensures that the staff stays within a timeline and accomplishes the goal of the war game
- **J-1/N-1**: Provides input as to how personnel support will be provided during an operation and how the operation may affect personnel status
- **J-2/N-2**: Has a dual role during the war game. First, role-plays the enemy commander and develops critical enemy decision points, projects enemy reactions to friendly actions, and determines enemy losses. Also is responsible for capturing the results of the enemy actions and counteractions in a war game worksheet. If the naval command is large enough, there may be a Red Cell that can assume the task of role-playing the enemy. Second, the J-2/N-2 provides input such as RFIs, named areas of interest (NAIs), target areas of interest (TAIs), high-value targets (HVTs) and high-payoff targets (HPTs). Also refines situation templates.
- **J-3/N-3**: Acts as the friendly force and ensures that the war game covers all of the operational aspects of the mission
- **J-4/N-4**: Provides analysis of logistics feasibility; identifies potential supply, transportation, and sustainment issues; and assesses the logistics functions that must be conducted in order to support the COA
- **Special staff**: In addition to the staff members listed above, the war game should include special staff personnel such as the Judge Advocate General or legal advisor to cover legal and ROE issues, the public affairs officer to handle questions about press guidance and themes, and a medical officer to determine if a COA may incur losses that would need additional medical support.

Role and Responsibility of the Red Cell
Ideally, the Red Cell consists of individuals of varied operational backgrounds and specialties. Combining their own operational experience with enemy tactics, weapons, and doctrine, the Red Cell provides enemy reactions to the friendly COAs during the COA war game. To be successful, the Red Cell must function as an extension of the J-2/N-2. The primary purpose of the Red Cell is to provide additional operational analysis of the enemy, tailored to the needs of the planning team. During the war game, the Red Cell employs ECOAs against the friendly COAs. Although the Red Cell is used principally at the JFMCC and NCC level and above, it can also be scaled for use by smaller units such as CSG, destroyer squadron, or air wing.

The objective of the Red Cell is not to defeat friendly COAs during the war game, but to assist the development and testing of friendly COAs. The Red Cell makes friendly COAs stronger and more viable for execution in battle. The J-2/N-2, in coordination with the J-3/N-3, determines the composition of the Red Cell and often provides a number of its analysts. The J-2/N-2 oversees the functioning of the Red Cell, as its analysis of the enemy must be coordinated with the J-2/N-2 staff. The J-2/N-2 provides the Red Cell with the initial detailed information on enemy location, weapons, tactics, doctrine, order of battle, and assessed COAs. Differences in analysis between the Red Cell and the J-2/N-2 must be identified and resolved. To be effective, the planning team and the Red Cell must exchange information and analysis continuously throughout the planning process.

Example Sketch & War Game Worksheets

Ref: NWP 5-01, Navy Planning (Jan '07), pp. 4-7 to 4-10.

Hasty War Game Worksheets (1 & 2)

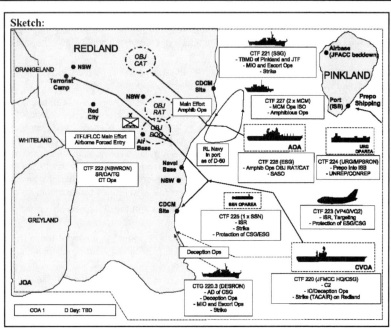

COA #	Critical Event: Establish Maritime Superiority					
Sequence Number	Friendly Action	Enemy Reaction	Friendly Counter Action	Time	Friendly Forces/ Assets	Remarks
1	Conduct offensive SUW, USW and strike warfare (STW) to achieve maritime superiority	Conduct coordinated attacks using CDCM, hit-and-run tactics by surface ships, maritime strike aircraft, and diesel subs. Conduct covert mining.	Maintain layered defenses; increase offensive ops as necessary. Conduct MCM in impacted area.	2–5 days	CSG (CVW & CGs, and DESRON) SUBRON (SSNs) SSG (CG, DDGs) PATRON (VP/AIP)	Shortfall in MCM assets. Supplement to ROE needed for preemptive strikes on Redland mine storage facilities. Adjust PIR to look for indications of mining.
2	Additional rows if needed to continue action/reaction for this critical event.				Identify any items of concern that become apparent during the war game: force shortfalls, ROE needs, coordination issues, impact upon commander's governing factors, etc.	
3						

Ref: NWP 5-01, Navy Planning, fig. 4-2, p. 4-7 to 4-8.

5-54 (Naval Planning) III. Course of Action Analysis (Wargaming)

War Game Worksheet - JFMCC, NCC or DTFC

SUBORDINATES

COMPONENT/FUNCTION	ACTION	REACTION	COUNTERACTION	REMARKS
CTF 220 (TRCSG)	Conduct offensive SUW against Redland naval forces. Conduct air strikes against Redland maritime strike support bases.	Conduct coordinated strikes by maritime air, ships, SS.	Increase SUW and strike operations.	
CTF 226 (NASESG)	Maintain miscellaneous operational details, local operations (MODLOC). B/P to conduct CSAR/tactical recovery of aircraft and personnel (TRAP).	Surveillance by MPA, combat support (CS) radars, and diesel subs. Increase CDCM activity-additional training at site and increased radar activity.	Move crisis action planning (CAP) station or CG to protect ESG. Move MODLOC outside CDCM range during daylight.	
CTF 221 (COWSSG)	Move to deception MODLOC and conduct IO deception plan. Conduct maritime interception operations (MIO) in JOA. Provide TBMD against Redland short-range ballistic missile (SRBM) launch on Pinkland.	Conduct ISR that penetrates/reveals the deception. Fortify hatches, doors; place obstacles/booby traps at access points to ships. Move missiles to protect against strikes and use as weapon of last resort.	Discontinue deception operations that adversely impact other operations. Use NSW and/or MEU (SOC) for visit, board, search, and seizure (VBSS). ISR to locate SRBMs.	Coordinate CTF 221 positioning for TBMD purposes with the area air defense commander (AADC).
CTF 227 (MCMRON)	Establish Q-routes in amphibious objective area (AOA).	Attack MCM units with CDCM and/or hit-and-run surface attacks.	CTF 220/221 provide defense and conduct counter strikes.	
CTF 223 (COMPATRON)	Conduct ISR in support of CTF-220 SUW and undersea warfare (USW) in coordination with CTF 225 and CTF 221.	Increased strip alert for fighters. Scramble for DEFPAT.	Continue ISR. Coordinate with JFMCC, CTF 220/221/226 for indications and warning (I&W) of Redland fighter activity.	
CTF 222 (NSWRON)	Insert into Redland to conduct SR and TG. Support CTF 221 for maritime interception operations (MIO)/expanded maritime interception operations (EMIO).	Increased patrols by Redland ground forces to locate NSW.	Continue SR and TG. Coordinate with JFMCC and ISR assets for I&W of Redland ground force movements. Shift ops as necessary. Extract as required.	Coordinate with JFSOCC.
CTF 225 (COMSUBRON)	Conduct offensive USW against Redland SS in the JOA. Conduct ISR.	Close littoral to evade SSNs.	Maintain aggressive ASW ops – ensure amphib landing area free of Redland SS.	

OPERATIONAL FUNCTIONS

COMPONENT/FUNCTION	ACTION	REACTION	COUNTERACTION	REMARKS
INTELLIGENCE	Provide I&W on Redland maritime strike forces & ISR in support of SUW & USW efforts.	Conduct deception to mask movement of strike force.	Activate named area of interest (NAI) 24.	
FIRES	Shaping TLAM and tactical aircraft (TACAIR) strikes against Redland C2, CDCM, naval facilities, and mine storage locations.	Conduct deception to mask movement of strike force. Move aircraft to bunkers or disperse. Move ships to dispersal bases. Move CDCMs.	Activate NAI 24.	Deconflict C2 strikes with IO plan.
LOGISTICS	Move MSF units to protected area (east) at onset of hostilities. B/P to conduct UNREP & CONREP as required.	Attack logistics shipping.	Provide additional protection for logistics and prepo shipping.	
COMMAND & CONTROL	CTF 220 given TACON of ESG and MCM units in order to coordinate antisurface warfare (ASUW) & ASW efforts.	CTF 225 given TACON of coalition SSN for USW.		
PROTECTION	Maintain CSG & ESG in ASW-sanitized MODLOCs. Establish air defense missile engagement zone (MEZ) and/or fighter engagement zone (FEZ).	Establish joint action area (JTAA).	Station SSG AEGIS ships for TBMD.	

COMPONENT/FUNCTION	ACTION	REACTION	COUNTERACTION	REMARKS
DECISION POINTS	Execution of amphibious ops into Redland.	Movement of MCM toward coast of Redland.	Insertion of NSW into Redland.	
CCIR	Location Redland SSs. Location Redland antiship cruise missile (ASCM) & CDCM units. Indications of Redland Maritime Strike activity. Evidence of Redland covert mining. Any mechanical/system/logistic discrepancy resulting in inability for unit self-defense. Any mechanical/system/logistic discrepancy resulting in inability for TLAM-capable unit to launch TLAM. Adverse weather conditions that can affect amphib ops or MIO/EMIO.			Establish a new NAI to determine movement of CDCM launchers.
BRANCHES	Terrorists found on board merchant during MIO/EMIO.	Redland mines in the vicinity of seaport of debarkation (SPOD).	Redland sub undetected.	

Ref: NWP 5-01, Navy Planning, fig. 4-3, p. 4-9 to 4-10.

may decide to withhold some assets for unforeseen tasks or opportunities, using this analysis to determine priorities of support. In this case, the commander may withhold one or more submarines to focus on the enemy submarine threat.

During the war game, the commander can modify the COA based on how the operation develops. When modifying the COA, the commander should validate the composition and location of decisive and shaping operations and reserve forces based on the factors of mission, enemy, terrain and weather, time, troops available, and civilian considerations (METT-TC), and adjust control measures as necessary. The commander may also identify combat situations or opportunities or additional critical events that require more analysis.

9. Refine Staff Estimates

The commander's staff continues to develop its estimates with the goal of capturing key support and execution considerations resulting from the war game. Staff members use these during the next step, COA comparison and decision. Issues for inclusion may include risk assessment and vulnerability assessment, casualty (to equipment or personnel) projections and/or limitations, personnel replacement requirements, projected enemy losses, ISR requirements and limitations, rules of engagement, high-value targets and high-payoff targets, METOC impacts, projected assets and resource requirements, requirements for prepositioned equipment and supplies, projected location of units and supplies for future operations, and C2 system requirements.

10. Update and Refine Intelligence Preparation of the Operational Environment (IPOE) Products

Intelligence preparation of the operational environment is a continuous, parallel process that takes place throughout all stages of planning. The planning team updates, refines, and prepares IPOE products to reflect the results of COA wargaming, facilitating the next step in the planning process, COA comparison and decision. By this point in the planning process, the following IPOE products and analysis must be well developed and detailed:

- Operational environment
- Analysis of the enemy
    1. Objectives
    2. Critical factors, DPs, and COGs
    3. High-value targets and HPTs, to include time-sensitive targets (TST) (Targets may and will change during the course of operations.)
- Enemy COAs: most likely and most dangerous

Additionally, PIRs have been identified, and a collection plan that focuses on NAIs has been built and refined.

See pp. 5-8 to 5-9 for an overview of the Intelligence Preparation of the Operational Environment (IPOE) process.

Target Examples

High-Payoff Targets (HPT)	High-Value Targets (HVT)	Time-Sensitive Targets (TST)
amphibious ships (for enemy amphibious assault) enemy petroleum, oils, and lubricants (POL) storage enemy commercial shipping bringing in logistics and military hardware enemy integrated air defense system (IADS) C2	enemy mine-laying assets enemy submarines enemy CDCMs along a SLOC	mobile surface-to-air missile (SAM) location of key enemy leader

Ref: NWP 5-01, Navy Planning, fig. 4-5, p. 4-15.

The Wargaming Process
Ref: NWP 5-01, Navy Planning (Jan '07), pp. 4-14 to 4-15.

Each game turn usually consists of three moves—two by the friendly force, one by the enemy force. The friendly force has two moves because the activity is intended to validate and refine the friendly force's COA, not the enemy's. If necessary, additional moves may be required to create desired effects.

When formally conducted, wargaming is a disciplined, interactive process that examines the execution of friendly COAs in relation to the enemy. When conducted informally, it may be as simple as a "what if" conversation between the naval commander, planning team, and staff. Whether formal or informal, wargaming attempts to foresee the friendly action, enemy reaction, and friendly counteraction dynamics of friendly versus enemy courses of action. In doing so, the commander and staff can visualize the flow of an operation and capture key elements that collectively define the synchronization of the operation. The analysis can focus on COA critical events selected by the commander or planning team or instead may wargame the COA over time (e.g., by phase, by day, etc.). To maintain the integrity of the wargaming process and strive for the best result to present to the commander:

- Evaluate each COA independently. Do not compare one COA against another
- Remain unbiased and avoid making premature conclusions
- Continually assess the suitability, feasibility, acceptability, distinguishability, and completeness of each COA
- Record the advantages and disadvantages of each COA
- Record data based on the commander's evaluation criteria for each COA
- Keep to the established timeline of the war game
- Identify possible branches and sequels for further planning

1. Friendly Actions
The war game begins with the first friendly action. The war game then proceeds as each war-fighting function representative gives the details of the friendly COA. Representatives explain how they would predict, preclude, and counter the enemy's action.

2. Enemy Reactions
Normally the N-2 or a selected Red Cell speaks for the enemy and responds to friendly actions. The N-2 or Red Cell uses an enemy synchronization matrix and event template to describe the enemy's activities. The event template is updated as new intelligence is received and as a result of the war game. These products depict the locations of NAI and when to collect information that will confirm or deny the adoption of a particular COA by the enemy; they also serve as a guide for collection planning. The N-2 describes enemy actions by warfighting function. The N-2 presents the enemy's CONOPS and ISR, including the intelligence collection assets of the enemy and how and when the enemy might employ them. Also, the N-2 describes how the enemy would organize its operational environment; identifies the location, composition, and expected strength of the enemy reserve; and determines the anticipated decision point and criteria that the enemy commander might use in committing reserves.

3. Counteractions
After the enemy reaction is executed, friendly forces provide a counteraction, and the various war-fighting functions' activities are discussed and recorded before advancing to the next series of events.

Development of an indications and warning matrix, showing key indicators of enemy actions, may be helpful in providing scope and direction to the collection plan. Modified combined obstacle overlays (MCOOs), doctrinal templates, and current enemy situation templates are updated.

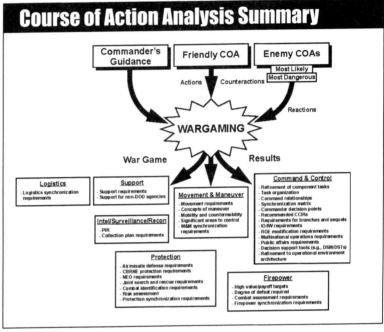

Ref: NWP 5-01, Navy Planning, fig. 4-6, p. 4-16.

Outputs

Course of action analysis steers the subsequent steps in the NPP. Figure 4-6 offers a summary of the wargaming processes. Required outputs of COA wargaming include the wargamed COA sketch and statement, as well as information on the commander's evaluation criteria. Additional outputs can include:

- War game results:
    1. Decision support matrix
    2. Refined task organization C2 requirements
    3. Identification of required assets and shortfalls
    4. Refined CCIRs
    5. Refined time-phased force and deployment data (TPFDD) inputs
    6. List of critical events and decision points
    7. Development of IO objectives and tasks
    8. Course of action war game worksheet
    9. War game synchronization matrix
- Updated IPOE products and refined staff estimates
- Subordinate commander's estimates of supportability
- Branches and sequels identified for further planning

The results of the war game are captured in a briefing to the commander and provide the contents and sequence for the COA war game brief.

Chap 5

IV. Course of Action Comparison & Decision

Ref: NWP 5-01, Navy Planning (Jan '07), chap. 5.

During the comparison step of the NPP, the planning team considers each retained COA for advantages and disadvantages. Each COA is evaluated in terms of the naval commander's previously established governing factors and, if modified, is tested a final time for feasibility and acceptability. This step is repeated until the naval commander selects the COA that offers the greatest prospect of accomplishing the mission.

IV. COA Comparison

Key Inputs	Key Outputs
• COA War Game Worksheet • Updated IPOE • Decision Support Matrix • Refined Staff Estimates • Governing Factors	• COA Decision • CONOPs • Updated IPOE • Refined Staff Estimates • WARNORD

1. Perform Course of Action Comparison
2. Perform Course of Action Evaluation
3. Make Final Tests of Feasibility and Acceptability
4. State Commander's Decision
5. Prepare Synchronization Matrix
6. Develop the Concept of Operations
7. Refine Intelligence Preparation of the Operational Environment

Ref: NWP 5-01, Navy Planning, fig. 5-1, p. 5-1.

Inputs

To help ensure that the COA comparison is thorough, the required input during this step is the COA war game worksheet, which includes the COA sketch, narrative, and war game results. The commander's governing factors are a critical input. Other information useful in this phase includes updated IPOE products, refined staff estimates, decision support matrix, and any branches and sequels identified for further planning. Additional information that may be useful includes: planning support tools,

(Naval Planning) IV. Course of Action Comparison and Decision 5-59

updated CCIRs, synchronization matrix, initial task organization, list of critical events and decision points, subordinate commander's estimates of supportability, identification of assets required, and shortfalls.

Process

1. Perform Course of Action Comparison

The actual comparison of the COAs is critical. A number of varying techniques exists for conducting the comparison, but each of them must assist the commander in reaching a sound decision. While many of the comparison techniques offer a numerical value, staffs must remember that these are simply decision aids. The greatest utility of these comparison techniques is not which COA has the highest score; rather, it is the insight into the strengths and weaknesses of each COA relative to a given governing factor.

The most common technique is to use a decision comparison matrix to facilitate the process. Four types of comparison decision matrices used are weighted numerical, nonweighted numerical, plus/minus/neutral, and advantages and disadvantages. Regardless of the comparison matrix used, each of the governing factors or evaluation criteria chosen must be clearly defined so that staff and commanders understand why a particular factor is weighted as high or low. This is vital to ensure that the process is completed correctly. For example, simplicity might be selected as a factor. The purpose of simplicity is to prepare clear, uncomplicated plans and concise orders to ensure thorough understanding. However, simplicity might be a negative factor if it will not facilitate mission execution.

1. Non-weighted Numerical

The non-weighted method numerically portrays subjectively chosen governing factors that are compared/evaluated on their individual merits and are all treated equally.

2. Weighted Numerical

The weighted technique numerically portrays subjectively chosen and subjectively weighted governing factors. Each staff member may use an individual matrix or recommend a personal choice of governing factors based on the respective functional area. The commander reviews this list and adds to or deletes from it. The list need not be a lengthy one; there should only be a few governing factors, though enough to differentiate between the COAs.

The primary consideration when assigning relative weights is to ensure that the selected factors are in balance and not artificially inflated. Comparing COAs by category is more accurate than attempting to aggregate a total score for each COA.

3. Plus/Minus/Neutral

The plus/minus/neutral matrix is used when credible quantitative (numeric) scores for how well each COA satisfies each governing factor are not relatively apparent, and instead qualitative scores must be used. The governing factors for each of the COAs are rated with a (+) for a positive influence, (-) for a negative influence, and (0) for a neutral influence in comparison to the baseline COA.

4. Advantages and Disadvantages

In completing the advantage and disadvantages matrix, staff members list the advantages and disadvantages of each COA retained. This is perhaps the most important comparison step and should be used in conjunction with one or more of the previous matrices. As the advantages and disadvantages are listed, the staff will begin to see where trade-offs or modifications will have to be made to a COA.'

Refer to NWP 5-01, Naval Planning, app. G for sample comparison matrices.

2. Perform Course of Action Evaluation

The trade-offs between the COAs become apparent as the advantages and disadvantages are tabulated for each retained COA, making this step the most valuable in the comparison. The advantages and disadvantages of any particular COA could be quite lengthy and detailed. Many advantages and disadvantages should be carried forward from the COA development and analysis steps (see Appendix G, Annex G-3). The basis of comparison can incorporate the performance relative to the measures of effectiveness (MOEs) (developed during the COA analysis phase) and the governing factor(s) established by the commander. When reviewing each COA, consider what additional actions, if any, could be taken to reduce or overcome any disadvantages made apparent by the analysis. For instance, analysis of a COA could reveal that the naval forces are placed at significant risk by an enemy's mine threat. However, the COA could potentially work if strikes were used to destroy the mines and/or their mine laying platforms or if a more robust ISR plan were implemented to determine if mines had been laid. If any changes are made to a COA, the planning team must war game the COA again to ensure that no new shortfalls have been introduced. To maintain an unbiased approach, any action proposed to overcome a disadvantage in one COA should be applied to all of the COAs, where appropriate.

3. Make Final Tests of Feasibility and Acceptability

If COAs are modified, the planning team applies the final tests for feasibility and acceptability based on the results of the analysis in the previous step. The planning team compares feasible COAs to identify the one that has the highest probability of success against the most likely ECOA and the most dangerous ECOA. As a result of the final tests for feasibility and acceptability, the planners may find none of the COAs analyzed to be valid; consequently, new COAs should be created. New comparisons, evaluations, and final feasibility and acceptability analyses must be completed on each new COA. These new COAs should then be analyzed against the ECOAs.

4. State Commander's Decision

After completing its analysis and comparisons, the planning team identifies a preferred COA and makes a recommendation to the commander. Subordinate commanders may be present but are not required; their participation either in person or via video teleconferencing (VTC) enhances the planning process. The decision brief includes:

- The intent of higher headquarters
- The approved mission statement
- The status of friendly forces
- An updated IPOE
- Analysis of ECOAs
- Friendly COAs, including assumptions used in planning, results of staff estimates, advantages and disadvantages (including risk) of each COA (with decision matrix showing COA comparison), and feasibility and acceptability estimates
- Recommended COA

The naval commander weighs the relative merits of the various COAs and selects the one that is most effective to accomplish the mission. The commander has many choices after the decision brief and may choose to:

- Select a COA without modification
- Select a COA with modification
- Select an entirely new COA by combining elements of multiple COAs
- Discard all COAs and resume mission analysis and COA development as required

(Naval Planning) IV. Course of Action Comparison and Decision 5-61

Commander/Staff Interaction During/After COA Comparison & Decision

Ref: NWC Maritime Component Commander's Handbook (Feb '10), p. 2-17.

The fourth step in the planning process is a comparison of the remaining COA's. The commander and staff develop and evaluate a list of important governing factors, consider each COA's advantages and disadvantages, identify actions to overcome disadvantages, make final tests for feasibility and acceptability, and weigh the relative merits of each. This step ends with the commander selecting a specific COA for further CONOPS development.

Selecting a specific COA is a major decision. In many ways, it is the commander's last chance to drastically change planning direction. After the decision, COA's not selected are discarded as the staff conducts detailed planning. Primary responsibility shifts from the OPT to the entire staff for completion of the remaining planning steps: prepare plans and orders and transition.

Commander's Preparation

A. Review previous briefs.

B. Update situation.

C. Gather close advisers and discuss each COA:

    (1) What are the key advantages and disadvantages?

    (2) How are the operational functions employed/impacted?

    (3) Have any assumptions been proved fact or invalid?

    (4) Which COAs depend on external support?

    (5) Review critical events: How does each COA approach the critical events?

    (6) Does each COA pass a timeliness test?

    (7) What is your subordinate commanders' understanding of the COAs?

    (8) Do the COAs adequately support any identified adjacent "supported" commanders?

D. Receive an executive summary from OPT lead. The operations planning group lead highlights any changes to the COAs as a result of the wargaming process:

    (1) Which COA is recommended?

    (2) What were significant advantages/disadvantages?

    (3) Were there any unique considerations?

    (4) Was there any significant disagreement on the recommended COA?

    (5) Has coordination between the staff and OPT been adequate? (The COA decision brief is not the forum for OPT or staff members to bring up new good ideas/concerns. Don't let the decision brief become the non-OPT staff's coordination vehicle.)

Actual Brief

A. After completing its analysis and comparison, the staff identifies its preferred COA and makes a recommendation. The staff then briefs the commander. Component commanders may be present, but are not required, for the decision brief; their participation, either in person or via video teleconferencing (VTC), enhances the planning process. The OPT lead should brief the commander and principal staff.

B. Does the COA comply with previous intent and guidance?

C. Does the COA protect the friendly COG and attack the enemy COG?

D. Are the mission statement and commander's intent still valid?

E. Possible decisions:

    (1) Select a presented COA.

    (2) Direct a hybrid COA.

    (3) Send the OPT back to COA development with additional guidance.

F. Actions. What immediate actions must take place to facilitate the selected COA?

    (1) Movement.

    (2) Logistics.

    (3) Rules of engagement.

    (4) Collection.

    (5) Shaping operations.

    (6) External coordination.

G. What branch planning needs to take place?

H. Assign responsibility for plans and order development.

Commander/Staff Interaction After Decision

Unless the commander selects a COA without modification, the new or modified COA should be analyzed fully to include wargaming. Upon COA decision, the commander should conduct a review of the COA with subordinate commanders. The mission statement must be reviewed to ensure all essential tasks are captured. The decision guides warning order development, the concept of operations, and orders development.

The staff and planning team need to elaborate and provide details to the COA; e.g., develop a CONOPS. First, the staff and planning team will complete the synchronization matrix which was initially created during COA analysis. The synchronization matrix lays out in a tabular format a description of each force, function focus and actions throughout the phases of the operation. This provides a straightforward method to show the linkages of a potentially complicated operation.

The CONOPS describes how the arrayed forces will accomplish the commander's intent. It is the central expression of the commander's operational design and governs development of supporting plans or annexes. During this step, units are converted from generic to specific units.

In a time-constrained environment, a CONOPS may be used in place of an OPORD to execute the operation. As such it needs to fully describe the operation. It needs to describe the decisive and supporting operations (with associated tasks and purposes) and the employment of naval and operational functions. The commander should consider reviewing the CONOPS after completion either in a brief or in text.

See p. 5-65 for further discussion of CONOPS.

Other than the first choice, any modified or new COA selected should be analyzed fully (to the maximum extent that planning time allows), to include wargaming. The commander may need to rely heavily on the planning team's professional judgment and experience; however, the ultimate decision is the commander's alone. The commander also bears the responsibility that goes along with the decision of selecting a COA.

The decision is a clear and concise statement by the commander setting forth the selected COA. The commander translates the COA selected into a brief statement of what the force as a whole is to do. The commander may amplify the statement with other elements of the mission as appropriate. Each of these elements should be explained in writing in relation to the physical environment in which the expected action is to take place. The wording of the decision is not bound by rigid form.

Lesson Learned

Observe two general rules in wording the commander's decision: Express it in terms of what is to be accomplished, and use simple language so that the meaning is unmistakable. The results of the COA comparison are briefed to the commander for his final approval on a COA.

Sample Decision Brief

BRIEFER	SUBJECT
J-5/N-5	Higher headquarters intent
	Restated mission
J-3/N-3	Status of own forces
J-2/N-2	Updated intelligence estimate
	Terrain analysis
	Weather analysis
	Enemy situation
J-3/N-3	Own COAs
J-3/N-3, J-2/N-2	Assumptions used in planning
J-1/N-1, J-4/N-4, J-6/N-6	Results of staff estimate
J-5/N-5	Advantages and disadvantages (including risk) of each COA (with decision matrix or table showing COA comparison)
	Recommended COA (may differ from other staff)
	COS Recommended COA

Ref: NWP 5-01, Navy Planning, fig. 5-2, p. 5-4.

5. Prepare Synchronization Matrix

Based on the commander's decision and final guidance, the decision-making portion of the NPP is completed. The staff and planning team now transition to operational planning, refine the COA, and prepare to issue the order. In order for the staff to issue the plan or order, it must first turn the selected COA into a clear, concise CONOPS. This is aided by completing a synchronization matrix, which is initially created during the war gaming step of the NPP and is now refined. This internal staff planning tool is used in much the same manner as the war gaming synchronization matrix. The commander can use the COA statement as the CONOPS statement. The COA sketch can become the basis for the operation overlay. In addition, the staff assists subordinate staffs with their planning and coordination as needed.

Refer to NWP 5-01, Naval Planning, app. H for further discussion of synchronization matrixes.

The Concept of Operations (CONOPS)

Ref: NWP 5-01, Navy Planning (Jan '07), pp. 5-5 to 5-6.

Using the synchronization matrix, the planning team expands and integrates the available information and provides the CONOPS—an elaboration of the selected COA. It should include the commander's vision of how major events are expected to occur in the forthcoming combat action and the commander's intent. The CONOPS must be developed quickly so that subordinate commanders have the time necessary to prepare their own plans and units for the impending action. Having already identified the risks associated with the selected COA, the naval commander refines what level of risk is acceptable to accomplish the mission and approves measures to reduce the risks. If there is time, there is a discussion of acceptable risks with lateral and senior commanders. However, the higher commander's approval must be obtained prior to accepting any risk that might imperil the higher commander's intent.

The CONOPS describes how arrayed forces will accomplish the commander's intent. It is the central expression of the commander's operational design and governs the development of supporting plans or annexes. Planners develop a scheme of maneuver by refining the initial array of forces and using graphic control measures to coordinate the operation and to show the relationship of friendly forces to one another, the enemy, and geography. During this step, units are converted from generic to specific units, such as the specific CSGs and ESGs. The CONOPS includes:

- The purpose of the operation
- A statement of where the commander will accept risk
- Identification of critical friendly events and phases of the operation (if phased)
- Designation of the decisive operation, along with its task and purpose
- Designation of shaping operations, along with their tasks and purposes, linked to how they support decisive operations
- Naval warfare functions (i.e., USW, SUW, AD, Strike, etc.)
- Intelligence, surveillance, and reconnaissance and protection operations
- An outline of the movements of the force
- Identification of options that may develop during an operation
- Location of engagement areas (surface, air, and subsurface) and objectives
- Responsibilities for AO and operating areas
- Concept of fires (i.e., employment of TLAM, carrier-based tactical aviation)
- Determined IO and/or deception concept of support and objectives
- Command and control attack priorities
- Prescribed formations or dispositions when necessary
- Priorities for logistics and sustaining operations
- Considerations of the potential effects of enemy WMD on the force.

6. Develop the Concept of Operations (CONOPS)

Using the synchronization matrix, the planning team expands and integrates the available information and provides the CONOPS—an elaboration of the selected COA. It should include the commander's vision of how major events are expected to occur in the forthcoming combat action and the commander's intent. The CONOPS must be developed quickly so that subordinate commanders have the time necessary to prepare their own plans and units for the impending action. Having already identified the risks associated with the selected COA, the naval commander refines what level of risk is acceptable to accomplish the mission and approves measures to reduce the risks. If there is time, there is a discussion of acceptable risks with lateral and senior commanders. However, the higher commander's approval must be obtained prior to accepting any risk that might imperil the higher commander's intent.

See previous page (p. 5-65) for discussion on developing the concept of operations.

7. Refine Intelligence Preparation of the Operational Environment (IPOE)

As the staff moves through the COA comparison process, it must remember to continually use and refine the IPOE as necessary.

See pp. 5-8 to 5-9 for an overview of the Intelligence Preparation of the Operational Environment (IPOE) process.

Outputs

The output of the COA comparison and decision provides the basis of orders development. The required output is the CONOPS. Additional outputs may include: updated IPOE products, planning support tools, updated CCIRs, refined staff estimates, and a WARNORD.

Key Points

- Keep the chain of command informed. Lower echelons of the planning team should keep their principal staff informed and, in turn, that staff should keep the major subordinate commanders informed of the planning progress.
- Members conducting COA comparison should be the same who were involved in the COA analysis; otherwise, much time is wasted in making new planners fully aware of the COAs.
- Use the governing factors and criteria to determine the best COA. Avoid the trap of comparing each COA to each other at this point. The best COA will become apparent as the process continues.

V. Plans and Orders Development

Ref: NWP 5-01, Navy Planning (Jan '07), chap. 6.

The plans and orders development step in the NPP communicates the commander's intent, guidance, and decisions in a clear, useful form that is easily understood by those executing the order. Operation plans are normally produced at the combatant command or JTF level with subordinate service or functional component commands (such as Navy component commands) producing supporting plans. In the case of a JFMCC or NCC, this would be the maritime supporting plan. Before proceeding, it is necessary to distinguish between plans and orders:

Plan
A plan is prepared in anticipation of operations and normally serves as the basis for an order. The procedures for producing a plan should therefore closely mirror the preparation of an order.

Order
An order is a written or oral communication that directs actions and focuses a subordinate's tasks and activities toward accomplishing the mission.

Various portions of the plan or order, such as the mission statement and staff estimates, have been prepared during previous steps of the NPP. The chief of staff or executive officer, as appropriate, directs plans or orders development. Plans or orders contain only critical or new information, not routine matters normally found in standing operating procedures. A good plan or order is judged on its usefulness, not its weight.

V. Plans & Orders Development

Key Inputs	Key Outputs
▪ Task organization ▪ Mission statement ▪ Commander's intent ▪ CONOPs ▪ Staff estimates	▪ OPORD/OPLAN ▪ Refined IPOE ▪ Planning support tools

1. Prepare Plans and Orders

2. Reconcile Plans and Orders

3. Backbrief and Crosswalk Plans and Orders

4. Commander Approves and Issues Plan or Order

Ref: NWP 5-01, Navy Planning, fig. 6-1, p. 6-1.

In the previous phase of the NPP, the planning team integrated the commander's selected COA with the staff estimates and planning support tools (developed in parallel) into a fully developed CONOPS. The planning team now translates the CONOPS into a clear, concise, and authoritative directive. This directive, whether it is a maritime supporting plan or an OPORD, is then back-briefed to the higher commander and cross-walked to other service and/or functional components to ensure that it is synchronized, understood, and meets the higher commander's intent. A well-written directive possesses important characteristics that help assure understanding of the directive and the accomplishment of the mission:

- **Clarity**: Each executing commander should be able to understand the directive thoroughly. Write in simple, understandable English and use proper military (doctrinal) terminology.
- **Brevity**: A good directive is concise. Avoid superfluous words and unnecessary details, but do not sacrifice clarity and completeness in the interest of brevity alone. State all major tasks of subordinates precisely but in a manner that allows each subordinate sufficient latitude to exercise initiative. Short sentences are more easily and quickly understood than longer ones.
- **Authoritativeness**: In the interest of simplicity and clarity, the affirmative form of expression should be used throughout all combat orders and plans.
- **Simplicity**: This requires that all elements are reduced to their simplest forms. All possibilities for misunderstanding must be eliminated.
- **Flexibility**: A good plan leaves room for adjustments that unexpected operating conditions might cause. Normally, the best plan provides the commander with the most flexibility.
- **Timeliness**: Plans and orders must be disseminated in enough time to allow adequate planning and preparation on the part of subordinate commands. Through the use of WARNORDs, subordinate units can commence their preparation before the receipt of the final plan or order. Concurrent planning saves time.
- **Completeness**: The plan or order must contain all the information necessary to coordinate and execute the forthcoming action. It also must provide control measures that are complete, understandable, and that maximize the subordinate commander's initiative. Only those details or methods of execution necessary to ensure that actions of the subordinate units concerned are synchronized with the CONOPS for the force as a whole should be prescribed.
- **Provides for the necessary organization**: A good plan clearly establishes command-and-support relationships and fixes responsibilities.

Inputs

The initial task organization, mission statement, commander's intent, CONOPS, staff estimates, specified tasks, and implied tasks are the required inputs for orders development. Other inputs may include:

- Updated intelligence and IPOE products
- Planning support tools
- Updated CCIRs
- Staff estimates
- Identified branches for further planning
- WARNORD
- Existing plans, standing operating procedures, and orders
- Chief of staff or executive officer orders development guidance

Commander/Staff Interaction During Plans and Orders Development

Ref: NWC Maritime Component Commander's Handbook (Feb '10), p. 2-23.

Orders development communicates the commander's intent, guidance, and decisions in a clear, useful form understandable to those executing the order. The commander normally is not closely involved in the administrative development of the plan or order. The MOC director, not the OPT lead, has responsibility for the writing and publishing of the order.

Plans and orders can come in many varieties, from the very detailed campaign plans and operations plans to simple verbal orders. They also include operation orders, warning orders, planning orders, alert orders, execute orders, and fragmentary orders. The more complex directives contain much of the amplifying information in appropriate annexes and appendices. However, the directive should always contain the essential information in the main body. The form may depend on the time available, the complexity of the operation, and the levels of command involved. However, in most cases, the directive should be standardized in the five-paragraph format.

- **Paragraph 1 — Situation.** The commander's summary of the general situation that ensures subordinates understand the background of the planned operations. Paragraph 1 often contains subparagraphs describing the higher commander's intent, friendly forces, and enemy forces.

- **Paragraph 2 — Mission.** The commander inserts his restated mission (containing essential tasks) developed during the MA.

- **Paragraph 3 — Execution.** This paragraph contains commander's intent, which will enable commanders two levels down to exercise initiative while keeping their actions aligned with the overall purpose of the mission. It also specifies objectives, tasks, and assignments for subordinates (by phase, as applicable, with clear criteria denoting phase completion).

- **Paragraph 4 — Administration and Logistics.** This paragraph describes the concept of support, logistics, personnel, public affairs, civil affairs, and medical services.

- **Paragraph 5 — Command and Control.** This paragraph specifies the command relationships, succession of command, and overall plan for communications.

Individual staff sections prepare appropriate annexes and appendices using staff estimates and the CONOPs as reference. Simultaneously, subordinate tactical organizations should conduct tactical planning to provide details to execute.

Orders development includes a two-step quality control process to ensure alignment and completeness. Reconciliation is an internal review within the fleet headquarters (HQ). A crosswalk is an external review conducted with higher, adjacent, and subordinate commanders and/or their staffs.

Orders reconciliation is the internal process in which the staff conducts a detailed review of the entire order. It ensures accuracy, agreement, coherency, and completeness and corrects any gaps. It compares commander's intent, the mission, and the CCIRs against the concept of operations and the supporting functional concepts (intelligence, logistics). It compares assigned tasks of the base order with the primary annexes to ensure linkage. The synchronization matrix initially developed in COA analysis can be expanded to accurately depict the linkage and alignment. Check the coordinating instructions to ensure completion and appropriateness. Ensure the PIR's and collection plan support CCIR(s). Identify and correct gaps and disagreements.

Orders crosswalk is the process of conducting the same detailed review that was done in reconciliation but executed with higher, adjacent, and subordinate staff representatives.

Process

1. Prepare Plans and Orders

Plans and orders are produced in a variety of forms. See Appendix L for a description of the purpose and format of WARNORDs, OPORDs, and fragmentary orders (FRAGORDs). Plans and orders can be detailed written documents with many supporting annexes, or orders may be simple verbal commands. Their form depends on the time available, complexity of the operation, and levels of command involved. Supporting portions of the plan or order, such as annexes and appendixes, are based on staff estimates, subordinate commander's estimates of supportability, and other planning documents.

A. Develop Base Paragraphs for Operation Plans and Orders

Directives most frequently use the standard five-paragraph format, briefly described below. In complex operations, much of the information required in the order is contained or amplified in the appropriate appendixes and annexes, such as synchronization and decision support matrices and logistics and sustainability analyses.

However, the essential form of the commander's CONOPS, including the commander's intent, command and control, task organization, and essential tasks and objectives, should be contained in the body of the order.

See facing page (p. 5-71) for further discussion and basic format.

B. Develop Appropriate Annexes, Appendixes, and Tabs

To keep the directive as simple and understandable as possible, details and amplifying information are placed in annexes. Typical annexes and their supporting appendixes and tabs include detailed plans for intelligence and operations (such as SAR/CSAR, movement, undersea warfare, etc.), instructions necessary for C2 (such as the communications plan), and information too complex to be covered completely in the basic plan (such as the detailed CONOPS, the logistics plan, and the complete task organization). Annexes may be referenced in the appropriate part of the body and should not include matters covered in SOPs. Annexes also allow for the selective distribution of certain information.

Lesson Learned

While it is not necessary to develop all annexes, it is important to quickly identify those to be developed and the responsible points of contact within each staff section. Additionally, someone should be designated to collect and review each annex for clarity, completeness, and consistency.

C. Confirm Time-Phased Force and Deployment Data (TPFDD)

A key aspect of each COA developed during the planning process is the availability and location of the forces required to execute the operation. This step represents the culmination of the deployment planning process. The planning team or appropriate staff section should conduct deployment planning in parallel with COA development since force availability, transportation, and sustainment are usually key determinants in the feasibility and acceptability of each COA considered. In conjunction with the development of each COA and its development into a detailed CONOPS, the TPFDD for the operation is reviewed, refined, confirmed, and prepared for execution.

If the plan or order is a supporting plan to a JTF OPLAN or OPORD, then the naval force's liaison on the JFC joint planning group (JPG) and/or joint deployment cell provides the required information for the deployment of naval forces during the development of the JFC's OPLAN/OPORD. The TPFDD is normally included as Appendix 1 to Annex A of the OPLAN/OPORD. CJSM 3122.02 JOPES Vol. III is the source document for TPFDD development and formatting.

Operation Plans & Orders (OPLANs/OPORDs)

Ref: NWP 5-01, Navy Planning (Jan '07), pp. 6-3 to 6-4.

Directives most frequently use the standard five-paragraph format, briefly described below. In complex operations, much of the information required in the order is contained or amplified in the appropriate appendixes and annexes, such as synchronization and decision support matrices and logistics and sustainability analyses.

However, the essential form of the commander's CONOPS, including the commander's intent, command and control, task organization, and essential tasks and objectives, should be contained in the body of the order. The format for OPLANs and OPORDs is contained in CJSM 3122.03A JOPES Vol. II. The five basic paragraphs for all plans and orders are:

Paragraph 1: Situation
This paragraph, the commander's summary of the general situation, ensures that subordinates understand the background for planned operations. It often contains sub-paragraphs describing enemy forces, friendly forces, and task organization, as well as higher headquarters guidance.

Paragraph 2: Mission
The commander inserts his own restated mission developed during mission analysis. This is derived from the mission analysis step and contains those tasks deemed essential to accomplish the mission.

Paragraph 3: Execution
This paragraph expresses the commander's intent for the operation, enabling subordinate commanders to better exercise initiative while keeping their actions aligned with the operation's overall purpose. It also specifies the objectives, tasks, and assignments for subordinate commanders. It should articulate not only the objective or task to be accomplished but also its purpose, so that subordinate commanders understand how their tasks and objectives contribute to the overall CONOPS.

Paragraph 4: Administration and Logistics
This paragraph describes the concepts of support, logistics, personnel, public affairs, civil affairs (CA), and medical services. The paragraph also addresses the levels of supply as they apply to the operation.

Paragraph 5: Command and Control
This paragraph specifies command relationships, succession of command, and the overall plan for communications and control.

Refer to NWP 5-01, Annex L for sample orders (warning order, basic operation order format, and basic fragmentary order format). The format for OPLANs and OPORDs is contained in CJSM 3122.03A JOPES Vol. II. concepts of support, logistics, personnel, public affairs, civil affairs (CA), and medical services.

Refer to The Joint Forces Operations & Doctrine SMARTbook (Guide to Joint, Multinational & Interagency Operations), chap. 3 for discussion of joint operation planning, to include operation plans and orders production (with sample formats).

2. Reconcile Plans and Orders

Orders reconciliation is an internal process in which the planning team conducts a detailed review of the entire order. This reconciliation ensures that the basic order and all the annexes, appendixes, etc., are complete and in agreement. It identifies discrepancies or gaps in the planning. If discrepancies or gaps are found, the planning team takes corrective action. Specifically, the planning team compares the commander's intent, the mission, and the CCIRs against the CONOPS and the supporting concepts. Also, the collection plan must support the CCIRs.

3. Backbrief and Crosswalk Plans and Orders

During the orders back brief and crosswalk, the planning team compares the order with the orders of higher commanders and other services and/or components to maximize synchronization and unity of effort and to ensure that the higher commander's intent is met. The crosswalk and back brief process can be done via telephone, VTC, scheduled meetings and briefings, or a combination of methods. This process is the culmination of a feedback process that has been occurring continuously during the planning process. The planning team should be in regular contact with the planning teams of the senior commander and the other services and/or components. Operation phasing, timing, and critical events decision points, as well as operational concepts, are compared among the components. The process identifies discrepancies or gaps in planning and, if any are found, enables the planning team to take corrective action.

Lesson Learned

New technologies in collaborative planning systems are extremely useful for both collaborative and parallel planning for U.S. military participants. However, all organizations involved in the planning process may not have access to these systems (e.g., other U.S. Government agencies and multinational components).

4. Commander Approves and Issues Plan or Order

The final action in orders development is the approval of the plan or order by the commander. While the commander does not have to sign every annex or appendix, it is critical that the commander review and sign the basic plan or order.

Outputs

The output of orders development is an approved plan or order. Additional outputs may include refined intelligence and IPOE products, planning support tools, and may also outline FRAGORDs for branches.

Key Points

1. Regularly scheduled meetings to share information among participants are critical, especially in a crisis situation.

2. When possible, all commanders should attend a single meeting/VTC for back briefs. Component commander feedback pays big dividends in cross-component understanding.

3. Crosswalks in multinational environments are absolutely essential due to different interpretations of languages and unique doctrines. The crosswalk should include Annex A, a task organization of the various plans/orders with the force structure listed in the TPFDD.

4. Centralized distribution of orders ensures that information is current and has been properly disseminated. The OPORD and supporting FRAGORDs must be the single source of authority. Briefings and slides may cause problems in dissemination of information because if not written in the plan, an order will not be executed.

Chap 5

VI. Transition

Ref: NWP 5-01, Navy Planning (Jan '07), chap. 7.

The purpose of transition is to ensure a successful shift from planning to execution. A good transition enhances the situational awareness of those who will execute the order, maintains the intent of the CONOPS, promotes unity of effort, and generates tempo. Transition facilitates the synchronization of plans between higher and subordinate commands and aids in integrated planning by ensuring the synchronization of the war-fighting functions. Transition requires free flow of information between commanders and staffs by all available means.

VI. Transition

Key Inputs	Key Outputs
• OPORD/OPLAN • Refined IPOE • Outline FRAGORDs for branches • Information for future missions/sequels	• Subordinate commanders and staffs • Ready to execute the order and possible branches • Prepared to plan sequels

1. Transition Briefing

2. Transition Drills

3. Confirmation Brief

Ref: NWP 5-01, Navy Planning, fig. 7-1, p. 7-1.

At higher echelons, where the planners may not be involved in plan execution, the commander may designate a representative as proponent for the plan or order. Normally this is the current operations representative to the planning team. After orders development, the proponent takes the approved plan or order forward to the staff charged with supervising execution. As a full participant in the development of the plan, the proponent can answer questions, aid in the use of the planning support tools, and assist the staff in determining necessary adjustments to the plan or order. Transition occurs at all levels of command. A formal transition normally occurs on staffs with separate planning and execution teams. Planning time and personnel may be limited at lower levels of command, or planners may be the same personnel as the executors.

Inputs

For transition to occur, an approved plan or order must exist. The approved plan or order, along with additional staff products, forms the input for transition. These inputs may include refined intelligence and IPOE products, planning support tools, outlined FRAGORDs for branches, information on possible future missions (sequels), and any outstanding issues.

Process

Successful transition ensures that those charged with executing the order have a thorough understanding of the plan. Regardless of the level of command, transition ensures that those who execute the order understand the commander's intent, the CONOPS, and NPP planning aids. Transition may be internal or external in the form of briefings or drills. Internally, transition occurs between future plans or future and current operations. Externally, transition occurs between the commander and his subordinate commands.

In the specific case of a naval command, the transition that occurs from planning to execution of an operation is done through a variety of formats. At an operational or operational-tactical level such as a JFMCC, NCC, or CTF, a commander may use such things as a daily intentions messages (DIMS), forums such as a warfare commanders' board or other meeting during the battle rhythm, or even voice communications with the force.

Further down at the tactical units such as a ship, the commander may use meetings, one-on-one communications, or night orders.

Role of the Planning Team

During transition, the planning team may perform the following:

- Conduct the internal transition briefing.
- Brief all tools (decision support matrix, synchronization matrix, execution checklist), enemy situation, CONOPS, and supporting concepts (intelligence, fires, logistics, maneuver) in detail. Current operations can conduct transition drills using this information.
- Assist the commander in the transition/execution drill.
- Coordinate with subordinate commanders on the confirmation briefing of their plan to the higher commander so that he/she can identify discrepancies between his subordinate commander's plans.
- Provide a transition proponent to current operations.

1. Transition Briefing

At the higher levels of command, transition may include a formal transition briefing to subordinate or adjacent commanders and to the staff supervising execution of the order. At lower levels, it might be less formal. The transition briefing provides an overview of the mission, commander's intent, task organization, and enemy and friendly situation. The briefing ensures that all actions necessary to implement the order are known and understood by those executing the order. The commander, deputy commander, or chief of staff may provide transition briefing guidance, including who will give the briefing, the briefing content, sequence, and who is required to attend. Time available dictates the level of detail possible in the transition briefing. Orders and supporting materials should be transmitted as early as possible before the transition briefing. The briefing may include items such as higher headquarters mission (tasks and intent), approved mission statement and commander's intent, CCIRs, EEFIs, task organization, situation (friendly and enemy), approved CONOPS (with supporting concepts), execution (including branches and potential sequels), coordinating instructions, decision points, and planning support tools (decision support template/matrix and synchronization matrix).

Planning Roles of the FPC, FOPS, COPS (and Internal/External Transition)

Ref: NWC Maritime Component Commander's Handbook (Feb '10), p. 2-25.

Future Plans Center (FPC)
The future plans center (FPC) conducts deliberate long-term operational planning; that is, planning that is focused on a time period beyond the scope covered by COPS and FOPS. Typically, the emphasis of the FPC is on planning the next phase of operations or sequels to the current operation. In a campaign, this could be planning the next major operation (the next phase of the campaign) or re-planning the initial effort based on assessments. The FPC is manned by personnel who are familiar with the Navy Planning Process and associated JOPES products. During an emergent crisis, FPC could be directed to lead the staff's effort to develop the JFMCC's OPLAN or OPORD.

Future Operations (FOPS)
FOPS conducts operational-level planning for near-term operations between those covered by the FPC and COPS. Typically, the emphasis of FOPS is on conducting planning in the current phase to include anticipated branch plans and crisis planning to deal with unanticipated circumstances. When it is assessed that the operation is not progressing as planned, it will fall to FOPS to adjust the plan to get back on track. Any operational plans developed by FOPS need to be synchronized and coordinated with the FPC and COPS. FOPS has primary responsibility for changing force allocation and resourcing approved plans. FOPS operates continuously and is composed of experts in various warfare areas who are assembled as the director of FOPS deems necessary to perform planning, commander's guidance development, orders preparation, and liaison with subordinates and other components.

Current Operations (COPS)
COPS' primary focus is on monitoring and assessing ongoing operations for compliance with the commander's intentions. COPS is responsible for overseeing execution of operations. COPS is the central point for all B2C2WG to forward and to receive information related to the execution of operations. COPS is responsible for monitoring the current situation and reflecting any changes to the execution of assigned orders by all subordinate forces. COPS must be capable of short-term operational planning, usually through a CAT and the development of associated FRAGORDS. COPS must also monitor the CCIRs. Lastly, COPS should be responsible for keeping track of the command relationships of subordinate maritime forces.

Transition
Transition is the final step of the NPP to ensure successful shift to execution. The commander's role is to ensure adequacy of the turnover of responsibility to the full staff and subordinate commands for execution. The commander needs to ensure that all aspects of the plan are discussed thoroughly. Possible branches and sequels and the status of their planning should be included. A methodical process considering each command's responsibilities is essential. There are two types of transition:

- **Internal Transition**. Internal transition moves the plan from **FPC to FOPS** or from **FOPS to COPS**. It is recommended to determine as much as possible how and when a plan will transition at the outset of plan development. The transition provides an overview of the mission, commander's intent, task organization, and enemy and friendly situation.

- **External Transition**. External transition is where FOPS briefs subordinate tactical commanders and staff. Subordinate tactical commanders may then have to provide a confirmation brief to the JFMCC, to ensure understanding and alignment with the JFMCC plan.

2. Transition Drills

Drills are important techniques used during transition to ensure the greatest possible understanding of the plan or order by those who must execute it. Drills improve the ability of the commander and staff to supervise operations.

A transition drill is a series of briefings, guided discussions, walk-throughs, or rehearsals used to facilitate understanding of the plan throughout all levels of the command. The commander and subordinate commanders conduct transition drills. Typically, a transition drill is the only drill used at lower levels of command, where the planning team both develops and executes the plan. Transition drills increase the situational awareness of the subordinate commanders and the staff and instill confidence and familiarity with the plan. Map exercises and rehearsals are all examples of transition drills.

See also p. 4-10, Rehearsals. NWP 5-01, Navy Planning, app. N contains an expanded discussion on the types of and techniques for drills and rehearsals.

3. Confirmation Brief

A confirmation brief is given by a subordinate commander once planning is complete. Subordinate commanders confirm the plan to their subordinates who will actually execute the mission, with the overall force commanders in attendance. The participants brief the execution portions of their subordinate plans, including the commander's intent, specific task and purpose, the relationship between their unit's mission and the other units in the operation, and their detailed operational plans, including actions on the objective, when applicable. The confirmation brief allows the higher commander to identify discrepancies between his order and the subordinates' plans and learn how the subordinate commanders intend to accomplish their mission.

Outputs

The outputs of a successful transition are subordinate commanders and staffs who understand the CONOPS and are ready to execute the order and possible branches and are prepared to plan sequels in priority.

Key Points

- Although a formal transition occurs on staffs with separate planning and executions teams, a similar process takes place at all levels of command. At the higher echelons, the commander may designate a representative as a proponent for the plan or order.

- Transitions can take the form of briefings, drills, exercises, or rehearsals. The level of understanding increases with time available to conduct the transition. As the completeness of the transition increases, additional preparation time and resources are required.

Chap 6
I. Fundamentals of Naval Logistics

Ref: NDP4, Naval Logistics (Feb '01), chap. 1.

I. The Mission of Naval Logistics

The mission of naval logistics is to provide and sustain the operational readiness of our naval forces, and to support the operational readiness of other forces as directed. In peace, operational readiness enables our naval forces to accomplish a wide variety of missions—independently or in conjunction with other services, agencies, allies, or coalition partners. In war, this same operational readiness is the root of war fighting effectiveness; it makes victory possible.

Effective logistics is a force multiplier, allowing the commander to maintain greater masses of power in harm's way for longer periods. This is accomplished through optimizing readiness at best value while providing responsive maintenance and sustainment. Naval logistics has historically provided the full range of logistics support to naval forces. Additionally, naval logistics forces provide sealift for the projection and sustainment of naval and non-naval forces.

Effective naval logistics enables us to carry out the Navy and Marine Corps' assigned roles. It supports our ability to conduct continuous forward presence, peacetime engagement, deterrence operations, and timely crisis response from the challenging maritime and littoral environment.

Through our logistics systems, Navy and Marine Corps striking power is always available, and always sustainable through an established support system. An extensive defense distribution system comprised of military bases at home and abroad, combat logistics force ships, and expeditionary support forces including airlift and sealift, as well as resources from sister Services, host nations, and commercial contractors provide the means for this projection power.

Sustained forward deployment of naval forces also allows our nation to pursue regional coalition-building and collective security efforts. Thus, naval logistics forces must be able to provide and receive support within a variety of organizational structures. Consequently, engagement in joint and multinational logistics efforts are increasingly critical to support mutual readiness and capability, enhancing the efficiency and effectiveness of our combat operations.

Naval logistics operations are conducted much the same in peace as they are in war. They support and sustain the war fighter whenever and wherever, differing mainly in the magnitude of the requirements placed on the logistics systems and the level and types of threat to which these systems are exposed. A viable, accessible, and ready reserve of trained personnel and effective equipment, and reliable sources of war materiel, must back active logistics forces. These resources must also include agreements and understandings that permit the sharing of logistics resources among other services, other nations, and the private sector of all engaged nations.

Refer to The Sustainment & Multifunctional Logistician's SMARTbook (Warfighter's Guide to Logistics, Personnel Services, & Health Services Support) -- updated with the latest doctrinal references (FM 4-0 Sustainment, FMI 4-93.2 Sustainment Brigade, JP 4-0 Joint Logistics, FM 3-35 Deployment & Redeployment, and more than 20 others joint and service publications) -- for complete discussion of strategic, operational, and tactical logistics.

(Naval Logistics) I. Fundamentals of Naval Logistics 6-1

II. Levels of Logistics Support

Ref: NDP4, Naval Logistics (Feb '01), pp. 5 to 6.

Logistics support is provided at the strategic, operational, and tactical levels, and involves interrelated and often overlapping functions and capabilities.

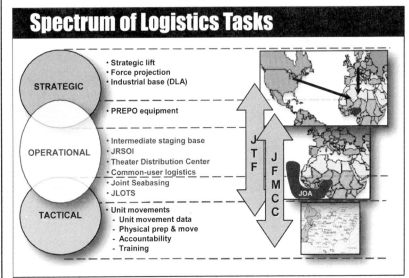

Ref: NWC Maritime Component Commander Handbook, fig. E-1, p. E-2.

A. Strategic Logistics

Strategic Logistics encompasses the ability to deploy and sustain forces executing the national military strategy whenever and wherever. It involves determination of requirements, personnel and materiel acquisition, and management of strategic airlift and sealift for the optimum levels of readiness at best value to the Navy. It also includes the role of pre-positioned equipment and materiel—both afloat and ashore—and our national ability to maintain the required support levels for the duration of operations.

B. Operational Logistics

The greater the scope or duration of anticipated military operations, the greater the impact of continuing effective strategic logistics operations. Operational Logistics involves coordinating and providing theater logistics resources to operating forces. It includes support activities to sustain campaigns and major operations within a theater and is the level at which joint logistics responsibilities and arrangements are coordinated. The unified combatant commanders and the supporting service component commanders are the main benefactors of this level of logistics.

C. Tactical Logistics

Navy tactical logistics encompasses the logistics support of forces within a battle group or amphibious readiness group and within Navy elements ashore, from both afloat platforms—including Combat Logistics Force (CLF) ships—and shore-based logistics support facilities. Tactical logistics support activities include maintenance, battle-damage repair, engineering, fueling, arming, moving, sustaining, material transshipment, personnel, and health service. Marine Corps tactical logistics, including combat service support (CSS), is provided by task-organized combat service support elements that complement the organic capabilities of the combat elements.

Scope of Logistics

Within the Navy and Marine Corps and throughout the Department of Defense, there continues to be pressure to reduce force levels and minimize system costs by rationalizing force constitution, projection, and sustainment around the world. From international and inter-Service acquisition programs to joint, multinational and interagency operations, cooperative activities have broadened both the resource base and the customer base for the naval logistician.

Whether for peacetime operations, war, or military operations other than war, logistics operations are conducted in support of forces, and are subject to the risks and uncertainties common to military missions. More broadly, logistics encompasses all of the processes, procedures, systems, and activities utilized to acquire, provide, maintain, and dispose of end products—equipment, supplies, facilities, services, and trained manpower—for military forces.

More than most components of military operations, logistics can be expressed mathematically. The quantification of requirements and capabilities demanded by the war fighters allow the logisticians to perform precise calculations and useful predictions. Projecting requirements for food or fuel in any operation confidently helps us project the outcome of maneuver or engagement. This predictive capability provides the baseline from which logisticians act in response to changing customers, customer locations, and support requirements.

Creative crisis response is another part of effective logistics; in spite of its scientific basis, logistics is also an art. Increased operating tempo and attrition of logistics capability through natural events, accidents, or enemy action combine to create shortfalls in support. These events reduce the reliability of previous projections, forcing the logistician to constantly monitor and adjust operations. Prediction, anticipation, innovation, and improvisation must be skillfully exercised as operations unfold. Logisticians must apply judgment and perception to the available information to ensure effective decision-making.

III. Process Elements

The activities of the logistics process may be reduced to four general elements—acquisition, distribution, sustainment, and disposition. Every logistics action may be expressed in terms of its contribution to one or more of these elements. These four elements make up our overall logistics process.

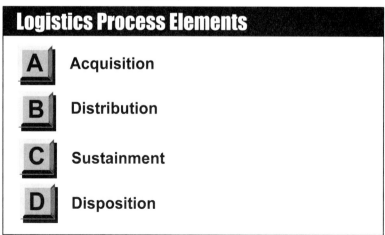

Ref: NDP-4 Naval Logistics, chap. 1.

A. Acquisition

The capability of naval forces rests on the investment in operational readiness. The principal acquisition organizations are the Navy and Marine Corps systems commands including Naval Sea Systems Command (NAVSEA), Naval Air Systems Command (NAVAIR), Space and Naval Warfare Systems Command (SPAWAR), Marine Corps Systems Command (MARCORSYSCOM), Naval Supply Systems Command (NAVSUP), Marine Corps Materiel Command, the Marine Corps Logistics Bases Command, Naval Medical Logistics Command (NAVMEDLOGCOM), the Naval Facilities Engineering Command (NAVFAC), the Defense Logistics Agency (DLA), and the General Services Administration (GSA). These organizations are responsible for procuring, producing, or constructing commodities, facilities, ordnance, and major weapon systems and end items. The Systems Commands are also responsible for life cycle management through a comprehensive systems support program known as Integrated Logistics Support (ILS). This program includes technical data, supply support, facilities, personnel, packaging, storage, handling and transportability, training and training support, maintenance planning, and design interface. This system also addresses environmental, safety, and health planning during acquisition.

Forward operations, geographically removed from much of the formal acquisition process, often demand time-sensitive reactions to support requirements. Local contracting can often support these requirements and reduce demand on the CONUS industrial base and may significantly reduce transportation requirements, while simultaneously reducing response time. NAVSUP coordinates the Navy Contingency Contracting Program through the Navy Regional Contracting Centers (NRCCs). The NRCCs provide a global network of field offices and deployable contracting capability.

NRCC contracting support may be augmented or supplemented by deploying additional reserve or other contracting support to theater. Additionally, DLA Contingency Support, Contract Administration Teams, and Fuels Management Teams can deploy to support CINC contracting needs. Within the engineering realm, NAVFAC administers the Construction Capabilities (CONCAP) contract, and also provides for the Navy timely real estate acquisition authority.

B. Distribution

Distribution refers to the processes used to get materiel, services, and personnel to the supported forces. It includes overall management, inventory control, and integration of information. Initiatives such as Direct Vendor Delivery have broadened the definition by moving distribution of selected items to the civilian sector. Increasingly, the logistics planner may incorporate non-military options into his mix of scarce distribution resources. Transportation decisions also depend upon what is being moved, its origin and destination, the lift assets available, and the urgency assigned. The transportation mode is based largely on the weight, size, urgency, and special handling requirements of the shipment. Airlift is normally reserved for passengers and high priority mail and cargo. Because a large proportion of naval operating forces are self-deploying, embarked on Navy ships, forward-deployed, or pre-positioned, distribution considerations during initial deployment are largely the concern of shore-based forces. Responsive distribution of sustainment is a monumental concern for all naval forces. High speed operational maneuver across broad areas of ocean, flexible reassignment of afloat units between task forces or groups, and operational movement of units in and out of theater (as in escort forces and shuttle ships) demand flexible distribution. Rapid embarkation and debarkation of Marine Corps forces, aircraft, staffs, and other units also challenge the distribution system by shifting customer locations.

The naval logistician must be adept at hitting constantly moving targets with critical sustainment, carefully monitoring ship and unit movements to anticipate the strategic and operational channels and modes most likely to put the support at the right place and time.

C. Sustainment

Sustainment is the provision of personnel, logistics, and other support required to maintain operations. This provision normally takes place at the operational level, where services and supplies processed through the distribution system actually reach the supported force. The term sustainment is also applied to specific materiel; in this usage, "sustainment" means those items planned or processed through the logistics system to fuel the sustainment element. Planners use this distinction to separate re-supply from forces in deployment planning. When national leaders call on naval forces, they expect both responsiveness and staying power.

Forward deployed naval forces carry with them initial sustainment stocks. Proper sustainment allows forces to remain on station as long as needed. Establishing and maintaining this reliable flow of materiel and services to operating forces is accomplished through the operation and management of logistics support activities. Sustainability depends on the effective participation of all providers across the functional areas of logistics.

D. Disposition

Disposition is the handling, stowage, retrograde, and disposal of materiel and resources released or returned by forces. Logistics economy, attainability, and sustainability are all dependent on the careful husbanding of limited resources. Similarly, efficient processing and shipment of excess materiel replenishes stocks available to other theaters, and can reduce the theater "footprint" needed by removing unessential stocks.

Disposition includes cleanup of environmental and other damage incident to operations. Minimizing environmental damage requires responsible and conscientious action at all levels. Naval commanders must also act to protect the environment during all phases of an operation. Noise, air and water pollution, waste disposal, hazardous materiel storage, and accidental discharge are examples where environmental damages potentially can occur.

All military forces are required to protect the environment to the extent operationally feasible through applicable DOD, local, national, and international environmental laws and regulations. The Navy, through the Supervisor of Salvage, has an oil spill-response capability including systems, equipment, materiel, and personnel. Working together with the Coast Guard, who has primary responsibility for oil-pollution response for U.S. waters including the Economic Exclusion Zone (EEZ), the Navy is committed to support cleanup actions in response to major oil and hazardous substance spills, accidental releases, and environmental terrorism.

IV. Principles of Logistics

Ref: NDP4, Naval Logistics (Feb '01), pp. 20 to 23.

Naval logistics—provided at the strategic, operational, and tactical levels; organized within the six major functional areas; and accomplished through application of the logistics process—is guided by a set of overarching principles. Each plan, action, organization, report, procedure, and piece of equipment may be defined and measured in terms of these principles. Each logistics decision is guided by the application of these principles. They are applicable to all military logistics, and provide the common foundation of joint and naval logistics doctrine. Both the operational commander, who needs to know the effective limits of the available logistics support, and the logistics planner, who has to ensure that all the essential elements of the logistics system are incorporated, must understand these principles. These principles of logistics include responsiveness, simplicity, flexibility, economy, attainability, sustainability, and survivability.

1. Responsiveness

Providing the right support at the right time and at the right place. This is the most important principle of logistics, because it addresses the effectiveness of the logistics effort, and in war an ineffective effort leads to defeat. Ensuring that adequate logistics resources are responsive to operational needs should be the focus of logistics planning. Such planning requires clear guidance from the commander to his planners.

It also requires clear communication between operational commanders and those who are responsible for providing logistics support. The operational commander's concept of operations must be thoroughly familiar to the supporting elements to ensure responsive, integrated support. Responsiveness is a product of logistics discipline, and commanders and logisticians who consistently overestimate their requirements—in quantity or priority—risk slowing the system's ability to respond.

2. Simplicity

Avoiding unnecessary complexity in preparing, planning, and conducting logistics operations. Providing logistics support is not simple, but plans that rely on basic systems and standardized procedures usually have the best chance for success. The operational commander can simplify the logistics task by maintaining cognizance of the available logistics capabilities, communicating clear priorities, and establishing support requirements based on current and accurate data.

3. Flexibility

Adapting logistics support to changing conditions. The dynamics of military operations are such that change is both inevitable and rapid. Logistics must be flexible enough to support changing missions; evolving concepts of operations; and shifting tactical, operational, and strategic conditions. A thorough understanding of the commander's intent enables logistics planners to support the fluid requirements of naval operations.

In striving for flexibility, the logistics commander considers such factors as alternative planning, anticipation, reserve capabilities, and redundancy. The task organization of shore-based support tailored from advanced base functional components is an example of flexible logistics response to anticipated operational requirements.

4. Economy

Effective employment of logistics support assets. Logistics assets are allocated on the basis of availability and the commander's objectives. Effective employment requires the operational commander to decide which resources must be committed and which should be kept in reserve. Additionally, the commander may need to allocate limited resources to support conflicting requirements. The prioritization of requirements in the face of limited forces, materiel, and lift capability is a key factor in determining the logistics feasibility of a plan. Common-user materiel, facilities and services may be sourced through joint, combined, or commercial providers at significant savings in transportation, stocks, and facilities. While certain redundancies may be necessary to responsiveness and survivability, the reduction in logistics "footprint" compounds savings by negating the requirement to support and protect larger logistics operations.

5. Attainability

The ability to acquire the minimum essential logistics support to begin operations. The difference between this minimum essential level of support and the commander's desired level of support determines the level of risk inherent in the operation from a logistics viewpoint.

The accurate determination of the minimum requirements, and the time it will take to reach that level given the available resources, allows the commander to determine the earliest possible date for the commencement of operations. The principle of attainability allows the commander to pursue a higher level of logistics confidence, but an operation undertaken without meeting the minimum needs determined under this principle is, by definition, destined to fail.

6. Sustainability

Ensuring adequate logistics support for the duration of the operation. Sustaining forces in an operation of undetermined duration and uncertain intensity is a tremendous challenge. Forces may operate with a diminished level of support for some time, but every means must be taken to maintain minimum essential support at all times. Sustainability derives from effective planning; accurate projections of requirements; careful application of the principles of economy, responsiveness, and flexibility to provide required support; and successful protection and maintenance of the lines of communication. Additionally, sustainability is dependent on discipline within the operating forces when establishing requirements and expending limited resources.

7. Survivability

Ensuring the functional effectiveness of the logistics infrastructure in spite of degradation and damage. Logistics forces, sites, transportation modes, lines of communication, and industrial centers are all high-value targets that must be protected. Logistics ships, aircraft, vehicles, and bases may be vulnerable to direct attack by enemy forces or terrorists.

Similarly, these assets and the systems that utilize them are subject to disruption by natural disaster, weather, communications failures, civil unrest, contract and labor disputes, legal challenges, and the political decisions of other nations. Survivability requires a robust and diverse logistics system capable of sustaining forces in the face of any obstacle. Dispersion of installations and materiel, maintenance of alternate modes of transportation and lines of communication, redundant logistics communication systems, adequate stock levels, reserves of equipment and personnel, phased delivery, effective use of deception operations, and alternate sources of supply can all support survivability. Force reconstitution and replacement, decontamination, reconstruction, re-equipment, repair, or relocation may restore the effectiveness of logistics systems degraded by battle damage or other events. Accordingly, the survivable logistics must include sufficient assets to support its own recovery as well as the operating forces.

The principles of logistics are always in evidence in a successful operation, but seldom have equal influence. At times the principles make conflicting demands. For instance, total responsiveness and survivability cannot be achieved with maximum economy. The operational commander, supported by his logistics planners, must weigh the relative importance of each principle to the specific operation. By carefully considering each principle in light of prevailing circumstances, the commander is guided toward an effective support plan that will be in consonance with operational requirements and the available logistics resources.

V. High Yield Logistics (HYL)

Ref: NDP4, Naval Logistics (Feb '01), pp. 83 to 86.

The Navy and Marine Corps have developed concepts designed to support military operations through a wide range of options. These will be increasingly expeditionary, forward positioned, sustainable, maneuverable, and streamlined. Many components of Focused Logistics are already impacting naval logistics operations. Current and projected initiatives are expanding electronic connectivity, real time access to the common operating picture, and a current global inventory of logistics assets and activity. High Yield Logistics (HYL) charts this course for the Navy's logistics strategy in the new millennium.

The plan's broad initiative is the reduction of the operating and support costs for fielded systems through technology insertion. By making funds available to purchase repair parts that are engineered for longer life and optimal performance through technology, the Navy is freeing up funds for modernizing weapon systems while maintaining readiness and sustainability. The "High Yield Logistics" strategy attacks all aspects of logistics. It focuses on three primary objectives, which describe how the vision will be achieved:

- Supporting the warfighter
- Outsourcing
- Optimizing

Precision

Precision is the watchword in the theater environment of the future. The right support must reach the customer when and where needed. Inventories must carry the needed items only. Logistics response time will decrease as inventories become more visible and management systems become more responsive. Depots must be sized and located to support the naval force without waste. Regionalization and consolidation process for depot and intermediate maintenance is already underway, and contractors and manufacturers will contribute through initiatives such as Life Cycle Support. Precision requires emphasis on joint operations and integration, with additional focus on the deliberate planning process to accurately determine and represent naval requirements and theater capabilities. Management of stocks, including WRM, must address sourcing, transportation, and positioning to maximize availability to naval or other supported forces.

Extensive and flexible sealift support must be available to place precise support where and when it is required. A distribution system relying on velocity to replace depth of stocks cannot afford a less-than-responsive transportation element. Strategic sealift enhancement through continuing addition of LMSR ships, and initiatives to ensure a successful National Defense Features program for the U.S. merchant fleet will support joint strategic sealift requirements. A follow-on CLF and naval integration into joint theater distribution will provide more efficient and faster distribution to the afloat forces. Enhanced logistics support of Marine Corps forces in the sea base will address improved support of operations from the sea.

Munitions management, including inventory reduction, revised positioning and storage, enhanced visibility, and uniform environmental protection will bring increased precision, economy, and efficiency. Modular organization of Fleet Hospitals, Naval Mobile Construction Battalions, Navy Cargo Handling Battalions, Navy Air Cargo Companies, and other expeditionary shore-based forces will allow precision responses to support requirements in the theater. The initiative for Fleet Hospital detachments responds directly to the JHSS for mobile distribution of essential care delivery.

Marine Corps Precision Logistics initiatives will measure logistics response times and repair cycle times for analysis and reduction, adapt commercial business practices to logistics operations, and improve Marine Corps logistics distribution and information systems. The Marine Corps Materiel Command will address Marine Corps connectivity to theater distribution systems, improving access to precise support through joint, common-user, and cross-Service capabilities.

Information
No single aspect of logistics has received more attention in the joint and multinational arenas than information. It is almost impossible to discuss any aspect of current or future operations without addressing the enabling power of information technology. Information will help identify and locate inventory, but it does not substitute for availability, delivery, or decision making.

The Naval Logistics Information Strategic Plan has been promulgated to streamline logistics processes, reduce life cycle costs, and create synergy of data. Naval logistics information systems will be reengineered and reorganized to reduce costs and cycle times. Outdated processes will be changed or eliminated to increase efficiency. Enhanced asset visibility programs will join with regionalization and inventory-sharing programs to minimize inventory requirements. New programs, such as Initial Requirements Determination/Readiness Based Sparing, will revolutionize parts planning and management. Advanced diagnostic and training systems will be employed to ensure the system and the operators or maintainers are at peak performance. New concepts like the Configuration Management Information System will provide more current and accurate systems data for each major systems installation.

Transformation
To effect change, the Navy and Marine Corps will reevaluate materiel, maintenance, and facilities management with the specific goal of identifying additional opportunities for outsourcing, consolidation, or regionalization. Express delivery services, Prime Vendor and Direct Vendor Delivery, common-user and cross-Service logistics, multinational support, and contingency contracting reflect the ongoing transformation of naval logistics. Tomorrow's naval logistics distribution system will be characterized by a concentration of expeditionary military logistics capabilities in theater, and a concentration of economical, capable private sector capabilities in support.

Partnership
Partnership focuses on integration of naval forces as good citizens. The global partnerships required to achieve theater success address both military and non-military issues. Key objectives of these partnerships include effective stewardship of our environment and human resources. Partnership involves every logistics function in timely environmental cleanup and comprehensive pollution prevention programs. It requires engineer support of environmental protection in facility and systems planning, integrated plans to protect the natural and cultural resources of naval installations, implementation of hazardous material control and management (HMC&M) at major bases, and inclusion of energy and environmental conservation factors in facility planning.

Sea-Based Logistics Concept
Navy and Marine Corps operational concepts such as "Forward...from the Sea" and "Operational Maneuver from the Sea" require bold departures in sustainment options. "Sea based Logistics" is a conceptual framework envisioning support of expeditionary Navy and Marine Corps shore-based forces from the sea base. The five primary tenets characterizing Sea based Logistics Concept are:

- **Primacy of the sea base** — over the horizon, reduced or eliminated footprint
- **Reduced demand** — seabased support, technology improvement, lighter force ashore
- **In-stride sustainment** — network-based, automated logistics for maneuver units
- **Adaptive response & joint operation**s — expanded missions, joint support
- **Force closure & reconstitution at sea** — building and restoring combat power

VI. Focused Logistics

Ref: NDP4, Naval Logistics (Feb '01), pp. 80 to 83.

Focused Logistics includes six tenets. These tenets combine to provide responsive support at any level and in any type of military operation, with reduced logistics response times, inventories, costs, infrastructure, and shortfalls.

1. Joint Theater Logistics Management

Synchronization and sharing of Service logistics capabilities can reduce the logistics presence required to support joint operations. Joint Theater Logistics C2 is one alternative offered to provide clear lines of authority by assigning responsibility for logistics support in joint operations to a single entity.

2. Joint Deployment/Rapid Distribution

Joint Deployment/Rapid Distribution is the process of moving forces to the operational area and providing them with accelerated delivery of logistics support. This requires improved transportation and information networks; visibility and accessibility will squeeze maximum support from limited assets. Navy and Marine Corps doctrine will continue to emphasize the unique characteristics and contributions of operations on and from the sea, and Navy and Marine forces may sometimes forego specific process steps, but the steps and claimants in the Service deployment processes are recognizable and definable in joint terms. It will focus at the strategic level on continuing improvements in core sealift and airlift capabilities, en route infrastructure agreements and upgrades, and increased utilization of commercial delivery.

3. Information Fusion

Information fusion will accomplish universal access to appropriate information through GCSS. GCSS will provide near-real-time logistics C2 and a common support picture through shared data and applications. The situational awareness and access to assets implied by network-centricity will extend to the logistics arena and ensure that reload and repair are as responsive and flexible as the operational maneuver they support.

4. Multinational Logistics

Continuing evolution of support relationships between the U.S. and allies and coalition partners will yield stronger regional contacts, more effective multinational operations, and equitable distribution of logistics tasking and responsibility. Efforts toward increased logistics cooperation amongst our allies are concentrated in four areas—common operational framework, expansion of bilateral agreements, interoperability through technology sharing, and leveraging the capabilities of multinational partnerships.

5. Joint Health Services Support

Focused Logistics directs a joint health service support strategy (JHSS) supporting Force Projection with essential care in theater, robust aeromedical evacuation, and definitive care in CONUS. This care is oriented to ensure a healthy and fit force, to prevent casualties, and to administer effective casualty care and management.

6. Agile Infrastructure

Agile infrastructure will improve joint logistics policies, structures, and functional processes to permit maximum economical application of these options. The result will be reduced logistics forces, equipment, supplies, and facilities—all achieved with the overriding objective of maintaining effective support. The actions necessary to reengineer infrastructure and achieve more economical logistics support will rely on outsourcing requirements where practical and effective, instituting commercial business practices, improved engineering and maintenance support, enhanced inventory management, and increased pre-positioning and war reserves.

II. Naval Logistics Planning

Ref: NDP4, Naval Logistics (Feb '01), chap. 2.

The dynamic process of providing logistics support to our operational forces is one characterized by the need to respond to continuous change: e.g., changes in support required because actual usage exceeds expected consumption; changes in user location to keep ahead of enemy moves; changes in quantities needed to replace losses in transit or at the theater depot. A responsive logistics planning system and integral information support allows naval logisticians to keep up with these necessary changes to maintain our operational war fighting readiness through uninterrupted logistic support whenever, wherever.

Naval logistics planning and information support is designed to answer these questions: What materials, facilities, and services are needed? Who is responsible for providing them? How, when, and where will they be provided? To find answers, we start with sources of logistics planning guidance, then apply a formal process that parallels operational planning procedures. The nature of the situation will determine whether we apply a deliberate or a crisis action planning process. Using one of these processes, a general plan is formulated that covers the organization, procedures, and policies of logistics support groups and the specific directives or instructions detailing the execution of support for a particular operation. These rules and tools are imperative for optimizing logistics systems responsiveness to the war fighter. Naval logistics information support systems keep the plan current, accurate, and adequate by providing data on the status of logistic resources, operational force needs, and the ability to meet those needs. Logistics planning and information support are thus complementary.

Information enables a commander to apply his experience and judgment to deviate from existing plans. Similarly, formal planning can organize and prioritize a commander's information needs, allowing him to select the best courses of action and adapt what he knows to the situation.

Operational Logistics (OPLOG)

Tactical, strategic, and operational logistical support and sustainment can be differentiated by the scale of military action. OPLOG extends from the theater's sustaining base or bases to the forward combat service support units and facilities organic to major tactical forces. Therefore, it links strategic logistics to tactical logistics. Its main purpose is to ensure that one's actions are continuous through all phases of an operation. Effective OPLOG must balance current consumption with the need to build up logistics support for subsequent operations. It must provide for lengthening the lines of communications and staging logistics support forward to maintain the desired operational tempo.

Operational logistics links tactical requirements to strategic capabilities to accomplish operational goals and objectives. It includes the support required to sustain operations. Operational logisticians assist in resolving tactical requirements and coordinating the allocation, apportionment, and distribution of resources within the AO. They coordinate with logisticians at the tactical level to identify theater shortfalls and communicate these shortfalls back to the strategic source. Operational-level logistics includes deployment, sustainment, and resource prioritization and allocation, and it identifies activities required to sustain the force. These fundamental decisions concerning force deployment and sustainment are key to providing successful logistical support.

I. The Logistics Planning Process

The planning guidance provided by joint, naval, and multinational doctrine forms a sound, consistent, and authoritative foundation for naval planning. Naval logistics planners should be ready to participate in joint and combined operations as partners in the planning process, and as spokespersons for Navy and Marine Corps interests and requirements. Joint Logistics Planning is part of the joint operation planning process. Operational planning, conducted simultaneously at the strategic, operational, and tactical levels, provides the framework for employment of military forces to achieve specified objectives during contingencies. Planners provide for five major activities of joint operations: mobilization, deployment, employment, sustainment, and redeployment.

The Deliberate Planning Process

Deliberate planning prepares for a possible contingency based on the best available information, using forces and resources apportioned by the CJCS in JSCP. Most deliberate planning is done in peacetime, based on assumptions regarding the political and military circumstances that may prevail when the plan is implemented. Deliberate planning is highly structured and occurs in regular cycles. It produces an operations plan (OPLAN), concept plan (CONPLAN), or functional plan.

Logistics planners prepare the staff logistics estimate during the concept development phase of deliberate planning. This provides the commander with the information to support courses of action (COA) selection, and is developed concurrently with the commander's estimate. Logistics also plays a major role in plan development as supported and supporting CINCs determine support requirements and resolve shortfalls. Finally, logistics planners at many levels prepare supporting plans to provide the mobilization, deployment, sustainment, reconstitution, and redeployment of forces and resources in the OPLAN.

Crisis Action Planning

Crisis Action Planning (CAP) is conducted in rapid response to actual circumstances. CAP follows the general pattern of deliberate planning, but adds flexibility for timely action. If an existing OPLAN is adaptable to the situation, CAP procedures are used to adapt an existing OPLAN to actual conditions or to develop and execute an operation order (OPORD).

II. Naval Logistics Planning

Naval planning occurs within the framework of the joint planning process. When naval forces are assigned, attached, or apportioned to unified or specified commanders, planning is done in support of the commander's intent. Naval planners provide input to the concept development (including the logistics estimate), plan development, and plan review phases. When a combined, joint, or naval task force is established, the component commander is directly responsible to the task force commander for development of supporting plans, including the logistics annex. Common planning processes allow products from each level or component to effectively support the overall plan.

For an OPLAN or OPORD, planners for Naval Component Commanders will receive the concept of operations, force apportionment, time-lines, and other pertinent information, and then promulgate the appropriate guidance, and task subordinate and supporting commanders to provide the additional information necessary to build the logistics plan. They will also work within the Joint Planning and Execution Community (JPEC) to resolve strategic and theater-wide planning issues. The Naval CINC logistics planners review and approve subordinate inputs, incorporate the data and requirements into the Navy plan, and represent the Navy CINC to resolve shortfalls, de-conflict issues, and develop comprehensive and feasible logistics annexes to the OPLAN or OPORD. In the event of a Navy OPORD or General Operation Order (OPGEN), the Navy CINC planners follow the same basic process, but without a requirement for JPEC refinement.

Navy CINC logistics planners often request and receive planning support from subordinate commands and supporting CINCs. Planners for the numbered fleets, or other command levels below the component commander, develop detailed logistics requirements. The numbered fleet commander considers the level of supported forces, the timing of their arrival, planned movements, projected operations tempo and the distance and capabilities of potential support sites and maintenance facilities.

Detailed support requirements and shortfalls are determined from these considerations. The CINC then incorporates these results in the final product, a detailed logistics concept of support. The Navy commander at this level—through his logistics staff and planners—may have responsibility for joint logistics coordination within theater, and will be concerned with both Navy and common-user theater stocks and services.

Transportation planning assumes a minor role for initial deployment of most ships. While sustainment requirements bring the afloat force into competition for limited strategic transportation assets, the transportation feasibility of ships' movement during the deployment phase of an operation is almost always assured. Initial requirements for every logistics function are transported with the supported force. Endurance loaded ships may operate for weeks with minimal external support.

Naval units at sea must also place planning emphasis on theater infrastructure and lines of communication (LOC) issues. Theater infrastructure concerns do influence planning with regard to ports and facilities for shore-based support, maritime prepositioning force (MPF), assault follow-on echelon (AFOE), and logistics over the shore (LOTS) operations; and other strategic sealift discharge. Forces afloat must also be concerned with characteristics of the theater regardless of land or sea.

Conversely, the support of ships at sea is complicated by specific environmental (wind/sea state) impediments to re-supply, and broader threat spectrums (including subsurface as well as surface and air). The U.S. Navy excels at underway replenishment (UNREP), but weather, threat, or operating conditions can render UNREP impossible at times.

Shore-based Naval logistics introduces additional planning requirements. Sites must be identified for the advanced logistics support sites (ALSSs) and forward logistics sites (FLSs). Capacities, layout, equipment, and competing requirements determine the throughput the sites can accommodate, and the logistics forces necessary to support that throughput. Plans are structured using ABFCs.

Logistics planning within the battle group generally addresses near-term operations and emergent requirements. The battle force logistics coordinator and staff will plan and develop operation tasks (OPTASKs) promulgating days of supply, logistics staff responsibilities, replenishment priorities, logistics reporting, and coordination of support within the battle force and with the underway replenishment group and shore-based logistics organizations.

Marine Corps logistics planning reflects the tradition of operations from the sea, focusing on its expeditionary nature. Expeditionary operations ashore generally require establishment of forward bases and creation of a theater logistics system. Expeditionary operations generally involve five phases of action: predeployment, deployment, entry, enabling/decisive actions, and redeployment. Planning for predeployment addresses logistics, interfaces, FSSG support to the deploying MAGTF, civilian support, mobilization personnel requirements, fiscal authority, facilities, and remain behind equipment. Entry planning includes amphibious operations, MPF operations, air contingency MAGTF operations, Marine Operations other than War, or any combination of the four. Enabling/decisive planning considers sustainment through lodgment for logistics and other support capabilities, sea basing requirements, the need for the sea echelon, and the potential to transition to sustained operations ashore. Finally, reconstitution planning provides for potential follow-on missions and redeployments.

(Naval Logistics) II. Naval Logistics Planning 6-13

III. Logistics Support to the Navy Planning Process (NPP)

Ref: NWC Maritime Component Commander Handbook (Feb '10), pp. E-6 to E-9.

The Logistics (Staff) Estimate

Commanders must be provided with an idea of cost, equipment, and manpower that it may take to sustain forces deployed. The LRC uses a logistic estimate to help with this process. Employment planning considerations directly impact the projection or deployment of forces. The concept of logistic support must be derived from the estimate of logistic supportability of one or more COAs developed during the commander's estimate phase of planning.

NWP 5-01, Navy Planning, annex K-3 provides a sample logistics estimate format.

The logistic estimate focuses on:

- Identify potential joint, common, and cross-service missions to be assigned to avoid duplication of effort and maximize efficiencies
- Identify stockage objective and accompanying supplies
- Identify availability of host-nation support (HNS), coalition support, pre-positioned stocks, and deficiencies
- Determine gross force closure times using existing ports, transportation infrastructure, and allocated transportation assets
- Identify bed-down/intermediate staging base requirements and environmental issues
- Analyze ability of the enemy to disrupt logistics operations
- Identify deployment and employment critical requirements
- Conduct initial logistic force structure analysis. This should include the availability of all required logistics assets and staging installations.
- Conduct initial logistic risk assessment based on availability of strategic lift, support forces, support alternatives, and results caused by loss of any support element or node

Intelligence support is critical to preparation of the logistic estimate and plan feasibility analysis. Hostile activities may impede forward movement, destroy logistic stockpiles, close airports and seaports, and destroy prime movers of critical logistic elements. Hostile actions may invalidate logistic support assumptions made during contingency planning.

There are multiple stages of preparing the logistics estimate: selecting a COA that is logistically supportable, developing a logistics concept that supports that COA, and creating a comprehensive logistics Annex D after extensive coordination and information gathering and analysis. Operational logistics is a complex, interdependent concept that if properly leveraged can serve as a significant combat power multiplier for the JFMCC and JTF. Logistics TTPs illustrate the linkage between future plans, future operations, and current operations:

- Understand the operating environment — geometry
- Monitor the fight and be proactive; anticipate change in main effort, priorities, and branch plan activation or requirements
- Understand logistics limitations (capabilities vice desires)
- Prioritize and coordinate the logistics effort at your level of command
- Understand higher and adjacent missions
- Leverage the littoral LOC
- Recognize that the JFMCC must operate ahead of the major subordinate commands

The problem for logisticians is to design a logistics system that will extend the operational reach of the force, increase the endurance of the force, and generate tempo of operations. The following outlines some key logistics activities associated with the NPP with emphasis on the front-loaded nature of required support and the continuous refinement of the logistics estimate.

Ref: NWC Maritime Component Commander Handbook, fig. E-3, p. E-5.

A. Mission Analysis

During mission analysis, the LRC will conduct requirements-capabilities analysis. Logistics capabilities are often initially a shortfall — so the design of a logistics system is critical to the overall success of the cooperation.

- Initiate JFMCC logistics estimate
- Understand the higher headquarters' situation (mission, intent, concept of operations, supporting concepts)
- Assign logistics representatives to B2C2WGs
- Determine logistical requirements
- Determine friendly force logistical capabilities
- Analyze the physical network and logistical infrastructure within the AO and AI
- Determine assumptions
- Understand friendly and enemy capabilities/vulnerabilities
- Give input to WARNORD
- Perform continuous collaboration with HHQ, adjacent components, and major subordinate commands

V. Logistics Support to the Navy Planning Process (CONT)

Logistics-Intelligence
Intelligence has a critical role in the logistical preparation of the OE. Operational logistics-intelligence can be defined as that information necessary to plan and conduct the deployment and sustainment during an operation. Among other things, it includes the physical network analysis. Logistics-intelligence also evaluates geography, efficiency of transportation, throughput capacity and enhancements, infrastructure protection, echelon of support, and assignment of responsibilities.

Logistics-related CCIRs
- What is our initial operational reach?
- When weather degrades our support by more than 20 percent for more than 24 hours.
- Any line of communications (LOC) interdiction that disrupts our distribution capability for more than 8 hours
- When readiness rates for the main effort fall below 80 percent
- When weather impacts on aviation and sea state conditions, disrupting sustainment operations for more than 12 hours
- Any imminent indications and warnings or actual terrorist/special operations forces (SOF) attacks on critical logistics nodes
- Use of chemical, biological, and radiological agents on coalition forces by enemy forces

B. Course-of-Action Development
The physical network analysis (PNA) that began with the identification of the physical network during MA continues during COA development and is updated as information becomes available throughout the planning steps. Results of the analysis of the physical network:

- Define the point of diminishing returns of resources (people and equipment)
- Help identify the logistics feasibility of COAs
- Characterize the risk inherent in the network

The PNA, along with the capabilities-versus-requirements analysis from the initial staff estimates, becomes the basis for developing logistics concept of support.

C. Course of Action Analysis (War-gaming)
War-gaming is a "what if" game played against the selected enemy COA. By this point, all COAs will work since they have been validated in the war game and the staff has provided supportability estimates. The intent of the war game is to improve the blue (friendly) COA by finding gaps and seams in the plan, identify branches and sequels, and validate/invalidate assumptions.

- Critically analyze all planning assumptions
- Document war game results; this will assist in development of decision support templates/matrices
- Develop general logistics CONOPS for each friendly COA
- Identify logistics governing factors to be used in COA comparison

- Continue to refine the logistics estimate
- Submit and respond to RFIs as required

D. Course-of-Action Comparison and Decision

The LRC advocates which friendly COA is most favorable from a logistics support perspective using governing factors developed in the preceding steps. Given the factors of time-space-force upon logistics resources, the LRC will likely prefer the COA that allows the most time to position assets and best positions the force to transition from one phase to the next.

The LRC validates and/or evaluates friendly COAs against established criteria.

- Advocate most favorable blue COA. Examples of logistics evaluation criteria include:
    - Which is the least vulnerable to weather?
    - Which places the most materiel forward at end state?
    - Which best prepares you for starting the next phase?
    - Which requires the least amount of external resources?
    - What can we control vice not control?
    - Which poses the biggest risk (effect success)?
- Refine the logistics functional section of the synchronization matrix
- Review major subordinate command estimates of supportability for each COA
- Refine logistics estimate
- Prepare logistics decision support template and matrix
- Submit and respond to RFIs as required
- Review major subordinate command estimates of supportability
- Prepare the CONOPS for logistics support

E. Plans and Orders Development

- Release the latest logistics estimate and publish the logistics CONOPS. The following JFMCC annexes use logistics products/information:
    - Annex D, Logistics
    - Annex E, Personnel
    - Annex P, Host Nation Support
    - Annex Q, Medical Services
    - Annex X, Execution Checklist
- Develop detailed logistics CONOPS and all reporting guidelines
- Conduct orders reconciliation and crosswalk

F. Transition

Participate in all rehearsals and brief-backs and focus on operational execution.

IV. Logistics Planning Considerations

Ref: NDP4, Naval Logistics (Feb '01), pp. 28 to 29; pp. 34-36.

Logistics is the responsibility of the operational commander, who must ensure that his operations and logistics experts integrate their operation and logistic plans. Overall feasibility of these plans will be determined by their ability to generate and move forces and materiel into the theater, then forward to our operating forces.

Logistics planning needs to shape, anticipate, innovate, and be conducted concurrently with operations planning. Logistics planning is performed in parallel with naval operations planning. Logistic planners identify and resolve support problems early by working concurrently with, and in support of operations planners. All planners must consider the overall support requirements and capabilities. The logistics concept of operations parallels the commander's concept of operations, permitting subsequent detailed, tactical-level, support planning. Detailed logistics planning should:

- Achieve optimum war fighter readiness
- Optimize logistics systems responsiveness to the warfighter
- Earmark significant time-phased support requirements necessary to maintain and sustain the warfighter whenever, wherever
- Identify personnel and cargo throughput at shore-based logistics sites
- Identify transportation requirements to support the movement of personnel, equipment, and supplies
- Outline the capabilities and limitations of ports, including the Logistics-Over-The-Shore (LOTS) capability to respond to normal and expanded requirements
- Recognize support methods and procedures required to meet the needs of the sea, air, and land lines of communications
- Coordinate and control movement into the contingency area
- Develop reasonable logistical assumptions
- Define the extent of needed host nation resources
- Designate alternative support sources for host nation support failure
- Identify the engineering and construction requirements for sustainability
- Identify the source of funding for logistics support
- Delineate contracting responsibilities and authority
- Consider the meteorologic and oceanographic limitations
- Identify health service support requirements
- Identify the service and maintenance support requirements for sustainability

Multinational Logistics Planning

Economic, military, environmental, and other crises seldom confine themselves to a single nation. The U.S. unified, sub-unified, or joint task force (JTF) commander normally acts within the U.S. chain of command to prepare both unilateral plans and joint plans in support of treaty or alliance commitments. Within the combined chain of command, the U.S. commander and component commanders coordinate these plans with coalition or alliance plans. The principles guiding operational and logistics planning within international organizations are much the same as those within joint and naval planning, and logistics remains a national responsibility under allied doctrine. However, certain planning considerations gain significance in multinational planning. The following are representative of areas the naval logistics planner must address chains of command, national security interests, logistics interoperability, and planning and reporting requirements, methods and formats.

III. Logistics Command and Control Systems

Ref: NDP4, Naval Logistics (Feb '01), chap. 3.

Naval commanders monitor and direct forces through command and control (C2) systems. C2 systems are bolstered by information systems offering reliable data and the organizational and analytical tools to manipulate that data in support of effective decision-making. The command and control and information systems used to monitor and direct naval forces and operations, including naval logistics forces and operations, function under national authority and within a joint command and control system to permit effective coordination and employment of forces.

Additionally, naval forces may be assigned to combined forces. In this mode, naval command and control systems connect to allied systems, and naval information systems share with allied systems that information necessary for combined operations, consistent with U.S. security considerations.

Effective logistics support requires commanders at every level of supported and supporting forces to understand the organization and associated information systems of naval, joint, and combined forces. Development of joint command and control, and information systems is progressing rapidly. Commonality and interoperability have assumed a higher priority than ever before. Advances in C2 have been made possible by advances in information and communications technology, and divergent approaches to command and control are drawn closer by computer processing power and electronic media. Advances in C2 extend to command and control of naval logistics (Log C2), and supporting information systems.

Adequate information about the availability and location of support—together with information on the physical and operational environment constraining distribution or execution of that support—allow maximum responsiveness and economy. By squeezing the most support from the available assets, effective command and control can positively influence overall efficiency of the logistics system. Likewise, a war fighter with reliable information on his logistics support can achieve the required level of confidence at a lower level of supply. Thus, excess stocks and requisitions are avoided, allowing the logistics information to serve as both an enabler and a product of logistics command and control. Naval logisticians will always depend on effective command and control to achieve maximum support from minimum resources, facilitating effective decision-making.

I. Naval Logistics System Organization

Naval logistics forces fall within the same overall command and control structure as operating forces. Forces in theater are assigned to the operational control (OPCON) of the supported CINC. The logistics organizations, systems, and forces are both components and customers of the overall logistics system. Regardless of location or employment, all logistics forces rely on supply and transportation systems to distribute their own support and to fuel the support they provide. Joint, naval, and multinational organizations often co-exist and interact to provide effective theater logistics command and control.

II. Theater Logistics Command and Control

Ref: NDP4, Naval Logistics (Feb '01), pp. 43 to 45; figs. 3-1 to 3-3.

Logistics command and control in theater is the responsibility of the combatant commander, while logistics support is a Service responsibility. The combatant commander normally accomplishes control of naval logistics operations through the naval component commander. Dependent on the size and nature of operations and assigned forces, there may be either Navy (NAVFOR) or Marine Corps (MARFOR) component commanders, or both. Naval forces afloat, including Marine forces, will likely be OPCON to NAVFOR. Marine Forces afloat may receive common-item logistics support via the NAVFOR. Conversely, Naval forces landed in support of ground operations, which may include Navy forces, will be likely OPCON to MARFOR and may receive common-item logistics support via the MARFOR. Also, a joint force commander may designate JTFs sourced entirely from a single Service, or from functional components of several Services.

Naval organization for effective logistics support is predicated on the nature of the forces supported, and may be tailored to specific theaters and operations. The joint force commander (JFC) will determine his appropriate Log C2 organization based on the mission, operating environment, and assigned assets. This organization will manage common-user and cross-Service logistics, monitor and report logistics operations and capabilities, advise the combatant commander on logistics matters, and represent the command to external logistics organizations. Regardless of what specific form the organization takes, it will generally constitute or include a Logistics Readiness Center (LRC). The logistics staff will focus its monitoring, advising, and internal and external coordinating activities within this center.

While the joint logistics C2 structure may take many forms, most can be categorized within three primary models:

A. Joint Logistics Augmented Organization

Augmented log organization utilizes the existing J4 organization as the theater Log C2 organization. This logistics staff, augmented as necessary by the relevant Services and Agencies personnel, extends its role beyond the internal staff logistics functions to provide coordination and tasking for joint force logistics.

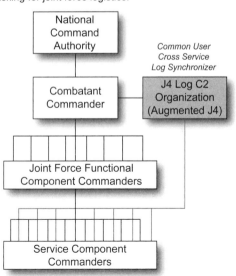

B. Joint Logistics Stand Alone Organization

In a second organizational form, a separate J4 focuses mainly on internal logistics, and a Log C2 tasking and coordinating position is created on the CINC's staff. Jointly staffed by the Services and Agencies, this organization can be activated and expanded as dictated by mission requirements.

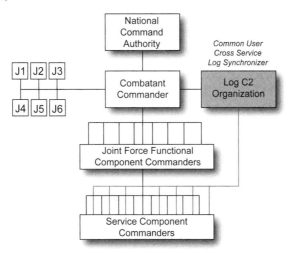

C. Joint Logistics Dominant Service Organization

For missions or areas where one Service represents the majority of the capabilities or requirements, the combatant commander may organize Log C2 by tasking the predominant involved Service's logistics agency with managing and coordinating joint requirements. Service and Agency liaison will be provided to represent component requirements.

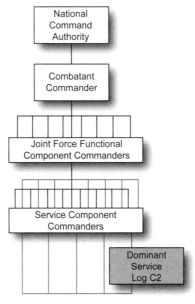

III. Naval Theater Logistics Command and Control

Ref: NDP4, Naval Logistics (Feb '01), pp. 44 to 48.

Naval Log C2 organization for forces afloat is often supported across multiple lines of communication from sites both within and without the operating area. These sites may be under control of different numbered fleet commanders. The forces afloat can also move in and out of the CINC's area of responsibility (AOR), drawing sustainment from changing CINCs as they transit the ocean. Shore-based forces in theater have different support requirements than forces afloat, and in some theaters may fall under different Service Log C2 organizations. Certain Navy forces ashore may be reassigned between Navy and Marine Corps (or other-Service) control, shifting support responsibilities. Finally, Marine Corps forces afloat shift substantial support requirements from Navy to Marine Corps logistics organizations when they go ashore. These challenges demand very flexible, but very well defined, Log C2. Like the unified CINC, the Navy fleet CINC has the three similar options for Log C2 organization.

The logistics forces of each numbered fleet overseas are organized into a standing task force. The commanders of these task forces act as the principal logistics agents for the Fleet Commander. They control assigned CLF shuttle ships; plan re-supply of ordnance, fuel, and repair parts; and plan and manage theater ship repairs in military and commercial yards. In some cases, the numbered fleet logistics task force commander may be "dual-hatted" on the CINC Fleet staff. In other cases, the Navy component commander may control logistics forces that are not assigned to a numbered fleet.

Logistics Command and Control Ashore

Navy expeditionary shore-based logistics forces include those assigned to the Navy component commander and those assigned to the Marine Corps forces. Navy expeditionary shore-based logistics forces not assigned to Marine Corps forces will normally be incorporated into logistics sites in theater. Commanders of these sites will report to the Navy component commander through the NCC's logistics C2 organization. Sites under NCC operational control include the ALSS and FLS central to Navy theater distribution.

While an ALSS or FLS will include airfields and seaports, Navy logistics forces will normally be tenants at these facilities, and will control only those forces and facilities specifically belonging to or given over to the Navy.

Operations conducted in the absence of ocean terminals include MPF operations, amphibious operations, and LOTS/JLOTS. MPF operations will be conducted through ports when available, but the organizational foundation of the operations remains the same. MPF operations require an aerial port of debarkation (APOD) for offload of two fly-in echelons (FIE). At the seaport or beach, the Naval Beach Group (NBG) commander becomes commander of the Naval Support Element and directs cargo offload operations. During amphibious operations, the NBG supports the landing.

A landing force support party (LFSP) is task organized from the NBG, Transportation Support Battalion, and other Navy organizations to provide initial combat service support. In JLOTS, the JFC will designate a JLOTS commander. Naval responsibilities will be as defined by joint doctrine and the JLOTS commander, and are generally influenced by the Service composition of the forces and sustainment being throughput. Naval logistics forces including medical battalions, dental battalions, medical logistics companies, and construction battalions also support MAGTF operations. These forces operate under the Marine Corps theater logistics organization.

The COMMARFOR may establish a MLC to facilitate reception, staging, onward movement, and integration (RSOI) and provide operational logistics to Marine forces. MLC is a task organization option, not a permanent organization. A FSSG may be assigned the resources and responsibility for MLC functions, based on the operational situation, geography, C2 (for both tactical operations and logistics), and infrastructure requirements. During deliberate planning the MLC supports the identification, preparation, and submission of host nation support, interservice support, and intertheater and intratheater requirements for the Marine Service component. The FSSG designated as the MLC deploys early to support arrival, assembly, and initial CSS missions to the MEF until its own CSSE can be established. The MLC then conducts general support and interfaces with other theater logistics agencies.

Marine Corps command and control of non-aviation logistics in the MAGTF is through the CSSE that may be a FSSG or subordinate element. All organizations in the MAGTF have limited logistics capability—when that capability requires augmentation, the CSSE provides combat service support. The CSSE commander takes direction from the MAGTF commander.

The Assistant Chief of Staff, Logistics (AC/S G-4) has staff cognizance for logistics, and identifies logistics requirements, recommends logistics priorities, and coordinates external support. As the MAGTF and its CSSE are task organized, the CSSE commander may use various C2 options. The FSSG commander may form a subordinate CSS Detachment (CSSD), centralizing control by giving the unit a general support mission, decentralizing control by giving the unit a direct support mission, or attaching the CSSD to the supported unit. The Assistant Chief of Staff, Aviation Logistics Department (AC/S ALD) coordinates aviation maintenance, aviation ordnance, aviation supply, and avionics for the MAGTF's ACE with the Marine Aviation Logistics Squadron (MALS). The ACE also possesses organic ground logistics capability in the Marine Wing Support Group (MWSG).

Logistics Command and Control Afloat

The commander of the afloat forces will exercise control of logistics through a Fleet Logistics Coordinator (FLC), Task Force Logistics Coordinator (TFLC), or Task Group Logistics Coordinator (TGLC). Guidance and direction for Navy logistics operations derives from OPGEN promulgated by the Navy operational commander to set general policies and procedures. An OPORD may be issued at various command levels to provide direction for specific operations. More specific guidance is provided by a series of OPORD appendixes or OPTASKs

IV. Multinational Theater Logistics Command and Control

Ref: NDP4, Naval Logistics (Feb '01), p. 49 to 52; fig. 3-4.

Command and control of logistics during combined operations is similar to joint command and control. Command and control of multinational logistics operations requires the commander and staff to be aware of the parallel national organizations involved, and to foster good relationships with national representatives at appropriate points within those "stovepipes." A few of the major factors in multinational operations follow:

- Combined operations can greatly multiply the overlapping organizations. As an example, defense of the Korean peninsula is entrusted to Service components reporting to a joint commander (CINCUSFK) working within a Republic of Korea/U.S. Combined Force Command (CFC) bilateral alliance that coexists with the United Nations Command (CINCUNC).

- Sovereign nations will not always give Multinational Force Commanders (MNFCs) operational control of their forces. When operational control is given, it may be accompanied by restrictive conditions that severely limit the commander's flexibility in employing the forces. This extends to logistics forces; OPCON may be extended to the MNFC, or limited directive authority may be granted.

- When OPCON over forces is granted, it does not automatically extend to logistics resources. Multinational operations do not provide directive authority over logistics unless specifically granted. Specified commanders within NATO are granted logistics redistribution authority to meet critical operational needs, but this is severely qualified.

- Forces are generally committed to multinational operations because of a community of interest; the military objectives of the force align substantially with the political objectives of the participants. Sovereign authority over forces, even those OPCON to the MNFC, supersedes any other.

The diversity of coalition or alliance members can be further complicated by the introduction of non-member nations into alliance operations. Ad hoc coalitions in response to emergent crises bring no C2 organization with them. It is assumed that when the U.S. is the coalition leader and dominant participant, U.S. joint or Service C2 organizations will prevail, but even willing coalition partners may sometimes lack the logistics robustness, interoperability, technology, or discipline to allow easy integration.

NATO Organization

There has been substantial success in developing combined command and control doctrine and procedures. NATO is the premier example of combined C2 for alliances involving many nations. As a standing organization that includes numerous members, operates in multiple theaters, and executes multiple types of missions, NATO reflects most of the challenges inherent in development of formal C2 for combined operations. NATO relies on a civil/military structure. Each member nation sends permanent ambassadors to the civil forums, and military representatives to the Military Committee. The highest civil forum is the North Atlantic Council (NAC).

The Military Committee (MC), which is one of several committees established under the authority of the NAC, is the highest military authority in NATO. NATO's upper level military structure is depicted in the NATO Logistics Handbook.

Forces assigned to NATO are task organized. Operational control passes to the Combined Joint Task Force (CJTF), but the United States does not relinquish command of its forces. The OPCON of forces does not entail control of logistics, which must be specified. When a CJTF is formed, the CJ4 is responsible for logistics coordination.

NATO Multinational Maritime Force (MNMF) doctrine requires member host nations to establish and operate necessary ALSSs or FLSs. Afloat support refers to logistics support ships providing sustainment, medical services, and repair support to MNMF ships underway or at anchor. Ashore support involves necessary sites, facilities, and forces to provide logistics support to the MNMF.

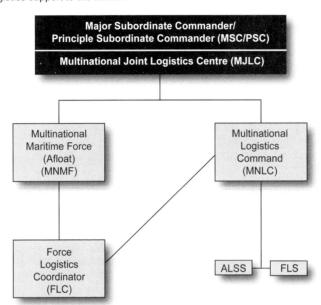

Afloat Support to the Multinational Maritime Force
Component forces may establish Multinational Logistics Centres or Commands (MNLCs). The MNLC for shore-based support of maritime operations is established as a command because of the requirement to command subordinate sites. Land or air MNLCs are established as centers, and are limited to coordinating authority.

Command and control of afloat support to the NATO MNMF is similar to U.S. Navy practices. The MNMF Commander assigns a Force or Group Logistics Coordinator (FLC/GLC) to ensure logistics readiness. The U.S. Navy commander within the MNMF will assign a Logistics Coordinator (LC) to provide coordination with the FLC/GLC. The Force Logistics Coordinator plans and executes MNMF afloat logistics policy, monitors inventory, and controls movement of sustainment to and within the task force or task group. The FLC is the Commander's direct liaison to the MNLC for shore-based support.

Shore-Based Support to the Multinational Maritime Force
Consistent with agreement between NATO Military Authorities and involved nations, the Multinational Logistics Commander (MNLC) controls and coordinates assigned shore-based logistics support forces. These include assets of the ALSS, FLSs, theater airlift, VOD, and shuttle tankers.

The MNLC will normally report to the NATO commander exercising OPCON of the Multinational Maritime Force. Shore-based theater distribution within NATO is very similar to the hub and spoke system operated by the U.S. Navy, but when an ALSS or FLS is established within a member nation, the commander of that site will be drawn from that nation. This facilitates cooperation and communication between NATO sites and the host nation, and often allows NATO to rely on existing capabilities with minimal startup delay.

V. Logistics Information Systems
Ref: NDP4, Naval Logistics (Feb '01), pp. 52 to 54.

Logistics command and control depend on the identification and communication of planned and actual support requirements, and the identification and application of logistics assets to meet those requirements. Conceptually, the quality of asset management determines whether scarce logistics resources can "stretch" to provide effective support. The commander, planner, or logistician is constrained by the accuracy and timeliness of available data. Such data is only useful when collected, analyzed, and refined into relevant information. Information systems enable every element, and support each principle of logistics. Information technologies have changed the way naval logisticians do business by fostering more efficient application of limited logistics resources. Increased emphasis on interoperability of equipment and standardization of procedures has allowed naval commanders and logisticians to lend to and gain from joint, other Service, and multinational logistics capabilities. The combined impact of the various systems in place and coming into service provides our naval forces with definitive advantages in planning, command and control, and operation of our logistics system.

Global Command and Control System (GCCS) and Other Primary Joint and Naval Logistics Information Systems

Global Command and Control System (GCCS) is the primary comprehensive automated command, control, communications, computers, and intelligence (C4I) system. It provides a worldwide network of military and commercial systems supporting information exchanges between the NCA, combatant commanders, and component commanders. Over 100 other major logistics information systems are in use by the Armed Forces, but as standardization continues, more systems feed common databases. The use of two emerging joint deployment information systems has been approved in an effort to enable the 72-Hour Time-Phased Force Deployment Data (TPFDD) time standard for deployment and provide operational capability in the near-term. The Transportation Coordinator's Automated Information For Movement System II (TC-AIMS II) will be the near-term joint single-source data system, and Joint Force Requirement Generator II (JFRG II) will be the near-term joint single-source feeder system for capturing and feeding unit movement requirements information into JOPES. TCAIMS II will exchange unclassified Organizational Equipment List (OEL), Unit Deployment List (UDL), and Unit Movement Data (UMD) files with the JFRG II. GCCS is the C2 migration system to bring Service systems together; in this sense, all legacy and migration systems that support or access the common data bases are "part of" GCCS. Many systems play some part in joint and naval logistics. Naval distribution is supported by numerous information and communication systems, offering management of inventories, movement, requirements, and other aspects of supply and transportation. These systems enhance distribution at every level of operations. Various systems at DLA, Service, and commercial locations provide the backbone of continental United States (CONUS) logistics, and support global distribution. Functionally specialized systems support disbursing, engineering, medical, repair, ordnance, fuel, and other operations. Together, these systems form an increasingly integrated network of information and decision support focused on effective logistics. NATO utilizes the Allied Command and Control Information System (ACCIS) for this function.

IV. Naval Theater Distribution

Ref: NDP4, Naval Logistics, chap. 4.

Expeditionary naval forces provide mission capabilities that can quickly reach and maintain station anywhere on the oceans, littorals, and airways. Exceptional mobility and sustainability mark the unique role of naval forces. Supply, engineering, transportation, maintenance, health services, and other services facilitate this readiness and thus the effectiveness of naval forces. A sound distribution system incorporating supply and transportation systems is critical for this sustainability.

I. The Defense Supply System

The greatest volume of materiel support is generated within CONUS through the defense supply system. This network of agency and service organizations includes the DLA, the GSA, Service supply systems, and miscellaneous DOD agencies. Primary naval components of this network are NAVSUP, Marine Corps Materiel Command (MARCORMATCOM), and NAVMEDLOGCOM.

Within the defense supply system, Integrated Materiel Managers (IMMs) are designated as the single point for acquisition and management of each item. This reduces redundancy and waste, and encourages economies of scale. DLA is the IMM for most items consumed by the Services. Exceptions generally fall into the areas of ordnance, major end items (ships, aircraft, and major equipment), repairables, cryptological material, and items with special circumstances dictating Service management.

Logistics support of operating forces is a Service responsibility, and each service maintains supply systems tailored to organic requirements. These systems are network components of the defense supply system. For the naval services, the primary supply systems are the Navy supply system and the Marine Corps supply system.

II. The Naval Supply System

The Naval supply system consists of NAVSUP, other naval organizations providing supply support, and organic supply capabilities of the operating forces. CINCPACFLT and CINCLANTFLT determine requisitioning channels for fleet units. These channels are coordinated with NAVSUP shore station channels, and are changed to reflect operational and distribution requirements.

NAVSUP conducts overall supply system management through an Inventory Control Point (ICP) and Fleet and Industrial Supply Centers (FISCs), and has responsibility for supply, disbursing, food services, postal services, and exchange services, as well as materiel transportation management. NAVSUP provides supply support to Navy forces, coordinates Navy participation in the defense supply system, establishes Navy supply methods and procedures, and provides certain contracting support. NAVSUP is organized functionally, with the following major components.

- Naval Inventory Control Point (NAVICP)
- Fleet and Industrial Supply Centers (FISC)
- Fleet Materiel Support Office (FMSO)
- Naval Transportation Support Center (NAVTRANS)
- Navy Petroleum Office (NAVPETOFF)
- Naval Ammunition Logistics Center (NALC)
- Navy Field Contracting Service (NFCS)

Other naval organizations providing significant supply support include:
- NAVMEDLOGCOM is the Navy and Marine Corps subject matter expert for medical materiel, and procures all medical and dental equipment, services, and supplies for naval forces.
- NAVFAC provides initial outfitting of chemical, biological, and radiological defense (CBR-D) material and equipment to overseas shore installations, and NCF and NBG units.
- SPAWAR provides software support for the fleet logistics programs that automate supply, inventory control, maintenance, and financial management.

III. The Marine Corps Supply System

The MARCORMATCOM has responsibility for materiel life cycle management of Marine Corps ground weapons systems, equipment, munitions and information systems. MARCORMATCOM exercises materiel support management through its two subordinate commands, Marine Corps Logistics Bases (MARCORLOGBASES) and Marine Corps Systems Command (MARCORSYSCOM). Together, these Commands plan, manage, and control the acquisition and sustainment of these systems. To properly sustain these systems, the Marine Corps executes its supply functions via wholesale and retail material management entities. At the wholesale level, MARCORLOGBASES performs traditional DOD inventory control point (ICP) functions for assigned items, as well as serving as the single service level manager for all Marine Corps ground weapons systems. At the retail level, Marine Expeditionary Forces (MEFs) operate intermediate stockpoints and process requisitions generated by the consumer level maintenance and supply systems. The Supply Battalions of the Force Service Support Groups (FSSG) operate these stockpoints and provide the primary source of supply for MEFs. The Navy provides support for Navy furnished material, ammunition, and equipment through cognizant SYSCOMs.

IV. The Naval Transportation System

Naval organic transportation assets are concentrated in sealift and airlift assets and with minimal land transportation assets. NAVSUP controls and oversees Navy materiel transportation through the Naval Transportation Support Center (NAVTRANS). This center provides Navy shippers with management guidance, provides limited mobile Navy overseas air cargo terminal, serves as the Navy shipper service representative to other transportation components, provides Navy airlift/sealift cargo requirement forecasts, and controls the Navy's Service-wide Transportation account.

The CNO and the CMC set policy for organic airlift. Navy organic transportation resources are heavily concentrated in the Naval Reserve, and the CNO has designated the Commander, Naval Air Reserve Force as executive agent for organic logistics aircraft. Limited aircraft are under the scheduling and administrative control of a variety of major Navy and Marine Corps claimants, providing direct support for major commands. Organic airlift assets provide a range of peacetime support in CONUS and overseas, but they are provided specifically to meet approved emergency or wartime requirements for organic support. All Navy and Marine Corps transport aircraft fall into the category of Operational Support Airlift (OSA).

The Joint Operational Support Airlift Center (JOSAC) uses data supplied through the Joint Air Logistics Information System (JALIS) to schedule theater support aircraft, including some Navy and Marine Corps assets. OSA includes operational support aircraft (such as those assigned to the major claimants), Navy-Unique Fleet Essential Aircraft (NUFEA), COD/VOD aircraft, Marine Corps helicopter and refueling aircraft operating in support of landing forces, and other miscellaneous aircraft. NUFEA and COD/VOD aircraft are assigned to Fleet CINCs to provide theater airlift support. Such support is not intended to replace common-user airlift; it is to provide specific support of fleet operations. Most commonly, fixed wing medium transport aircraft will

The Defense Transportation System (DTS)
Ref: NDP4, Naval Logistics (Feb '01), pp. 60 to 61.

The Defense Transportation System (DTS) provides global transportation. The DTS includes military and commercial assets, systems, and services of the Department of Defense, including those contracted or controlled by DOD. The DTS does not include Service-unique assets or those assigned to a theater CINC. The Commander in Chief of the U.S. Transportation Command (USTRANSCOM) is the unified commander designated as the DOD single manager for common-user transportation. USTRANSCOM manages military transportation through three component commands:

Air Mobility Command (AMC), Military Traffic Management Command (MTMC), and Military Sealift Command (MSC)
In addition to their roles as component commands of USTRANSCOM, these are major commands of the Air Force, Army, and Navy respectively. Assets controlled or operated by USTRANSCOM components include a wide range of military, domestic commercial, and foreign commercial. The ability to readily access commercial capacity for continuing operations and surge requirements permits economical deployment of a responsive and flexible transportation system.

See pp. 1-40 to 1-41.

As the naval component of USTRANSCOM, MSC operates the Strategic Sealift Force to provide surge and sustainment shipping, and pre-positioning. Organic common-user sealift ships are part of this command and operate in reduced operating status, and can be activated on four days to full status. MSC responsibilities include negotiation and procurement of sealift ships, and activation and oversight of Ready-Reserve Fleet (RRF) ships (in coordination with Maritime Administration (MARAD)). MSC also schedules DOD controlled shipping, coordinates required ship services with port authorities, and maintains availability and status data on MSC-controlled ships. MSC supports joint deployments with Afloat Prepositioning Ships, stocked with materiel and supplies for all Services. Army rapid deployment requirements are addressed by the MSC Large Medium Speed Roll-on/Roll-off (LMSR) ships and Fast Sealift Ships (FSS).

In addition to USTRANSCOM components, the DTS includes other government agencies that manage or administer civil transportation assets. These include:

Maritime Administration (MARAD)
The Maritime Administration (MARAD) supports and oversees the U.S. Merchant Marine. In addition it owns and manages the Ready Reserve Force (RRF) ships. These ships are available for activation and employment in strategic sealift operations. RRF ships in active service are under the operational control of the Military Sealift Command. MARAD also requisitions ocean shipping and coordinates activities with the NATO Defense Shipping Authority for allocation of NATO sealift assets to meet U.S. requirements during a NATO contingency.

See pp. 1-42 to 1-43.

Coast Guard
The Coast Guard provides safety and security of shipping, waterways, harbors, and ports. The USCG has civil law enforcement authority to ensure water safety, navigational safety, and vessel inspections, maintains aids to navigation, and licenses merchant mariners. The USCG is unique in that it is a military service, that upon declaration of war or presidential direction, changes operational control from the Department of Transportation (DOT), to the Department of the Navy.

See pp. 1-27 to 1-32.

V. The Hub and Spoke Concept of Navy Theater Distribution

Ref: NWC Maritime Component Commander's Handbook (Feb '10), pp. 6-11 to 6-17.

Navy theater distribution is accomplished through a hub and spoke system. Non-self-deploying forces transit this system on their way to employment, and all Navy forces receive sustainment through this system.

Unlike the CONUS and shipboard systems at each end of the pipeline, the theater distribution system may not pre-exist. It may form around existing theater structures, or stand up from scratch. It may incorporate existing bases, but will probably be expeditionary. It will be manned by forces drawn from some combination of active, reserve, joint, combined, and civilian sources. It will deploy quickly and commence operations with a mix of facilities and equipment drawn from many sources. Theater distribution to naval forces works because highly mobile and forward-focused Navy and Marine Corps forces have the backing of a deliberate, responsive, and robust infrastructure that operates under proven procedures to provide responsive support under all conditions.

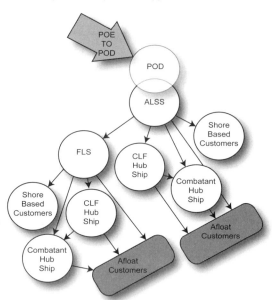

A. Primary Hubs Ashore – The Advanced Logistics Support Site

The primary theater distribution hub is called an ALSS. The ALSS centers on the availability of an aerial port and seaport in relative proximity. An ALSS normally possesses the capabilities to receive, store, consolidate, and transfer the full range of required support for forward-deployed Navy forces. An ALSS is stood up at military or civilian sites using a mix of active and reserve units augmented by contract, host nation, and allied and coalition support. These units cover required logistics functions, as well as administration and support of the ALSS itself. Tenders and hospital ships can also augment an ALSS.

Onward movement from the ALSS may be via organic or contracted local delivery; organic, common-user, or combined theater airlift; CLF shuttle ship; common-user or combined ground or water transportation; customer pickup; COD; or organic, contracted, or VOD. This movement is directed to the next transshipment point or to the end user.

B. Secondary Hubs Ashore – The Forward Logistics Site

The final transshipment point ashore in theater is the FLS. A FLS is usually closer to the operating forces than the ALSS, and capabilities may range from very austere to nearly as capable as an ALSS. Naval expeditionary logistics units and equipment, host nation, and contracted resources are task organized and assigned based on established and anticipated support requirements. FLSs normally include both a seaport and airport, but may have only one or the other when appropriate to the support requirement or site availability. While most logistics flow is from the ALSS to the FLS, FLSs may receive direct shipments into the theater in response to operational emergencies. FLSs are expeditionary and are established, moved, and disestablished readily in response to movement of the supported forces.

The final spoke in distribution can be via CLF ship to afloat forces by local issue within the ALSS/FLS, or by surface delivery from the ALSS or FLS for nearby customers and ships in port. Similarly, customer pickup—usually by ship's organic helicopters or by shore based customers—may complete operational distribution. For forces at sea, distribution may proceed through direct delivery or through an afloat hub.

Navy COD aircraft provide direct links from the closest FLS or ALSS to carriers at sea. COD is usually reserved for the highest priority passengers, mail, and cargo (PMC). Weight, size, and cube are strictly limited on COD aircraft, so both the volume and nature of support via COD is very restricted.

VOD aircraft offer another option for direct delivery of high priority PMC from the FLS or ALSS to ships with adequate helicopter facilities. VOD has commonly been accomplished by Navy heavy helicopters, but other-Service, allied, and commercial aircraft have proven capable of operating to and from Navy ships with proper aircraft and crew certification and standardization of procedures. Considerations in VOD are generally similar to those involved with COD.

C. Hubs and Spokes Afloat – Replenishment at Sea

To accomplish effective support at sea, the Navy establishes hubs afloat. These hubs carry or receive PMC for transfer to other ships at sea. Replenishment at sea, primarily through UNREP, is done by moving materiel across rigging between two ships (CONREP) or by military or commercial helicopter (VERTREP).

The primary hub for UNREP is the CLF ship. These ships are configured and equipped for cargo transfer to other ships underway. Other military or commercial vessels can be pressed into service by embarking cargo afloat rig teams (CARTs) to provide and operate temporary rigging for limited ship-to-ship transfers. Primary CLF ships involved in UNREP of supplies today are the T-AFS, T-AO, T-AE, and the AOE. CLF hub ships primarily receive materiel in port at the FLS, ALSS, or other supply point.

The T-AFS is stocked to a specific plan—called a fleet issue load list (FILL)—for issue to requisitioning ships. In addition, the CLF ship will pick up freight for ships it will be replenishing. Dependent on operations, materiel availability, and logistics replenishment (LOGREP) schedules, the T-AFS may also receive materiel from other CLF ships at sea. T-AOs and AOEs load petroleum products at defense energy supply points (DESPs) at the FLS, ALSS, or other locations in or near the theater. T-AOs and AOEs at sea also receive transfers from other oilers and point to point tankers; this is termed consolidation (CONSOL). T-AOs and AOEs also load ordnance at in theater U.S. ordnance facilities.

Combatant ships also serve as hub ships for specific support. Many ships cannot accept COD aircraft or heavy VOD aircraft. Delivery of high priority air-transportable PMC to these ships may be via COD/VOD to a carrier or VOD to another large-deck air capable ship (LHD/LHA). For VOD delivery, a CLF ship in company will normally perform as hub ship when available. When a CLF ship is not available, or when most of the VOD delivery is destined for the large deck combatant, the combatant will normally accept delivery of items for ships in company, with final transfer to be subject to operational considerations.

(Naval Logistics) IV. Naval Theater Distribution 6-31

operate between the ALSSs and FLSs, supporting the COD/VOD operations from those sites to the fleet at sea.

NUFEA aircraft also support deployment, redeployment, and sustainment of shore-based naval forces. CLF ships provide strategic transportation during initial deployment, and are capable of providing additional emergency strategic lift. Similarly, hospital ships and pre positioning ships act as defense stock points, strategic transporters, theater transporters, and combat service support providers. With these various assets, a full range of strategic and theater distribution functions is possible with limited or no theater shore-based support. While forward basing, fixed or expeditionary, is critical to support maneuver and provide economy of operations and throughput capacity, naval forces afloat are able to maintain station anywhere.

VI. The Logistics Pipeline

The flow of logistics support to the operating forces has often been depicted as a flow through a pipeline channeling support from sources (most commonly CONUS-based acquisition), through nodes (bases, stock points, fixed and expeditionary sites, etc.), to the end user (forces). Personnel and materiel flow from seaports and airports of embarkation (SPOE/APOE) via strategic lift. This strategic phase of distribution ends at the aerial port or seaport of debarkation (APOD/SPOD) in theater. RSOI of forces commences at these nodes.

Theater distribution entails both operational and tactical logistics. Once at the end user, logistic resources, unless expended, must reenter the logistics pipeline in the reverse flow during redeployment, disposal, or other retrograde actions.

Acquisition decisions have traditionally limited overseas purchases. Consequently, overseas purchases for afloat forces were limited to consumable supplies and port and intermediate maintenance services. Planners and operational commanders now place greater emphasis on the use of host nation, allied, coalition partner, or other foreign support reducing throughput through the naval logistics pipeline. Civilian contractors—domestic and foreign—directly provide support previously accomplished by the Services.

Transportation modes for the pipeline are selected based on mission need, capability, transportation priority, regulatory restrictions, and available capacity. Regulatory restrictions include transportation and storage issues such as hazardous material regulations, and security and custody issues such as registered mail regulations. Strategic transportation choices include a range of military and commercial options, both foreign and domestic.

The combatant commander's options for operational (theater) lift extend across services, modes, allied services, and host nation or other nation capabilities. Selection of the service support provider at point of issue in theater is straightforward for the naval forces afloat, but vary significantly for shore based forces. Tactical service support is normally under Service control. The majority of common-user items will be requisitioned by an organic supply organization for issue within the operating unit. However, as directed by the CINC, shore based naval forces may utilize common-user, other-Service, combined, contract, host or other-nation, or small purchase options to effect tactical delivery of a commodity or service.

Tactical Distribution Within the Battle Group

Delivery to the final hub ends the operational level of theater distribution. Afloat units link directly to the supply system; they are individual customers that requisition directly, maintaining their own inventories and operating budgets. While different types and classes of ships have widely differing capabilities, basic combat service support is organic—self-contained and self-deploying with the ship; larger ships enjoy organic logistics capabilities ranging up to some intermediate level maintenance. Consolidated Shipboard Allowance Lists (COSALs), Aviation Consolidated Allowance Lists (AVCALs), and carefully developed and tested planning factors for

VII. Force Projection

Ref: NDP4, Naval Logistics (Feb '01), pp. G-4 to G-6. See also pp. 1-41 and 1-47.

Force projection is the doctrine under which U.S. armed forces are employed in overseas missions. Joint doctrine identifies five activities in force projection. They are Mobilization, Deployment, Employment, Sustainment, and Redeployment. These generally parallel the four logistics process elements that support and enable the employment of forces. **Deployment and redeployment** are of critical interest to the operational commander or logistician in theater. **Mobilization**—concerned with the acquisition, assembly, and organization of assets—is primarily a strategic activity that the operational commander will be able to influence only in indirect or incidental manners. **Sustainment** is received at the tactical level out of services and supplies distributed to the end user.

Phases of Deployment

The first phase is pre-deployment activities that begin at the point of origin, and include planning and preparations required to prepare units and materiel for deployment. For deployments requiring strategic movement, this normally means movement of materiel or forces to a seaport or airport, and preparation for loading. CONUS, MTMC will coordinate commercial movement requirements with Service transportation authorities. Substantial Navy and Marine Corps forces including reserves and war reserves will deploy through this phase. For OPLAN execution, the POE for each movement of sustainment or forces is designated in the TPFDD.

The second phase is the movement from the POE in CONUS or elsewhere to the theater POD. This is the strategic transportation phase of deployment. Most Navy operating forces are self-deploying, as are the War Reserve Material (WRM) and ships of the MPF. Shore based naval forces and sustainment requiring strategic movement will rely on DTS for this phase. Normally, movement from the POE to the POD involves sealift or airlift, but movement of materiel and forces between theaters can be by land or inland waterway. TPFDD development and refinement in the planning process is critical to apportioning available lift. Material, forces, and personnel are loaded at the POE in accordance with established transportation priorities, and load and storage limitations. When the strategic lift arrives at the theater POD and is downloaded, strategic movement is complete and the final phase known as RSOI begins.

The third phase, theater distribution, commences when forces or sustainment arrive at a POD. This "arrival" can occur administratively when WR prepositioned ashore in theater is broken out. For terminals with significant throughput for naval forces, the Navy and/or Marine Corps will normally assign appropriate liaison or forces to ensure accurate identification and rapid handling of their respective resources. At aerial ports, this will often entail the deployment of an Air Cargo Company (ACC), element-sized and configured to the projected throughput with appropriate terminal operation capabilities present. The MSC will establish offices at ocean terminals to support MSC controlled ships and operations. The Navy Cargo Handling and Port Group (NAVCHAPGRU), Navy Cargo Handling Battalions (NCHBs) and other expeditionary units may be assigned to support aerial and sea port operations. Common-user SPOD operations will normally be under MTMC control, and will be operated by Army, civilian, host nation, joint, combined or Navy forces.

RSOI is the final phase of the deployment process in force projection. RSOI of Navy and Marine Corps forces and sustainment may occur through service, joint, or combined organizations. The relevance of RSOI is more evident for forces ashore than for forces afloat or items of supply.

Refer to The Sustainment & Multifunctional Logistician's SMARTbook for complete discussion of Mobilization, Deployment, Employment, Sustainment, and Redeployment.

endurance loading allow the efficiency of supply planning necessitated by finite storage limits. U.S. Navy afloat supply operations permit tailored, focused throughput of precise requirements. This, in turn, allows streamlined distribution featuring reduced logistics footprint in theater, minimal intermediate inventories, and negligible movement of superfluous supply to forward areas.

CLF ships within a battle force or battle group conduct tactical distribution during replenishment at sea, and other hub ships distribute PMC to ships in company as possible. LOGREP cycles are determined by the operational commander in response to operational requirements, unit locations, elapsed time since replenishment, and urgency of requirements. By minimizing forward inventories and shore-based infrastructure, and providing the means to rapidly move sustainment to and between units in direct response to precise requirements, Navy tactical distribution and shipboard supply have predicted and practiced the future direction of joint logistics.

Tactical Distribution Ashore

Naval forces ashore rely on a combination of unit-organic, Navy fixed base, Naval Expeditionary Logistics Support Force (NELSF), contract, common-user, host nation, cross-Service, and multinational sustainment. As most Navy shore-based forces—other than those assigned to Marine Corps forces—will be within the ALSS or FLS, tactical distribution is largely confined to immediate issue or local delivery. Thus, while sourcing of sustainment may be very flexible and innovative, tactical distribution is generally simple and direct. Exceptions arise in areas where U.S. Navy shore-based operations are not in proximity to adequate support. An example is the remote and austere FLS with very limited organic capability; tactical distribution of support as basic as disbursing payments for local contracts can require periodic movement of support either down the operational distribution channels from the ALSS or back from supported forces.

It is not uncommon for afloat forces to provide critical tactical support of remote FLSs. Such distribution is accomplished through local coordination between the FLS and either the ALSS or the afloat forces. The NCF, the medical force, and other Navy forces assigned to support Marine Corps units derive tactical distribution through Marine Corps channels. Marine Corps tactical distribution ashore is accomplished through organic unit capabilities and units of the CSSE of the MAGTF. The Marine Corps identifies tactical logistics as the tactical-level execution of logistics functions by either CSS units or unit organic actions. The Marine Corps consider combat service support as intermediate support provided to units lacking organic capability.

Naval Theater Distribution in Multinational Operations

NATO has developed distribution procedures and policies allowing combined support. NATO naval operational logistics are similar to that of the U.S. and readily understood by U.S. naval logisticians. The information given above on hub and spoke theater distribution ashore and within the battle force generally applies to NATO maritime logistics operations. The principles and policies of NATO establish logistics support as a collective responsibility, effected by the cooperation of the nations and the transfer of sufficient authority over logistics resources to enable effective employment and sustainment of forces. Implicit in this is an understanding that transfer of authority, or even transfer of a repair part, is voluntary an may be prevented or limited by national laws and interests.

For U.S. forces, this translates to strategic distribution. National supply systems inform MNLC of all PMC en route. Once forces or materiel reach the ALSS, RSOI are the responsibility of the MNLC organization. Host nation and multinational agreements for specific support will often result in substantial savings in distribution. Shared resources and shortened transportation legs made possible by these agreements allow a distribution system that is at once more responsive and more economical.

I. Navy Theater Security Cooperation

Ref: Tactical Commander's Handbook for Theater Security Cooperation, chap. 1.

The recent centennial of the Great White Fleet highlights the Navy's sustained involvement in security cooperation and related activities over many decades. In a globally connected world bridged by cyberspace, we are now threatened by adversarial non-state actors and irregular challenges, as well as traditional threats, which demands increased efforts toward building partnership capacity. Theater security cooperation (TSC) is a key mission area that has become increasingly important in the 21st century.

Thus, the Navy has invigorated its efforts supporting TSC's primary objective — to build international partnerships toward a more secure and stable global environment. In today's Navy, nearly every officer, chief, and sailor will contribute to the TSC mission in one capacity or another. In order to ensure our service contributions meet our nation's objectives, it is imperative to understand what TSC is and how best to plan and execute assigned tasking. Lessons learned and post-deployment briefs have continued to highlight the need for clear, concise TSC guidance for the operator.

This chapter addresses that requirement by providing the basic tools and knowledge to accomplish the TSC mission at the tactical level. It discusses the TSC concept and terminology to ensure the fleet has a common understanding of the mission. It also explores considerations common to all tactical TSC activities, such as cultural awareness training and multinational coordination. We also delineate three categories of TSC activities and discuss more specific planning factors associated with the process of building relationships, partner capability, and partner capacity through security cooperation activities.

I. Guidance

In 2007, the naval services released *A Cooperative Strategy for 21st Century Seapower*. This strategy emphasizes the importance of "building confidence and trust among nations through collective security efforts." These efforts occur across the full range of military operations, contribute to the building of partnership capacity, and establish conditions favorable to U.S. security. The Chief of Naval Operations (CNO) reiterated the importance of TSC in his Guidance for 2010.

Navy component commanders (NCCs) blend this service approach to TSC with geographic combatant commanders' theater campaign plans in order to provide administrative and operational direction to the fleet. This fleet guidance is promulgated as the Maritime Security Cooperation Plan (MSCP) and is issued by each NCC. This plan contains the Navy's theater, regional, and country-specific security cooperation objectives, which guide tactical-level activities with foreign nations. It also serves as the starting point for tactical commanders assigned to execute the TSC mission.

Refer to The Stability, Peace & Counterinsurgency SMARTbook, chap. 6 for complete discussion of theater security-related topics to include: military engagement and deterrence, security cooperation and assistance, and security force assistance.

II. Security Cooperation and Assistance - Planning and Activities

Ref: JP 3-0 (Change 1), Joint Operations (Feb '08), chap. VII; JP 5-0, Joint Operation Planning (Dec '06), pp. I-3 to I-4 and CJCSM 3113.01A, Responsibilities for the Coordination and Review of Security Cooperation Strategies.

In an era of persistent conflict, the United States supports the internal defense and development of international partners, regardless of whether those partners are highly developed and stable or less developed and emerging. While many of these partners are nations, they can also include alliances, coalitions, and regional organizations. U.S. support to these partners ranges from providing humanitarian assistance to major combat operations. U.S. support includes conducting conflict transformation, bolstering partner legitimacy, and building partner capacity. A vital part of these three aspects of U.S. support is assisting partner security forces.

A. Security Cooperation

Security cooperation is the means by which Department of Defense (DOD) encourages and enables countries and organizations to work with us to achieve strategic objectives. Security cooperation consists of a focused program of bilateral and multilateral defense activities conducted with foreign countries to serve mutual security interests and build defense partnerships. Security cooperation efforts also should be aligned with and support strategic communication themes, messages, and actions. The SecDef identifies security cooperation objectives, assesses the effectiveness of security cooperation activities, and revises goals as required to ensure continued support for US interests abroad.

DOD's senior civilian and military leadership — in conjunction with CCDRs, Service Chiefs, and support agencies — focus their activities on achieving the security cooperation objectives identified by the SecDef. Security cooperation planning links these activities with security cooperation objectives by identifying, prioritizing, and integrating them to optimize their overall contribution to specified US security interests. Security cooperation activities are grouped into six categories:

- Military contacts, including senior official visits, port visits, counterpart visits, conferences, staff talks, and personnel and unit exchange programs.
- Nation assistance, including foreign internal defense, security assistance programs, and planned humanitarian and civic assistance activities.
- Multinational training.
- Multinational exercises, including those in support of the Partnership for Peace Program.
- Multinational education for US personnel and personnel from other nations, both overseas and in the United States.
- Arms control and treaty monitoring activities.

In response to direction in the DOD Security Cooperation Guidance (SCG), CCDRs, Service Chiefs, and combat support agencies' directors prepare security cooperation strategies in accordance with SCG objectives for CJCS review and SecDef approval, with the GCCs as the supported entities. These strategies serve as the basis for security cooperation planning. Collaboration among the combatant commands, Services, and combat support agencies is essential. Equally important is the close coordination with US agencies that represent other instruments of national power, and particularly with the US chiefs of mission (ambassadors) in the GCCs' AORs. The functional combatant commands, Services, and DOD agencies communicate their intended security cooperation activities to the responsible GCCs, execute their activities in support of approved security cooperation strategies, and assist in the annual assessment of the effectiveness of their security cooperation activities.

CJCSM 3113.01A, Responsibilities for the Coordination and Review of Security Cooperation Activities

CJCSM 3113.01A establishes the responsibilities and procedures for the coordination and review of security cooperation strategies submitted to the Chairman of the Joint Chiefs of Staff. Security cooperation strategies are published by all combatant commanders and Service Chiefs annually as assigned. The Security Cooperation Guidance (reference a) was completely rewritten in 2005. The instruction was also rewritten to replace previous guidance pertaining to theater engagement planning.

B. Security Assistance

Security assistance is a group of programs authorized by the Foreign Assistance Act of 1961, as amended, and the Arms Export Control Act of 1976, as amended, or other related statutes by which the United States provides defense articles, military training, and other defense-related services by grant, loan, credit, or cash sales in furtherance of national policies and objectives (JP 3-57). Security assistance is a specific subset of security cooperation and may focus on external or internal threats.

DOD 5105.38-M describes the scope of security assistance programs in detail. Security assistance allows the transfer of military articles and services to friendly foreign governments. These transfers may be carried out via sales, grants, leases, or loans. If these transfers are essential to the security and economic well-being of allied governments and international organizations, they are equally vital to the security and economic well-being of the United States. Security assistance cannot be conducted using Title 10 or exercise funds. Security assistance can also include funding peace operations.

C. Theater Security Cooperation (TSC)

Joint Publication 3.07-1, Joint Tactics, Techniques, and Procedures for Foreign Internal Defense, states that security cooperation involves all DOD interactions with foreign defense establishments to build defense relationships in order to accomplish three main objectives:

- Promote specific United States (U.S.) security interests
- Develop allied and friendly military capabilities for self-defense and multinational operations
- Provide U.S. forces with peacetime and contingency access to a host nation

This definition describes security cooperation as the sum of all military-to-military activities, which enhances our ties to and interoperability with foreign nations; however, this does not exclude those DOD interactions with nongovernmental organizations (NGOs), intergovernmental organizations (IGOs), or foreign populations, as these vital activities contribute to broader military security cooperation objectives.

Joint Publication 1-02, DOD Dictionary of Military and Associated Terms, defines a security cooperation activity as a "military activity that involves other nations and is intended to shape the operational environment in peacetime." This means that nearly every interaction we have with a foreign nation can contribute to security cooperation objectives, framing this all-important mission that directly supports the combatant commander's theater strategy.

With these definitions in mind, it becomes evident that theater security cooperation is a mission that consists of the set of military activities undertaken with other nations to build relationships and contribute to strategic theater objectives. Theater security cooperation is not an event. It is a mission set that encompasses a range of activities, each contributing in specific ways to security cooperation objectives.

III. Theater Security Cooperation Activities
Ref: Tactical Commander's Handbook for Theater Security Cooperation, pp. 1:4 to 1:5.

TSC utilizes a number of tools to accomplish security cooperation objectives. These tools are outlined in the Secretary of Defense's Guidance for Employment of the Force and categorize every security cooperation activity. Some tools, like foreign military sales, are utilized only by the most senior levels of DOD; however, many TSC tools are utilized primarily by deployed Navy forces. A list of these tools, with official definitions or descriptions, follows:

Combined/Multinational Exercises and Training
Military maneuvers or simulated wartime operations involving planning, preparation, and execution for training and evaluation with allies (combined) or foreign nations/coalition partners (multinational).

Counter-proliferation Operations
Defeat the threat and/or use of weapons of mass destruction against the United States and our forces, friends, allies, and partners.

Counterdrug Operations
Active measures taken to Detect, monitor, and counter the production, trafficking, and use of illegal drugs.

Defense Support to Public Diplomacy
Activities and measures taken by DOD components to support and facilitate public diplomacy efforts of the U.S. Government (USG). Public diplomacy refers to those overt international public information activities of the USG designed to promote U.S. foreign policy objectives by seeking to understand, inform, and influence foreign audiences and opinion makers and by broadening the dialogue between American citizens and institutions and their counterparts abroad.

Facilities and Infrastructure Support Projects
Military construction investments and cooperative infrastructure development with allies and international partners at host-nation installations (including main operating bases, forward-operating sites, cooperative security locations, pre-positioned equipment and materials, and other en route and support infrastructure).

Humanitarian Assistance
Programs conducted to relieve or reduce the results of natural or man-made disasters or other endemic conditions, such as human pain, disease, hunger, or privation, which might present a serious threat to life or which can result in great damage to or loss of property. The assistance provided is designed to supplement or complement the efforts of host-nation civil authorities or agencies that have primary responsibility for providing humanitarian assistance.

Information Sharing/Intelligence Cooperation
Activities that increase partnership capacity to collect and disseminate information. It seeks to improve multinational/combined interoperability and awareness.•

Defense and Military Contacts
These are focused and tailored interactions to support defense reform; staff interoperability; senior defense official visits (civilians, officers, noncommissioned officers [NCOs]); bilateral and multilateral planning events; exchanges; staff talks; conferences; and officer, NCO, and unit exchange programs.

II. Security Cooperation Activity Planning

Ref: Tactical Commander's Handbook for Theater Security Cooperation, chap. 2

Security cooperation activities are military activities that involve other nations and are intended to shape the operational environment in peacetime. As with any mission, there are unique considerations for planning TSC activities. They include inputs and outputs, intelligence preparation of the operational environment (IPOE) geared toward learning about a partner nation, and the integration of civilian personnel into activity execution.

Whether your activity is a multinational live-fire exercise or a shipboard distinguished visitor reception, there are common factors that must be considered during TSC planning and preparation.

The Tactical Commander's Handbook for Theater Security Cooperation provides a number of references that will provide further information on the following topics. An electronic version of this handbook and the associated reference materials is also accessible online via the NWDC command Web site as follows: https://ndls.nwdc.navy.mil/assets/commandfolders.aspx (CAC required).

I. Planning

The first step upon receipt of the TSC mission or single activity tasking is to conduct the Navy planning process (NPP). While the process focuses on the operational level, it is equally applicable to the tactical level, with the understanding that some of the steps may have been addressed by higher headquarters. This process, however, is critical, as it outlines the commander's necessary steps to properly prepare for the TSC activity.

Though Navy and joint doctrine are invaluable resources that should be read for fuller context, this handbook contains an activity checklist that covers many of the items that factor into the Navy planning process. This checklist will walk you through the process and primary considerations for TSC activities.

A. Lessons Learned

Lessons learned must be a vital component in a commander's mission analysis. The Navy Lessons Learned Information System (NLLIS) is one of the most significant resources you have at your disposal for planning a mission. It is absolutely imperative that it be utilized in order to gain knowledge from past operations, which will benefit your mission.

NLLIS may be accessed at https://www.jllis.mil/navy (unclassified) and http://www.jllis.smil.mil/navy/ (classified). You may contact the NWDC Lessons Learned Directorate at nwdcnavylessons@nwdc.navy.mil and request all relevant lessons for your mission. NWDC will compile the results and e-mail them to your command.

B. Security Cooperation Guidance

For every theater, region, and country where the United States engages in security cooperation activities, there exists guidance outlining specific objectives and end states. The country-specific information resides in appendixes of the combatant commander's theater campaign plan and is usually further defined in the annexes of the NCC's Maritime Security Cooperation Plan.

C. Strategic Communication

JP 5-0, Joint Operation Planning, defines strategic communication as "focused United States Government efforts to understand and engage key audiences to create, strengthen, or preserve conditions favorable for the advancement of United States Government interests, policies, and objectives through the use of coordinated programs, plans, themes, messages, and products synchronized with the actions of all instruments of national power." Strategic communication guidance will be provided to tactical commanders prior to execution of the TSC mission and activities.

D. Cultural Awareness

All of a command's work toward mission success may be largely negated if cultural training does not occur. Security cooperation is primarily about building relationships and, to do so, you must understand your partners. Cultural awareness training is a key component of the TSC mission.

E. American Embassy Coordination

The American Embassy (AMEMB) is an invaluable and critical resource for units operating in a given host nation. There are varying guidelines for the proper means to establish communication with the relevant AMEMB. Higher headquarters guidance will detail the steps you should take to begin AMEMB coordination. The most common point of entry will be the Military Group (MILGRU) at the AMEMB.

Contact information for the MILGRU at the respective AMEMB is available from the NCC and numbered fleet staff.

F. Liaison Officers

The partner nation will most likely embark or attach one or more of its military and/or security force officers to assist in planning and carrying out TSC activities. Your approach should be inclusive and to make these officers part of the planning and execution effort. When presenting in public, the visual of the liaison officer alongside U.S. leadership is a powerful one; therefore, efforts should be made in this regard. Accommodations should be made to ensure cultural sensitivities (e.g., dietary restrictions, religious considerations, etc.) are addressed when lodging partner nation representatives. Coordination to get these officers on staff will occur via the NCC and AMEMB MILGRU.

G. Force Protection

Navy forces operating in foreign waters and in host nations must be prepared to face threats to the safety and security of personnel and platforms. Force protection consists of preventative measures taken to mitigate hostile actions. Tactical commanders oversee the development of force protection plans for each security cooperation activity as directed by the pertinent NCC.

H. Public Affairs

JP 3-61, Public Affairs (Draft), states that "public affairs (PA) are those public information, command information, and community relations activities directed toward both the external and internal publics with interest in the DOD." This is one of the most critical aspects of security cooperation due to its impact on perception and influence. Most staffs will have a professional PA officer (PAO).

I. Communications

Communications paths to enable the sharing of information between U.S. and multinational forces must be established early to facilitate coordination prior to and during security cooperation activities. Local communications options and availability must be investigated to ensure a smooth transition once operations begin. Considerations are vast and range from local allocated transmission frequencies and power output restrictions to addressing security concerns in allowing multinational partners access to our networks and necessary throughout Web sites.

J. Legal

Legal constraints and restraints are paramount considerations in preparing for security cooperation activities. Navy forces will comply with applicable national and international laws during the course of activities. The staff judge advocate or legal officer will ensure that U.S. activities and associated requirements are in compliance with U.S. and international law.

K. Environmental Considerations

Environmental considerations range from physical conditions to legal constraints. Planners and operators must account for environmental factors to minimize adverse environmental impact, ensure the safety and health of personnel, and reduce post-deployment environmental cleanup.

Refer to DOD Instruction 4715.8, Environmental Remediation for DOD Activities Overseas, for considerations that may be helpful for planning.

L. Funding

There are several sources of funding for TSC activities, each dependent on the specific nature of the activity. The fiscal landscape can be hard to navigate because of the numerous resources available, but it is important to ensure undertaken activities utilize the correct funding line and remain within legal constraints. These resources will be requested by the NCC on your behalf; however, as the tactical unit in place, you must be familiar with these funding lines and which activities they cover. The legal restrictions on how the funds may be used are some of the most critical information to your operation to ensure our activities are within the bounds of American law.

For more information, refer to appendix A of NTTP 3.07-15, Navy Component Commander Support to Theater Security Cooperation (Draft) and appendix E of JP 1-06, Financial Management Support in Joint Operations.

M. Personnel

There are varied requirements that need to be met when personnel are ashore in foreign nations. Commanders must ensure that personnel and any partnering NGO representatives have proper documentation, such as passports and visas. Additionally, DOD theater requirements must be met, such as isolated personnel reports (ISOPREPs), immunizations, and other medical requirements necessary to operate in areas with potential infectious disease and other endemic conditions. These are just a few examples of the personnel considerations that must be taken into account.

For more information, refer to JP 1-0, Personnel Support to Joint Operations, specifically Appendixes C, E, and U.

N. Data Collection and Reporting

Your presence in-country or through direct engagement with foreign nationals aboard a U.S. vessel allows for the opportunity to collect data and information that will assist in the assessment process. As has been mentioned above, surveys and questionnaires may be the means for some TSC activities. Other activities will require a more thorough understanding of the TSC objectives for the foreign nation in order to collect the right information.

For more information on data collection, measures of effectiveness, and measures of performance, refer to TACMEMO 3-32.2-09, Operational Assessment.

O. Information and Knowledge Management

The importance of information and knowledge management cannot be overstated. These tools provide timely, relevant, and prioritized information (e.g., processed data) in an organized manner and usable format to the commander. They aid in storing information and providing the commander the knowledge necessary to make decisions. You should ensure that procedures and processes are in place to collect relevant information, store it, and analyze it to facilitate mission success.

II. General Security Cooperation Activity Checklist

Ref: Tactical Commander's Handbook for Theater Security Cooperation, App. A

1. Read security cooperation guidance (most likely on SIPRNET), Navy component commander guidance, numbered fleet guidance (where different from above), and country-specific guidance for relevant host nation.
2. Identify and read relevant operational doctrine and directives for the TSC activity.
3. Obtain higher headquarters activity guidance: activity objectives, measures of effectiveness and measures of performance, strategic communication guidance, and public affairs guidance.
4. Review Navy lessons learned database.
5. Submit and obtain unit diplomatic/country clearance.
6. Conduct mission analysis to answer the questions.
   - What tasks must the command do for mission accomplishment?
   - What is the purpose of the activity?
   - What limitations have been placed on own force's actions?
   - What forces/assets are available to support the operation?
   - What additional assets are needed?
7. Develop mission analysis outputs.
   - Determine specified, implied, and essential tasks
   - Identify constraints and restraints
   - Determine critical factors, centers of gravity, and decisive points
   - Conduct risk assessment
   - Develop commander's intent
   - Develop commander's critical information requirements
8. Identify security activity funding source.
9. Address personnel requirements: passport/visa/military ID and customs policy, required medical screening and immunizations, complete isolated personnel report (as required), ensure government charge card authorization and policies of use are promulgated, and ensure proper uniforms for geography, weather, and activity.
10. Request staff judge advocate assistance in identifying memorandums of understanding/status of forces agreements/visiting forces agreements that exist with host nation.
11. Identify host-nation, interagency, intergovernmental organizations, and nongovernmental organization points of contact for coordination.
12. Identify logistics path and anticipated requirements.
13. Arrange for cultural awareness training.
14. Identify means of ship-to-shore communications.
15. Obtain format of situational reports to higher headquarters, frequency of reporting, and circuit to make report.
16. Depending on time restraints, identify pre-deployment site survey team or advance command element (arrives prior to activity execution) to prepare for activity.
17. Draft lessons learned during the course of activity planning and execution. Submit upon activity completion or as directed.
18. Draft after-action briefing per higher headquarters' direction.

III. Building Relationships, Capabilities & Capacity

Ref: Tactical Commander's Handbook for Theater Security Cooperation, chap. 3 - 5.

Mission Commander

The mission commander is responsible for all resources and activities associated with extended deployments. These deployments encompass a variety of TSC activities conducted with a number of foreign nations over the course of several months. The mission commander is usually assigned to execute humanitarian and civic assistance and foreign humanitarian assistance contingency operations. The commander must ensure the crew is properly trained and equipped to accomplish the mission in support of theater objectives. The commander also assumes responsibility for the support and integration of multinational, NGO, and IGO liaisons attached to U.S. TSC forces.

Security Cooperation Progression

 Building Relationships

 Building Capabilities

 Building Capacity

Ref: Tactical Cdr's Handbook for Theater Security Cooperation, fig. 1, p. 1-6.

A primary challenge of the mission commander is the enormity of the several activities taking place simultaneously and in serial. The commander must ensure safe and successful mission execution while continuing to remain forward-looking for upcoming activities and contributing to the assessment of activities already executed. It is imperative for the commander to provide daily guidance on assigned tasks while not losing sight of the strategic communication and TSC objectives that guide the mission.

Mission success squarely hinges on the commander's ability to build relationships with interagency partners and host-nation leadership, as well as to synchronize actions with messages in a legal and efficient manner by, with, and through the professionalism of his command.

Social Norms and Customs

In small-scale, direct engagements with groups of a foreign populace, it is especially important to be observant of social norms and customs. This goes beyond cultural awareness and is a more specific approach to the particular people in the town, city, and region of the country where you are operating. Efforts should be made to determine social nuances, such as topics to be avoided, proper greetings, food restrictions, hygiene, etc. While cultural awareness training serves as an introduction to a nation, social norms and customs training should be geared toward etiquette and personal interactions. The best source of this information is the American Embassy and foreign nation liaison officers.

I. Building Relationships

Ref: Tactical Commander's Handbook for Theater Security Cooperation, chap. 3.

"Building relationships" is a broad term that characterizes any security cooperation activity where the primary goals are engagement and personal interaction between DOD representatives and other nations. This personal interaction is not limited to obvious events such as receptions and photo opportunities. It includes all the activities that build relationships with a foreign nation's government officials and populace. These are activities that may set the stage and serve as a precursor to further contact with foreign defense establishments and work toward accomplishment of military objectives. Building relationships with a nation increases the opportunity for, and likelihood of, partnerships being built between foreign defense forces and DOD.

The topics outlined below are starting points for planning security cooperation activities that specifically facilitate building relationships with foreign nation officials and the population at large. Naturally, every possible interaction or activity could not be covered, but the information and associated references provide a solid foundation to begin detailed planning for the activity. In each instance, it is important to remember that TSC objectives cannot be met without the groundwork that is laid in the course of formal and casual interactions with a foreign nation.

Humanitarian Assistance

When it comes to TSC, humanitarian assistance encompasses a range of activities that are not necessarily limited by its doctrinal definition. These activities include HCA, FHA, disaster relief, and COMREL projects.

The foreign defense establishment and AMEMB are your primary points of contact for any of these areas of humanitarian assistance; however, the foreign populace serves as your primary interaction point. The respect and compassion shown to the people of the affected nation are essential, as such goodwill can enable trust and deeper relationships with the host nation.

Again, though humanitarian assistance efforts may not always have a foreign military face as the immediate benefactor, your contribution to the humanitarian assistance mission extends beyond alleviating human suffering to forming strong partnerships. For these types of operations, it is necessary to collaborate with NGOs, IGOs, and interagency representatives. Due to the large lift capacity the military brings, many of these representatives may rely on your command to serve as a hub for planning and coordination, as well as transportation to landing zones and activity sites. As a result, it is imperative that you consider the legal and security factors in transporting and networking civilians on military crafts and information systems.

Distinguished Visitors Embarks/Receptions

Distinguished visitors (DV's) embarks and official receptions are activities that support interactions with foreign leaders and government representatives. The strategic objectives of these events are building foundations for long relationships and establishing trust and confidence. The public affairs visuals that result from these activities are powerful indicators of burgeoning or ongoing partnerships between nations. The tactical objective is normally to expose distinguished visitors, media organizations, and the foreign population to Navy personnel and assets. Commanders should be prepared with strategic communication and public affairs guidance prior to engaging DV's to ensure the message is consistent; however, just as important are the nonverbal communicators, such as cultural awareness and understanding of local customs, that speak to the sincerity of the gestures.

Media liaisons may be assigned to escort and monitor the activities of the foreign media onboard. This is done to ensure the visitors remain safe and that their needs, such as communication, power, and translation needs, are met where feasible. Receptions absolutely must conform to local customs and protocols as they pertain to dietary restrictions, rendering of proper honors, and translator support. If there are gift or slide presentations, ensure they are translated into the languages of all primary nations in attendance. Ensure that logistics support (e.g., food, gifts, tour capacity, etc.) is flexible, as last minute guests and media requests are common occurrences.

Port Visits

Port visits, whether conducted for quality of life, to take on supplies, or to maintain a forward presence, are a major staple in the Navy TSC activity portfolio; therefore, each port visit should be examined for potential opportunities to contribute to country-specific TSC objectives. In the instances where specific TSC objectives may not be overtly tied to liberty port visits, it is important to remember that future relations with a country can be impacted, positively or negatively, by Navy personnel conduct ashore. Commanders will be provided with points for emphasis through the strategic communication and public affairs guidance for port visits, which will ensure our relationships with the foreign nation are strengthened.

Commanders should also realize the TSC implications inherent in simply accessing a nation's port facilities. Observations and relationships built with port authorities are often captured in after-action port visit messages but should also be examined for TSC contributions.

Subject Matter Expert Exchanges/Ship Rider Programs

Subject matter expert exchanges (SME) and ship rider programs are important elements of combatant commanders' security cooperation endeavors. While the SME exchange is executed by the services and at all levels of command, the ship rider program is unique to the naval services. These two activities serve as a means to build stronger relationships and as a venue for the professional exchange of information toward partnerships and interoperability. The procedures for conducting SME exchanges or hosting ship riders are determined by the combatant commander and carried out by the NCC. Guidance for this process is available from the NCC staff and MSCP; however, basic considerations apply, such as protection of classified information, foreign disclosure agreements, and cultural and dietary requirements. In all cases, ensure the foreign nation representative is provided with a meaningful and contributory role. The respect shown to SME's and ship riders often contributes to interoperability and partnerships as much as the military information exchanged.

Defense Support to Public Diplomacy

JP 1-02 defines defense support to public diplomacy (DSPD) as "those activities and measures taken by the Department of Defense components to support and facilitate public diplomacy efforts of the United States Government." In security cooperation, DSPD is most often those activities that contribute to Department of State objectives that apply directly or indirectly to TSC objectives. In some cases, the primary goal for DSPD with respect to TSC is to further relationships with a foreign nation's government and populace. These interactions lay the foundation for TSC specific activities.

(Theater Security Coop) III. Building Relationships, Capabilities & Capacity 7-11

II. Building Capabilities

Ref: Tactical Commander's Handbook for Theater Security Cooperation, chap. 4.

While building relationships will always be a part of security cooperation activities, there are many instances when an activity's main objective is to increase a partner nation's capabilities. Building a partner's capability is mostly about increasing its ability to execute a specified course of action in support of multinational operations. Exercises and training at various levels are the primary vehicles to building this capability.

While each exercise or training event will have detailed objectives, it is critical that commanders make certain not to lose sight of the operational and strategic TSC objectives that the activity supports. The major multinational exercises will be planned several months out through a series of planning conferences and/or staff talks. These meetings will produce tasking orders that will guide your participation in the exercise. Your responsibilities to the TSC mission are covered in large part during the preparation and execution of the exercise. In fact, the planning conferences themselves can be seen as TSC activities, and your unit will benefit from employing the considerations outlined earlier.

A. Exercises

JP 1-02 defines exercise as "a military maneuver or simulated wartime operation involving planning, preparation, and execution. It is carried out for the purpose of training and evaluation." These activities may be conducted at sea or in the field, using modeling and simulation software, or by tabletop planning and war gaming. For TSC, the most important aspect is building the partner's capability to execute courses of action on its own as well as in an interoperable fashion with our forces. Exercises range in size and scope based on the objectives and participants. However, no matter the scope or size, the objective of the exercise is to make our forces more interoperable with foreign nations in order to facilitate execution of real-world missions when necessary. The following paragraphs explore these different exercises in more detail.

Joint Chiefs of Staff Multinational/Combined Exercises

These exercises are part of the Joint Exercise Programs and are a principal means for combatant commanders to maintain ready forces, exercise security cooperation plans, and achieve joint and multinational training. While tactical commanders will not lead or plan these exercises, they may often serve as vital contributors during preparation and execution. Exercise guidance will be promulgated prior to commencement outlining the exercise objectives as well as the country-specific TSC objectives. Assessment most often occurs using the joint mission-essential task list from which the Navy develops its mission-essential task list (NMETL). The NMETL will provide information for the post-exercise assessment and certifications that may supplement the TSC assessment and analysis.

Bilateral Exercises

These exercises are usually arranged at the theater level and are not part of the JCS-approved list of exercises. These exercises may be field exercises or command post exercises, but the interaction between forces remains the major thrust for strengthening relationships and building a partner's capability. Tactical commanders will most often be a part of the planning process but will not lead it. The NCC will coordinate with the foreign partner to determine events and objectives that tactical commanders will execute.

Unit-Level Exercises

These exercises are small-scale but provide valuable opportunities to increase tactical interoperability. Activities that fall into this category include passing exercises (PASSEX) that may include rendering honors or semaphore drills; mine and explosive ordnance disposal exercises (MINEX/ EODEX); and tracking exercises (TRACKEX). Obviously, there are many more exercises that qualify as unit-level. However, regardless of the subject matter, in addition to exercise objectives, commanders must be familiar with the TSC objectives for the nations involved in these exercises to ensure activities support the broader and long-term goals.

Exercise-Related Construction

Exercise-related construction refers to military construction investments and cooperative infrastructure development with allies and international partners at host-nation installations (including main operating bases, forward operating sites, cooperative security locations, pre-positioned equipment and materials, and other en route and support infrastructure).

Tactical commanders and detachment officers-in-charge will execute these activities that have been arranged and planned at the combatant commander and component levels. This activity does not include events such as community relations construction projects. Instead, these projects usually occur during the course of an exercise and focus on projects that will increase a partner's military capability to operate.

B. Training

Multinational Training

Multinational training refers to those activities where DOD forces and foreign nations endeavor to make stronger operational ties and capabilities to facilitate interoperability during real-world operations. These training activities may range from shipboard firefighting and line-handling to complex information technology applications. This sort of training is normally done in concert with exercises or even community relations projects. The goal is to increase proficiency in a given task to bolster the ability to execute specific courses of action toward security cooperation objectives. The training should be progressive and build on skills that have been previously taught. To do this, commanders will need a thorough understanding of a country's capabilities, previous instruction, and a nation's ability to train in the absence of U.S. Navy forces.

Security Force Assistance (SFA) Training

Security force assistance (SFA) is a burgeoning term encompassed by the security cooperation mission. Specifically, SFA refers to DOD activities that contribute to the development of the increased capacity and capability of foreign security forces and their support mechanisms. SFA training specifically refers to actions taken to prepare a foreign security force for operations. SFA activities are not limited to foreign military forces; they may include police, intelligence services, coast guard forces, or customs officials.

C. Information Sharing and Intelligence Cooperation

The exchange of information and intelligence is a primary benefit of working with foreign nations. These activities make for stronger alliances and partnerships toward the mutual benefit of all involved. The resulting maritime security fosters a global environment of cooperation and security. The sharing of information extends from personal interaction to information technology networks that enable rapid exchange of data.

Intelligence cooperation takes advantage of participating nations' areas of expertise to facilitate a more complete picture of the operational environment. Disclosure agreements and policies must be reviewed and adhered to before embarking on these activities to ensure U.S. interests are protected.

However, the more information that is shared within allowed limits, the more prosperous the multinational partnerships and operations will be. Commanders should take care to remain focused on security cooperation objectives while enabling the sharing of information and intelligence.

III. Building Capacity

Ref: Tactical Commander's Handbook for Theater Security Cooperation, chap. 5.

Building partnership capacity is a further extension of increasing a foreign nation's ability to execute specified courses of action. Whereas building capabilities facilitates a partner's ability to be interoperable with DOD toward specific objectives, building capacity refers to enhancing a partner's ability to carry out real-world operations with U.S. forces or in absence of them. The Office of the Secretary of Defense for Policy states that building capacity enables our partners to disrupt internal terrorist, insurgent, and criminal activity, meet trans-border challenges, and address maritime security challenges. To accomplish these goals, DOD contributes to a partner's specialized skills and equipment.

The tactical commander's efforts in this regard are primarily focused on executing multinational operations. Building capacity is the culmination of significant investment in building relationships and building capabilities. In terms of security cooperation, building capacity applies specifically to counterdrug, counterterrorism, counter-piracy, and counter-proliferation operations. While these missions are clearly stated, the tactical commander must take the extra step to ensure actual activities are tied to country and theater security cooperation objectives.

Maritime Security

Maritime security operations help maintain security on the seas. They are one of the most important Navy efforts used to combat sea-based terrorism and other illegal activities, such as hijacking, piracy, and human trafficking. They also include those activities that provide port security and protection of sea-based assets, such as oil platforms. Many of the Navy's operations are elements that contribute to the global security mission to make international waters peaceful and safe. Maritime security is inherently a global mission, as the vast expanse of the maritime domain can only be made safer through multinational cooperation. The following paragraphs provide more detail on specific tasks that make up maritime security.

Counter Drug (CD)

Counterdrug (CD) operations are those civil or military actions taken to reduce or eliminate illicit drug trafficking. DOD's principal counterdrug mission is detection and monitoring (D&M), and the desired end result of successful D&M is interdiction and apprehension by law enforcement agencies. CD operations are inherently interagency and/or multinational in nature. Coordination and collaboration can be accomplished by integrating the efforts of military, civilian agency, and multinational planners early in the planning process. Planners must understand that some of the agencies and multinational organizations that lead or might become involved in CD operations will have different goals, capabilities, limitations (such as policy and resource restraints), standards, and operational philosophies.

Counter Proliferation (CP)

Counter-proliferation (CP) is defined as those actions (e.g., detect and monitor, prepare to conduct counter-proliferation operations, offensive operations, weapons of mass destruction, active defense, and passive defense) taken to defeat the threat and/or use of weapons of mass destruction against the United States and our military forces, friends, and allies. The objective of CP operations is to deter, interdict, attack, and defend against the full range of possible WMD acquisition, development, and employment scenarios. CP operations are intended to reduce the WMD threat and require a balanced and integrated concept of operations to defeat hostile WMD threats. CP operations require timely intelligence and thorough planning throughout the range of military operations. There are a number of supporting roles for which coalition partners and allies are well suited: site and team security, transportation, medical support, language support, and intelligence. The presence of international players increases the legitimacy of WMD elimination efforts and fosters greater cooperation in the overarching CP challenge. A systematic process must be implemented to determine classification and release ability guidance for coalition partners and allies.

Counterterrorism/Counterinsurgency

Counterterrorism refers to operations that include the offensive measures taken to prevent, deter, preempt, and respond to terrorism. **Counterinsurgency** is those military, paramilitary, political, economic, psychological, and civic actions taken by a government to defeat insurgency. The targets of these operations are usually non-state actors who pose threats to numerous nations. Thus, these operations are almost always multinational. Tactical commanders at sea and ashore will execute these operations on host-nation soil and most likely with host-nation defense and/or security forces. Whereas these activities are usually centered on capturing terrorists and defeating an insurgency, security cooperation objectives in this regard are most likely focused on bolstering the partner nation's capability and capacity. Commanders should seek to meet both of these objectives.

Security Force Assistance (SFA) Operations

SFA operations are those activities wherein DOD forces are supporting a foreign security force in real-world security operations. These operations are the result of an evolutionary process of SFA training and other security cooperation activities. Tactical commanders will execute SFA operations by augmenting foreign security force operations ashore, at port facilities, and in the littoral environment. These operations are conducted at the behest of the foreign nation and may range from information sharing in peacetime to security patrols during a contingency or time of combat.

Planning
Command and Control

Although nations will often participate in multinational operations, they rarely, if ever, relinquish national command of their forces. Forces participating in a multinational operation will always have at least two distinct chains of command: a national chain of command and a multinational chain of command. In some multinational environments, it might be prudent or advantageous to place appropriate U.S. forces under the operational control of a foreign commander to achieve specified military objectives. In general, the more centralized the command structure, the greater the multinational forces' ability to achieve unity of effort. Integrated command structures, operating within their alliance framework, afford the greatest degree of control. A parallel structure, with its separate lines of command, typically offers the least control and ability to achieve unity of effort. Lead nation structures can exhibit a wide range of control depending on the command relationships assigned.

Refer to The Joint Forces Operations & Doctrine SMARTbook for further discussion of multinational operations.

Rules of Engagement (ROE)

It is essential that multinational forces understand each other's rules of engagement (ROE), as it cannot be assumed that each will react in an identical fashion to a given situation. Without this understanding, events could result in misperceptions, confusion, and even fratricide. In many cases, commanders of deployed member forces may lack the authority to speak on behalf of their nation in the ROE development process. U.S. forces assigned operational control will follow the ROE of the multinational force unless otherwise directed by the President or Secretary of Defense. U.S. forces will be assigned and remain under operational control to a foreign commander only if the U.S. Combatant Commander and higher authority determine that the ROE for that multinational force are consistent with U.S. policy guidance on individual and unit self-defense as contained in the standing rules of engagement (SROE).

Pre-Deployment Site Survey (PDSS)

The pre-deployment site survey (PDSS) is a critical part of any TSC activity that requires forces going ashore. Ideally, the PDSS occurs several months prior to execution; however, this is a luxury not always afforded to emergent and time-sensitive activities, such as foreign humanitarian assistance (FHA) or disaster relief. In either case, it is imperative that these surveys occur, as they facilitate planning, safety, and force protection issues, among others. The PDSS team comprises several members (normally from four to eight) who serve as the initial coordination agents with the foreign nation and other participating organizations, examine local geographic conditions, and identify potential site locations that will best contribute to the desired effects. The data collected from the PDSS contributes to the spectrum of information that commanders need to assess the operational environment and prepare for operations.

Interagency, Nongovernmental Organizations (NGOs), and Intergovernmental Organizations (IGOs)

NGOs are private, self-governing, not-for-profit organizations dedicated to alleviating human suffering; promoting education, health care, economic development, environmental protection, human rights, and conflict resolution; and/or encouraging the establishment of democratic institutions and civil society. IGOs are organizations created by a formal agreement (e.g., a treaty) between two or more governments. They may be established on a global, regional, or functional basis for wide-ranging or narrowly defined purposes and are formed to protect and promote national interests shared by member states. "Interagency" refers to the USG agencies and departments, including DOD. For TSC, interagency is the coordination that occurs between elements of DOD and engaged USG agencies for the purpose of achieving an objective.

Successful interagency, IGO, and NGO coordination enables the USG to build international support, conserve resources, and conduct coherent operations that efficiently achieve shared international goals; however, coordinating and integrating efforts between your unit and other government agencies, IGO's, and NGOs should not be equated with the command and control of a military operation. The various USG agencies' different, and sometimes conflicting, goals, policies, procedures, and decision-making techniques make unity of effort a challenge. The key premise is to ensure interagency/NGO/IGO inclusion in the relevant planning process and:

- Forge a collective definition of the problem in clear and unambiguous terms
- Understand the overall USG strategic goal in addition to the security cooperation objectives
- Understand the differences between U.S. objectives and those of IGO's and NGOs
- Establish a common frame of reference

Refer to The Stability, Peace & Counterinsurgency SMARTbook, chap. 6 for complete discussion of these topics to include military engagement, security cooperation, and security force cooperation. Chap. 7 covers multinational operations, and chap. 8 discusses interagency, IGO and NGO coordination.

Index

A
Acquisition, 6-4
Aegis, 2-16
Afloat Support, 6-25
Air Mobility Command (AMC), 1-41, 6-29
Air Superiority, 2-22
Air Tasking Order (ATO), 2-57
Aircraft Carriers, 1-8
Airspace Coordinating Order (ACO), 2-57
Airspace Management, 4-9
Alliance, 2-71
Allocated, 2-39
American Embassy, 7-7
Amphibious Operations, 2-52, 2-61
Amphibious Ready Group (ARG), 1-4
Amphibious Warfare, 1-14
Analyze Available Forces and Assets, 5-26
Analyze the Higher Commander's Mission, 5-20
Antisubmarine Warfare (ASW), 2-54
Apportioned, 2-39
Area of Operations (AO), 2-7, 2-46
Assess, 4-15
Assigned, 2-39
Assumptions, 5-50
Attached, 2-39
Attack on Maritime Trade, 2-30
Availability of Forces, 2-39, 5-26
Aviation Combat Element (ACE), 1-21

B
B2C2WG, 3-6
Backbrief, 5-72
Ballistic Missile Defense (BMD), 2-16
Basing/Deployment Area Control, 2-22
Battle Damage Assessment (BDA), 4-22
Battle Rhythm, 3-8
Battlefield, 2-7
Battlespace, 2-7
Battlespace Management, 4-10
Bilateral Exercises, 7-12
Boards, Bureaus, Centers, Cells, and Working Groups (B2C2WG), 3-6
Building Capabilities, 7-9, 7-12
Building Capacity, 7-9, 7-15
Building Relationships, 7-9, 7-10

C
Campaigns, 2-13
Carrier Air Wing (CVN), 1-8
Carrier Strike Group (CSG), 1-4, 3-28
Centers of Gravity, 5-26
Choke-Point Control, 2-22
Civil Reserve Air Fleet (CRAF), 1-44
Civil-Military Coordination, 2-74
Civil Support Operations, 2-12
Coalition, 2-71
Collateral Damage, 3-33
Combat Assessment (CA), 2-57, 4-22
Combatant Command (COCOM/CCDR), 2-34, 2-37
Command and Control, 3-11
Command and Control Relationships, 2-34, 5-39
Command Authority 2-36
Command Element (CE), 1-21
Command Structure, 1-2, 2-68
Commander's Critical Information Requirements (CCIR), 4-12, 5-30
Commander's Decision Cycle, 4-14
Commander's Initial Guidance, 5-19
Commander's Intent, 5-21
Commander's Planning Guidance, 5-32
Commercial Blockade, 2-31
Communications, 4-15
Composite Warfare Commander (CWC), 2-65, 3-27
Concept of Operation (CONOPS), 5-66
Confirmation Brief, 5-76
Contiguous Zone, 3-31
Continuing Activities, 4-8
Control Areas, 3-24
Counter Drug, 7-14
Counter Proliferation, 7-14
Counterinsurgency, 2-12, 7-15
Counterterrorism, 2-12, 7-15
Course of Action Analysis (Wargaming), 5-47
Course of Action Comparison, 5-60
Course of Action Development, 5-35
Course of Action Evaluation, 5-61
Course of Action Options, 5-38
Crisis Response Operations, 2-13
Critical Events, 5-50
Critical Factors, 5-26
Crosswalk, 5-72
Cultural Awareness, 7-7
Current Operations (COPS), 5-75
Customs, 7-9
Cyber Warfare, 2-16

Index-1

D

Decision Brief, 5-64
Decisive Points, 5-26
Defense of Maritime Trade, 2-31
Defense Supply System, 6-27
Defense Support to Public Diplomacy, 7-11
Defense Transportation System (DTS), 6-29
Defensive Posture, 2-19
Design, 5-14
Deterrence, 2-13
Direct, 4-15
Disposition, 6-5
Distribution, 6-4
Doctrine, 2-1

E

Effects, 5-4
End State, 5-25, 5-28
Environmental Law, 3-36
Essential Tasks, 5-22
Establishing Authority, 2-41
Evaluation Guidance, 5-41
Event Horizons, 4-18, 5-14
Exclusive Economic Zone (EEZ), 3-31
Execution, 4-11
Exercises, 7-12
Exercise-Related Construction, 7-13
Expeditionary Strike Force, 3-28
Expeditionary Strike Group (ESG), 3-28
Externally Imposed Limitations, 5-23

F

Fires, 2-56, 3-11
Focused Logistics, 6-10
Force Projection, 1-39, 6-33
Foreign Flag Ships, 1-43
Strategic Deployment, 1-44
Foreign Humanitarian Assistance (FHA), 2-12
Forward Logistics Site, 6-31
Framing the Problem, 5-14, 5-15

Friendly Force Information Requirements (FFIRs), 5-20
Friendly Forces, 5-50
Full Command, 2-35
Functional Component Commander, 2-44
Functional Coordinators, 2-70
Future Operations (FOPS), 5-75
Future Plans Center (FPC), 5-75

G

Global Command and Control System (GCCS), 6-26
Global Force Management (GFM), 1-2, 2-39
Global Response Forces (GRF), 1-25
Governing Factors, 5-51
Ground Command Element (GCE), 1-21

H

High Seas, 3-31
High Yield Logistics (HYL), 6-8
Higher Headquarters, 5-18
Homeland Defense, 2-12
Hub and Spoke Concept, 6-30
Humanitarian Assistance/Disaster Relief (HA/DR), 3-36, 7-10

I

Implied Tasks, 5-22
Information Sharing, 7-13
Initial Commander's Intent, 5-28
Initial Risk Assessment, 5-27
Integrated Command Structure, 2-69
Integrating Processes, 4-8
Intelligence, 3-12, 7-13
Intelligence Preparation of the Operational Environment (IPOE), 4-8, 5-8, 5-56, 5-66

Intelligence, Surveillance, and Reconnaissance (ISR), 2-53, 4-9
Internal Waters, 3-31

J

Joint Deployment Process, 1-45
Joint Force Commander (JFC), 2-38
Joint Force Components, 2-37
Joint Force Maritime Component Commander (JFMCC), 2-44, 2-45
Joint Maritime Operations Targeting, 2-58, 3-33
Joint Operation Planning and Execution System (JOPES), 5-5
Joint Operations Area (JOA), 2-7
Joint Operational Areas, 2-6
Joint Operations Rehearsals, 4-10
Joint Task Force (JTF), 2-38

L

Law of Armed Conflict, 3-29
Lawful Combatants, 3-33
Lead Nation, 2-69
Legal Regimes of Oceans and Airspace, 3-27
Level of Control, 3-21
Levels of Maritime Command, 2-33
Levels of War, 2-9
Liaison, 5-7
Limitations, 5-23
Limited Contingency Operations, 2-13
Logistics, 6-1
Logistics Combat Element (LCE), 1-21
Logistics Command and Control Systems, 6-19
Logistics Information Systems, 6-26
Logistics Pipeline, 6-32
Logistics Planning Considerations, 6-18
Logistics Planning Process, 6-12

M

Major Naval Operations, 2-14
Major Operations, 2-13
Marine Air/Ground Task Force (MAGTF), 1-21
Marine Corps Reserve, 1-20
Marine Corps Supply System, 6-28
Marine Expeditionary Brigade (MEB), 1-23
Marine Expeditionary Force (MEF), 1-22
Marine Expeditionary Unit (MEU), 1-4, 1-21, 1-24
Marine Special Operations Forces, 1-38
Maritime Administration (MARAD), 1-42, 6-29
Maritime Battlefield, 2-7
Maritime Battlespace, 2-7
Maritime Campaign, 2-14
Maritime Control Area, 2-7
Maritime Domain, 2-4
Maritime Domain Awareness (MDA), 1-30, 3-22
Maritime Headquarters (MHQ), 3-1
Maritime Interception Operations (MIO), 2-52
Maritime Law, 3-25
Maritime Operational Command, 3-19
Maritime Operational Threat Response (MOTR), 2-17
Maritime Operations, 2-1, 2-9
Maritime Operations Center (MOC), 2-51, 3-1
Maritime Prepositioning Force (MPF), 1-25
Maritime Security, 7-14
Maritime Security Regimes, 1-30
Maritime Superiority, 2-1
Maritime Supremacy, 2-1
Maritime Theater Structure, 2-7
Maritime Trade, 2-15
Maritime Trade, 2-30, 2-32
Maritime Warning Zones, 3-32
Measure of Effectiveness (MOE), 4-19
Measure of Performance (MOP), 4-19
Method, 5-25, 5-28
Military Engagement, 2-13
Military Sealift Command (MSC), 1-41, 6-29
Military Surface Deployment and Distribution (SDDC), 1-41
Military Traffic Management Command (MTMC), 6-29
Mine Warfare, 1-6
Mission Analysis, 5-17
Mission Commander, 7-9
Mission Statement, 5-21
Monitor, 4-15
Movement and Maneuver, 2-52, 3-13
MSC Support Ships, 1-16
Multinational Command, 2-72
Multinational Operations, 2-71
Multinational Training, 7-14
Munitions Effectiveness Assessment (MEA), 4-22

N

Narrow Seas, 2-23
National Command, 2-71
National Defense Reserve Fleet (NDRF), 1-42
NATO Organization, 6-24
Naval Attack, 2-14
Naval Aviation, 1-8
Naval Battle, 2-14
Naval Fleet Auxiliary Force (NFAF), 1-5, 1-16
Naval Logistics, 6-1
Naval Operations, 2-19
Naval Raid, 2-14
Naval Special Warfare (NSW), 1-33
Naval Strike, 2-14
Naval Supply System, 6-27
Naval Theater Distribution, 6-27
Naval Transportation System, 6-28
Navy Command, 2-8
Navy Planning, 5-1
Navy Planning Process (NPP), 4-5, 5-2
Navy Rehearsals, 4-10
Navy Reserve, 1-18
Non Time-Sensitive Planning, 5-2
Interagency, Intergovernmental & NGOs, 2-70, 7-16
Numbered Fleets, 1-2

O

Objectives, Naval, 2-19
Offensive Mining, 2-31
Officer in Tactical Command (OTC), 2-66, 3-27
Open Ocean, 2-23
Operating Forces, 1-2, 1-20
Operation Plans & Orders, 5-71
Operational Approach, 5-15
Operational Area, 2-7
Operational Art, 5-10
Operational Command (OPCOM), 2-35
Operational Control (OPCON), 2-34, 2-35
Operational Design, 5-11
Operational Factors, 2-20
Operational Law (OPLAW), 3-25
Operational Level of Command (OLC), 2-8, 3-20
Operational Logistics (OPLOG), 6-2, 6-11
Operational Protection, 4-9
Operational-Level Functions, 3-10, 3-20
Operations Process, 4-1
Orders, 5-67

P

Peace Operations, 2-12
Piracy, 3-31
Plan, 4-15
Planning, 4-5
Planning Horizons, 5-14
Planning Assumptions, 5-27
Planning Team, 5-74
Planning Team, 5-6
Plans and Orders, 5-67
Port Visits, 7-11
Ports, Waterways & Coastal Security (PCWS), 1-30
Preparation Activities, 4-9

Index-3

Preparation, 4-7
Prepositioning Ships, 1-16
Primary Hubs Ashore, 6-30
Principal Headquarters, 1-3
Principle Warfare Commanders (PWC), 2-68
Priority Information Requirements (PIRs), 5-20
Projecting Power Ashore, 2-26
Protection, 3-13
Protection of Maritime Trade, 2-31
Purpose, 5-23, 5-25, 5-28

R

Ready Reserve Force (RFF), 1-42
Reattack, 4-23
Red Cell, 5-53
Reframe the Problem, 5-16
Rehearsals, 4-10
Relative Combat Power, 5-36
Replenishment at Sea, 6-31
Right of Visit, 3-34
Risk Assessment, 4-16
Risk Management, 4-17
Rules For the Use of Force (RUF), 3-26
Rules of Engagement (ROE), 3-26, 7-15

S

Sea Control, 2-19, 2-24
Sea Denial, 2-27
Sea-Air-Land (SEAL) Teams, 1-36
Sea-Based Logistics, 6-9
Seaborne Nuclear Deterrent Forces, 2-15
Sealift Program, 1-16
Secondary Hubs Ashore, 6-31
Security, 4-9
Security Assistance, 7-3
Security Cooperation, 2-13, 7-2
Security Force Assistance (SFA), 7-13, 7-15
Self-Defense, 3-35
Service Component Commander, 2-42

Seven-Minute Drills, 3-17
Shore Forces, 1-2
Shore-Based Support, 6-25
Single Port Manager (SPM), 1-46
Situational Understanding, 4-12
Small Vessel Security Strategy (SVSS), 1-31
Special Marine Air/Ground Task Force (SPMAGTF), 1-24
Special Mission Ships (SMS), 1-16
Specified Tasks, 5-22
Staff Estimates, 5-7, 5-19, 5-56
Staff Transition, 3-16
Strategic Communication, 7-6
Strategic Lift, 1-39
Strategic Offensive Posture, 2-19
Subject Matter Expert Exchanges, 7-11
Submarine Forces, 1-12
Submarine Operating Authority (SUBOPAUTH), 2-55
Subordinate Unified Command, 2-37
Support Situations (SUPSITs), 3-27
Supporting Establishment, 1-20
Surface Combatants, 1-10
Surface Strike Group (SSG), 1-5, 3-28
Sustainment, 3-14, 6-5
Synchronization Matrix, 5-64

T

Tactical Command (TACOM), 2-35
Tactical Control (TACON), 2-34, 2-35
Tactical Distribution, 6-32, 6-34
Tactical Headquarters, 2-64
Tactical Operations, 2-14
Targeting, 2-58, 3-33, 4-9
Task Elements, 3-26

Task Forces, 1-3, 3-26
Task Groups, 3-26
Task Organization, 2-62
Task Units, 3-26
Task List Libraries, 5-23
Tenth Fleet, 2-18
Terrain Management, 4-9
Territorial Sea, 3-31
Terrorism, 1-31
Theater ASW Commander (TASWC), 2-55
Theater Distribution, 6-27
Theater of Operations, 2-7
Theater Security Cooperation (TSC), 7-1, 7-3
Theater Structure, 2-7
Time-Sensitive Planning, 5-2
Training, 7-13
Transition, 5-73
Transition Drills, 5-76

U

U.S. Coast Guard, 1-27
U.S. Cyber Command (USCYBERCOM), 2-16
U.S. Flag Fleet, 1-42
U.S. Marine Corps, 1-19
U.S. Marine Corps Special Operations Forces, 1-38
U.S. Maritime Administration (MARAD), 1-42, 6-29
U.S. Navy, 1-1
U.S. Special Operations Command (USSOCOM), 1-34
U.S. Tenth Fleet, 2-18
U.S. Transportation Command (USTRANSCOM), 1-40
Undersea/Antisubmarine Warfare (USW/ASW), 2-54
Unified Action, 2-11
Unified Command, 2-37
Unit-Level Exercises, 7-12

W

Wargaming, 5-47
Wargaming Methods, 5-51
Warning Order, 5-34
Waterspace Management (WSM), 2-53, 4-9

The Essentials of Warfighting
Military SMARTbooks

SMARTbooks - The Essentials of Warfighting! Recognized as a doctrinal reference standard by military professionals around the world, SMARTbooks are designed with all levels of Soldiers, Sailors, Airmen, Marines and Civilians in mind.

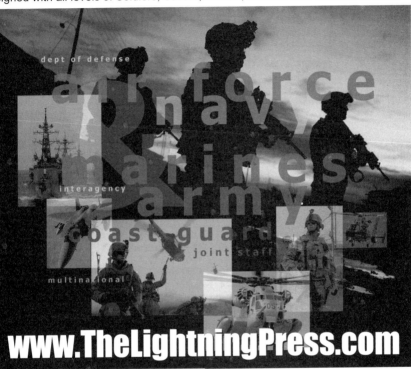

www.TheLightningPress.com

SMARTbooks can be used as quick reference guides during actual tactical combat operations, as lesson plans in support of training exercises and as study guides at military education and professional development courses. Serving a generation of warfighters, military reference SMARTbooks have become "mission-essential" around the world:

- Military education and professional development courses/schools: officer and noncommissioned officer basic and advanced courses, NCO Academy, Command & General Staff College (CGSC), Intermediate Level Education (ILE), Joint Forces Staff College (JFSC), War College, West Point and ROTC
- National Training Center (NTC), Joint Readiness Training Center (JRTC) and Battle Command Training Program (BCTP)
- Active, Reserve and Guard units across the full-spectrum of operations
- Global War on Terrorism operations in Iraq, Afghanistan and the Asia-Pacific
- Combatant Command (COCOM) and JTF Headquarters around the world
- Allied, coalition and multinational partner support and training to include NATO, Iraq and the Afghanistan National Army

SMARTpurchase!

View, download samples and purchase SMARTbooks online at: **www.TheLightningPress.com**. Register your SMARTbooks to receive email notification of SMARTupdates, new titles & revisions

The Joint Forces Operations & Doctrine SMARTbook (2nd Rev. Ed.)
Guide to Joint, Multinational & Interagency Operations

The Army Operations & Doctrine SMARTbook (4th Rev. Ed.)
Guide to FM 3-0 Operations & the Six Warfighting Functions

The Naval Operations & Planning SMARTbook
Guide to Designing, Planning & Conducting Maritime Operations

The Stability, Peace & Counterinsurgency SMARTbook
Nontraditional Approaches in a Dynamic Security Environment

The Battle Staff SMARTbook (3rd Rev. Ed.)
Guide to Designing, Planning & Conducting Military Operations

The Small Unit Tactics SMARTbook (First Edition)
Leader's Reference Guide to Conducting Tactical Operations

The Leader's SMARTbook (3rd Rev. Ed.) With Change 1 (FM 7-0 SMARTupdate)
Military Leadership & Training for Full Spectrum Operations

The Sustainment & Multifunctional Logistician's SMARTbook (2nd Rev. Ed.)
Warfighter's Guide to Logistics, Personnel Services, & Health Services Support

Web: www.TheLightningPress.com
E-mail: SMARTbooks@TheLightningPress.com

24-hour Voicemail/Fax/Order:
Record or fax your order toll-free at 1-800-997-8827

Mail, Check & Money Order:
2227 Arrowhead Blvd., Lakeland, FL 33813